Computational Methods for Coupled Problems in Science and Engineering

Computational Methods for Coupled Problems in Science and Engineering

Editors

Simona Perotto
Gianluigi Rozza
Antonia Larese

Basel • Beijing • Wuhan • Barcelona • Belgrade • Novi Sad • Cluj • Manchester

Editors

Simona Perotto
MOX—Dipartimento
di Matematica
Politecnico di Milano
Milano
Italy

Gianluigi Rozza
SISSA mathLab
International School for
Advanced Studies
Trieste
Italy

Antonia Larese
Department of Mathematics
Università degli Studi
di Padova
Padova
Italy

Editorial Office
MDPI
St. Alban-Anlage 66
4052 Basel, Switzerland

This is a reprint of articles from the Special Issue published online in the open access journal *Mathematical and Computational Applications* (ISSN 2297-8747) (available at: https://www.mdpi.com/journal/mca/special_issues/COUPLED2021).

For citation purposes, cite each article independently as indicated on the article page online and as indicated below:

Lastname, A.A.; Lastname, B.B. Article Title. *Journal Name* **Year**, *Volume Number*, Page Range.

ISBN 978-3-7258-1137-3 (Hbk)
ISBN 978-3-7258-1138-0 (PDF)
doi.org/10.3390/books978-3-7258-1138-0

Cover image courtesy of Simona Perotto

© 2024 by the authors. Articles in this book are Open Access and distributed under the Creative Commons Attribution (CC BY) license. The book as a whole is distributed by MDPI under the terms and conditions of the Creative Commons Attribution-NonCommercial-NoDerivs (CC BY-NC-ND) license.

Contents

Ruo Li and Wei Zhong
Towards Building the OP-Mapped WENO Schemes: A General Methodology
Reprinted from: *Math. Comput. Appl.* **2021**, *26*, 67, doi:10.3390/mca26040067 1

Camille Carvalho
Modified Representations for the Close Evaluation Problem
Reprinted from: *Math. Comput. Appl.* **2021**, *26*, 69, doi:10.3390/mca26040069 47

Matteo Gavazzoni, Nicola Ferro, Simona Perotto and Stefano Foletti
Multi-Physics Inverse Homogenization for the Design of Innovative Cellular Materials:
Application to Thermo-Elastic Problems
Reprinted from: *Math. Comput. Appl.* **2022**, *27*, 15, doi:10.3390/mca27010015 68

Sophie Valcke, Andrea Piacentini and Gabriel Jonville
Benchmarking Regridding Libraries Used in Earth System Modelling
Reprinted from: *Math. Comput. Appl.* **2022**, *27*, 31, doi:10.3390/mca27020031 89

Swadesh Pal and Roderick Melnik
Coupled Neural–Glial Dynamics and the Role of Astrocytes in Alzheimer's Disease
Reprinted from: *Math. Comput. Appl.* **2022**, *27*, 33, doi:10.3390/mca27030033 115

Sourav Dutta, Peter Rivera-Casillas, Brent Styles and Matthew Farthing
Reduced Order Modeling Using Advection-Aware Autoencoders
Reprinted from: *Math. Comput. Appl.* **2022**, *27*, 34, doi:10.3390/mca27030034 129

Yeram Lim, Tamara Chambers, Christine Walck, Safeer Siddicky, Erin Mannen and Victor Huayamave
Challenges in Kinetic-Kinematic Driven Musculoskeletal Subject-Specific Infant Modeling
Reprinted from: *Math. Comput. Appl.* **2022**, *27*, 36, doi:10.3390/mca27030036 157

Stefan Banholzer, Luca Mechelli and Stefan Volkwein
A Trust Region Reduced Basis Pascoletti-Serafini Algorithm for Multi-Objective
PDE-Constrained Parameter Optimization
Reprinted from: *Math. Comput. Appl.* **2022**, *27*, 39, doi:10.3390/mca27030039 168

Kyle Davis, Miriam Schulte and Benjamin Uekermann
Enhancing Quasi-Newton Acceleration for Fluid-Structure Interaction
Reprinted from: *Math. Comput. Appl.* **2022**, *27*, 40, doi:10.3390/mca27030040 197

Rohan Singla, Shubham Gupta and Arnab Chanda
A Computational Fluid Dynamics-Based Model for Assessing Rupture Risk in Cerebral Arteries
with Varying Aneurysm Sizes
Reprinted from: *Math. Comput. Appl.* **2023**, *28*, 90, doi:10.3390/mca28040090 215

 **Mathematical and Computational Applications**

Article

Towards Building the OP-Mapped WENO Schemes: A General Methodology

Ruo Li [1,†] and Wei Zhong [2,3,*,†]

1. CAPT, LMAM and School of Mathematical Sciences, Peking University, Beijing 100871, China; rli@math.pku.edu.cn
2. School of Mathematical Sciences, Peking University, Beijing 100871, China
3. Northwest Institute of Nuclear Technology, Xi'an 710024, China
* Correspondence: zhongwei2016@pku.edu.cn or zhongwei@ac.nint.cn
† Both authors contributed equally to this work.

Abstract: A serious and ubiquitous issue in existing mapped WENO schemes is that most of them can hardly preserve high resolutions, but in the meantime prevent spurious oscillations in the solving of hyperbolic conservation laws with long output times. Our goal for this article was to address this widely known problem. In our previous work, the *order-preserving (OP)* criterion was originally introduced and carefully used to devise a new mapped WENO scheme that performs satisfactorily in long simulations, and hence it was indicated that the *OP* criterion plays a critical role in the maintenance of low-dissipation and robustness for mapped WENO schemes. Thus, in our present work, we firstly defined the family of mapped WENO schemes, whose mappings meet the *OP* criterion, as OP-Mapped WENO. Next, we attentively took a closer look at the mappings of various existing mapped WENO schemes and devised a general formula for them. That helped us to extend the *OP* criterion to the design of improved mappings. Then, we created a generalized implementation of obtaining a group of OP-Mapped WENO schemes, named MOP-WENO-X, as they are developed from the existing mapped WENO-X schemes, where the notation "X" is used to identify the version of the existing mapped WENO scheme. Finally, extensive numerical experiments and comparisons with competing schemes were conducted to demonstrate the enhanced performances of the MOP-WENO-X schemes.

Keywords: *order-preserving* mapping; OP-Mapped WENO; hyperbolic conservation laws

1. Introduction

The essentially non-oscillatory (ENO) schemes [1–4] and the weighted ENO (WENO) schemes [5–8] have been developed quite successfully in recent decades to solve the hyperbolic conservation problems, especially those that may generate discontinuities and smooth small-scale structures as time evolves in their solutions, even if the initial condition is smooth. The main purpose of this study was to find a general method to introduce the *order-preserving (OP)* mapping proposed in our previous work [9] for improving the existing mapped WENO schemes for the approximation of the hyperbolic conservation laws in the form

$$\frac{\partial \mathbf{u}}{\partial t} + \nabla \cdot \mathbf{F}(\mathbf{u}) = 0, \tag{1}$$

where $\mathbf{u} = (u_1, \cdots, u_m) \in \mathbb{R}^m$ is the vector of the conserved variables and $\mathbf{F}(\mathbf{u})$ is the vector of the Cartesian components of flux. In recent years, there have been many works by Dumbser [10], Boscheri [11–13], Tsoutsanis [14,15], Titarev and Toro [16–19], Semplice [20,21], Puppo [22], Russo [23,24], and others on WENO approaches. These researches embraced a wide range of issues, e.g., the ADER-WENO finite volume schemes, the Cool WENO schemes, the unstructured WENO schemes, the Compact central WENO schemes, and so on. However, because of space limitations, it is very difficult to provide detailed descriptions of them here, and we refer the reader to our references for more details. In the present study, our main concern was to improve the performances of the $(2r-1)$th-order mapped WENO schemes, so we briefly review recent developments in this field in the following.

Harten et al. [1] introduced the ENO schemes. They used the smoothest stencil from r possible candidate stencils based on the local smoothness to perform a polynomial reconstruction such that it yielded high-order accuracy in smooth regions but avoided spurious oscillations at or near discontinuities. Liu, Osher, and Chan [7] introduced the first WENO scheme, an improved version of the ENO methodology with a cell-averaged approach, by using a nonlinear convex combination of all the candidate stencils to achieve a higher order of accuracy than the ENO schemes, while retaining the essential non-oscillatory property at or near discontinuities. In other words, it achieves $(r+1)$th-order accuracy from the rth-order ENO schemes [1–3] in the smooth regions while behaving similarly to the rth-order ENO schemes in regions including discontinuities. In [8], Jiang and Shu proposed the classic WENO-JS scheme, along with a new measurement for the smoothness of the numerical solutions on substencils (hereafter, denoted by smoothness indicator), by using the sum of the normalized squares of the scaled L_2-norms of all the derivatives of r local interpolating polynomials, to obtain $(2r-1)$th-order accuracy from the rth-order ENO schemes.

The WENO-JS scheme has become a very popular and quite successful methodology for solving compressible flows modeled through hyperbolic conservation laws in the form of Equation (1). However, it was less than fifth-order for many cases, such as at or near critical points of order $n_{cp}=1$ in the smooth regions. Here, we refer to n_{cp} as the order of the critical point; e.g., $n_{cp}=1$ corresponds to $f'=0, f''\neq 0$ and $n_{cp}=2$ corresponds to $f'=0, f''=0, f'''\neq 0$. In particular, Henrick et al. [25] identified that the fifth-order WENO-JS scheme fails to yield the optimal convergence order at or near critical points where the first derivative vanishes but the third derivative does not. Then, in the same article, they derived the necessary and sufficient conditions on the nonlinear weights for optimality of the convergence rate of the fifth-order WENO schemes and these conditions were reduced to a simpler sufficient condition [26] which could be easily extended to the $(2r-1)$th-order WENO schemes [27]. Moreover, also in [25], Henrick et al. devised the original mapped WENO scheme, named WENO-M hereafter, by constructing a mapping function that satisfies the sufficient condition to achieve the optimal order of accuracy.

Later, following the idea of incorporating a mapping procedure to keep the nonlinear weights of the convex combination of stencils as near as possible to the ideal weights of optimal order accuracy, various versions of mapped WENO schemes have been successfully proposed. In [27], Feng et al. rewrote the mapping function of the WENO-M scheme in a simple and more meaningful form and then extended it to a general class of improved mapping functions, leading to the family of the WENO-IM(k, A) schemes, where k is a positive even integer and A a positive real number. It was indicated that by taking $k=2$ and $A=0.1$ in the WENO-IM(k, A) scheme, far better numerical solutions with less dissipation and higher resolution could be obtained than that of the WENO-M scheme. Unfortunately, the numerical experiments in [28] showed that the seventh and ninth-order WENO-IM(2, 0.1) schemes generated evident spurious oscillations near discontinuities when the output time was large. In addition, our numerical experiments, as shown in Figures 10, 12 and 14, indicate that, even for the fifth-order WENO-IM(2, 0.1) scheme, the spurious oscillations are also produced when the grid number increases or a different initial condition is used. Recently, Feng et al. [29] pointed out that, when the WENO-M scheme is used for solving the problems with discontinuities for long output times, its mapping function may amplify the effect from the non-smooth stencils, leading to a potential loss of accuracy near discontinuities. To amend this drawback, a piecewise polynomial mapping function with two additional requirements, that is, $g'(0)=0$ and $g'(1)=0$ ($g(x)$ denotes the mapping function), to the original criteria in [25] was proposed. The recommended WENO-PM6 scheme [29] achieved significantly higher resolution than the WENO-M scheme when computing the one-dimensional linear advection problem with long output times. However, it may generate spurious oscillations near the discontinuities, as shown in Figure 8 of [27] and Figures 3–8 of [28].

More mapped WENO schemes, such as the WENO-PPMn [30], WENO-RM($mn0$) [28], WENO-MAIMi [31], WENO-ACM [32] schemes, and others, have been successfully developed to enhance the performances of the classic WENO-JS scheme in some ways, such as letting it achieve optimal convergence orders near critical points in smooth regions; having less numerical dissipation; letting it achieve higher resolutions near discontinuities; or reducing the computational costs. See the references for more details. However, as mentioned in previously published literature [27,28], most of the existing improved mapped WENO schemes could not prevent the spurious oscillations near discontinuities, especially for long-output-time simulations. Moreover, when simulating the two-dimensional problems with strong shock waves, the post-shock oscillations, which were systematically studied for WENO schemes by Zhang et al. [33], become very severe in the solutions of most of the existing improved mapped WENO schemes [32].

In our previous study [9], we studied the nonlinear weights of the existing mapped WENO schemes by taking the ones developed in [9,27,29,31] as examples. It was found that the order of the nonlinear weights for the substencils of the same global stencil has been changed at many points in the mapping processes of all these considered mapped WENO schemes. The order-change of nonlinear weights is caused by weight increasing of non-smooth substencils and weight decreasing of smooth substencils. It was revealed that this is the essential cause of the potential loss of accuracy of the WENO-M scheme and the spurious oscillation generation of the existing improved mapped WENO schemes, through theoretical analysis and extensive numerical tests. In the same article, the definition of the *order-preserving (OP)* mapping was given and suggested as an additional criterion in the design of the mapping function. Then a new mapped WENO scheme with its mapping function satisfying the additional criterion was proposed. Extensive numerical experiments showed that this scheme can achieve the optimal convergence order of accuracy, even at critical points. It also can decrease the numerical dissipation and obtain high resolution, but does not generate spurious oscillation near discontinuities, even if the output time is large. Moreover, it was observed clearly that it exhibits a significant advantage in reducing the post-shock oscillations when calculating the problems with strong shock waves in two dimensions.

In this article, the idea of introducing the *OP* criterion into the design of the mapping functions proposed in [9] is extended to various existing mapped WENO schemes. First of all, we give the common name of *OP-Mapped WENO* to the family of the mapped WENO schemes whose mappings are *OP*. A general formula for the mapping functions of various existing mapped WENO schemes is presented, which allows the extension of the *OP* criterion to all existing mapped WENO schemes. The notation MOP-WENO-X is used to denote the improved mapped WENO scheme considering the *OP* criterion based on the existing WENO-X scheme. A new function named **minDist** is defined (see Definition 4 in Section 3.3 below). A general algorithm to construct *OP* mappings through the existing mapping functions by using the **minDist** function is proposed.

Extensive numerical tests were conducted to demonstrate the performances of the MOP-WENO-X schemes: (1) A series of accuracy tests show the abilities of the MOP-WENO-X schemes to achieve optimal convergence order in smooth regions with first-order critical points and their advantages in long-output-time simulations of problems with very high-order critical points. (2) The one-dimensional linear advection equation with two kinds of initial conditions for long output times are presented to demonstrate that the MOP-WENO-X schemes can obtain high resolution, and meanwhile avoid spurious oscillation near discontinuities. (3) Some benchmark tests with strong shock waves modeled via the two-dimensional Euler equations were computed. It is clear that the MOP-WENO-X schemes also enjoy a significant advantage in reducing the post-shock oscillations.

The remainder of this paper is organized as follows. In Section 2, we briefly review the preliminaries to understand the finite volume method and the procedures of the WENO-JS [8], WENO-M [25], and other versions of mapped WENO schemes. Section 3 presents a general method to introduce the OP mapping for improving the existing mapped WENO schemes. Some numerical results are provided in Section 4 to illustrate the advantages of the proposed WENO schemes. Finally, concluding remarks are given in Section 5 to close this paper.

2. Brief Review of the WENO Schemes

For simplicity of presentation but without loss of generality, we denote our topic with the following one-dimensional linear hyperbolic conservation equation:

$$\frac{\partial u}{\partial t} + \frac{\partial f(u)}{\partial x} = 0, \quad x_l < x < x_r, t > 0, \tag{2}$$

with the initial condition $u(x,0) = u_0(x)$. We confine our attention to the uniform meshes in this paper, and for the WENO method with non-uniform meshes, one can refer to [34,35]. Throughout this paper, we assume that the given domain $[x_l, x_r]$ is discretized into the set of uniform cells $I_j := [x_{j-1/2}, x_{j+1/2}], j = 1, \cdots, N$ with the cell size $\Delta x = \frac{x_r - x_l}{N}$. The associated cell centers and cell boundaries are denoted by $x_j = x_l + (j - 1/2)\Delta x$ and $x_{j\pm 1/2} = x_j \pm \Delta x/2$, respectively. The notation $\bar{u}(x_j, t) = \frac{1}{\Delta x}\int_{x_{j-1/2}}^{x_{j+1/2}} u(\xi, t)\mathrm{d}\xi$ indicates the cell average of I_j. The one-dimensional linear hyperbolic conservation equation in Equation (2) can be approximated by a system of ordinary differential equations, yielding the semi-discrete finite volume form:

$$\begin{aligned}\frac{\mathrm{d}\bar{u}_j(t)}{\mathrm{d}t} &\approx \mathcal{L}(u_j), \\ \mathcal{L}(u_j) &= -\frac{1}{\Delta x}\left(\hat{f}_{j+1/2} - \hat{f}_{j-1/2}\right),\end{aligned} \tag{3}$$

where $\bar{u}_j(t)$ is the numerical approximation of the cell average $\bar{u}(x_j, t)$, and the numerical flux $\hat{f}_{j\pm 1/2}$ is a replacement of the physical flux function $f(u)$ at the cell boundaries $x_{j\pm 1/2}$ and it is defined by $\hat{f}_{j\pm 1/2} = \hat{f}(u_{j\pm 1/2}^-, u_{j\pm 1/2}^+)$. $u_{j\pm 1/2}^\pm$ refer to the limits of u, and their values of $u_{j\pm 1/2}^\pm$ can be obtained by reconstruction, for instance, the WENO reconstruction procedures shown later. In this paper, we use the global Lax–Friedrichs flux:

$$\hat{f}(a,b) = \frac{1}{2}\left[f(a) + f(b) - \alpha(b - a)\right],$$

where $\alpha = \max_u |f'(u)|$ is a constant and the maximum is taken over the whole range of u.

2.1. The WENO-JS Reconstruction

Firstly, we review the process of the classic fifth-order WENO-JS reconstruction [8]. For brevity, we describe only the reconstruction procedure of the left-biased $u_{j+1/2}^-$, and the right-biased one $u_{j+1/2}^+$ can trivially be computed by mirror symmetry with respect to the location $x_{j+1/2}$ of $u_{j+1/2}^-$. We drop the subscript "−" below just for simplicity of notation.

To construct the values of $u_{j+1/2}$ from known cell average values $\bar{u}_j$, a 5-point global stencil $S^5 = \{I_{j-2}, I_{j-1}, I_j, I_{j+1}, I_{j+2}\}$ is used in the fifth-order WENO-JS scheme. It is subdivided into three 3-point substencils $S_s = \{I_{j+s-2}, I_{j+s-1}, I_{j+s}\}$ with $s = 0, 1, 2$. It is known that the third-order approximations of $u(x_{j+1/2}, t)$ associated with these substencils are explicitly given by

$$u^0_{j+1/2} = \frac{1}{6}(2\bar{u}_{j-2} - 7\bar{u}_{j-1} + 11\bar{u}_j),$$
$$u^1_{j+1/2} = \frac{1}{6}(-\bar{u}_{j-1} + 5\bar{u}_j + 2\bar{u}_{j+1}), \qquad (4)$$
$$u^2_{j+1/2} = \frac{1}{6}(2\bar{u}_j + 5\bar{u}_{j+1} - \bar{u}_{j+2}).$$

Then the $u_{j+1/2}$ of global stencil S^5 is computed by a weighted average of those third-order approximations of substencils, taking the form

$$u_{j+1/2} = \sum_{s=0}^{2} \omega_s u^s_{j+1/2}. \qquad (5)$$

The nonlinear weights ω_s in the classic WENO-JS scheme are defined as

$$\omega_s^{JS} = \frac{\alpha_s^{JS}}{\sum_{l=0}^{2} \alpha_l^{JS}}, \quad \alpha_s^{JS} = \frac{d_s}{(\epsilon + \beta_s)^2}, \quad s = 0, 1, 2, \qquad (6)$$

where d_0, d_1, d_2 are called the ideal weights of ω_s since they generate the central upstream fifth-order scheme for the global stencil S^5. It is known that $d_0 = 0.1, d_1 = 0.6, d_2 = 0.3$ and in smooth regions we can get $\sum_{s=0}^{2} d_s u^s_{j+1/2} = u(x_{j+1/2}, t) + O(\Delta x^5)$. ϵ is a small positive number introduced to prevent the denominator from becoming zero. The parameters β_s are the smoothness indicators for the third-order approximations $u^s_{j+1/2}$ and their explicit formulas can be obtained from [8], taking the form

$$\beta_0 = \frac{13}{12}(\bar{u}_{j-2} - 2\bar{u}_{j-1} + \bar{u}_j)^2 + \frac{1}{4}(\bar{u}_{j-2} - 4\bar{u}_{j-1} + 3\bar{u}_j)^2,$$
$$\beta_1 = \frac{13}{12}(\bar{u}_{j-1} - 2\bar{u}_j + \bar{u}_{j+1})^2 + \frac{1}{4}(\bar{u}_{j-1} - \bar{u}_{j+1})^2,$$
$$\beta_2 = \frac{13}{12}(\bar{u}_j - 2\bar{u}_{j+1} + \bar{u}_{j+2})^2 + \frac{1}{4}(3\bar{u}_j - 4\bar{u}_{j+1} + \bar{u}_{j+2})^2.$$

In general, the fifth-order WENO-JS scheme is able to recover the optimal convergence rate of accuracy in smooth regions. However, when at or near critical points where the first derivative vanishes but the third derivative does not simultaneously, it loses accuracy and its convergence rate of accuracy decreases to third-order or even less. We refer to [25] for more details.

2.2. The Mapped WENO Reconstructions

To address the issue of the WENO-JS scheme mentioned above, Henrick et al. [25] made a systematic truncation error analysis of Equation (3) in its corresponding finite difference version by using the Taylor series expansions of the Equation (4), and hence they derived the necessary and sufficient conditions on the weights for the fifth-order WENO scheme to achieve the formal fifth-order of convergence at smooth regions of the solution, taking the form

$$\sum_{s=0}^{2}(\omega_s^{\pm} - d_s) = O(\Delta x^6), \quad \sum_{s=0}^{2} A_s(\omega_s^+ - \omega_s^-) = O(\Delta x^3), \quad \omega_s^{\pm} - d_s = O(\Delta x^2), \qquad (7)$$

where the superscripts "+" and "−" on ω_s correspond to their use in either $u^s_{j+1/2}$ and $u^s_{j-1/2}$ stencils respectively, and the parameter A_s is independent of Δx and it is given explicitly in Equation (16) in [25] for the fifth-order version WENO-JS scheme. Since the first equation in Equation (7) always holds due to the normalization, a simpler sufficient condition for the fifth-order convergence is given as [26]

$$\omega_s^{\pm} - d_s = O(\Delta x^3), \quad s = 0, 1, 2. \qquad (8)$$

The conditions Equation (7) or Equation (8) may not hold in the case of smooth extrema or at critical points when the fifth-order WENO-JS scheme is used. An innovative idea of fixing this deficiency, originally proposed by Henrick in [25], is to design a mapping function to make ω_s approximating the ideal weights d_s at critical points to the required third order $O(\Delta x^3)$. The first mapping function devised by Henrick et al. in [25] is given as

$$\left(g^M\right)_s(\omega) = \frac{\omega\left(d_s + d_s^2 - 3d_s\omega + \omega^2\right)}{d_s^2 + (1 - 2d_s)\omega}, \qquad s = 0, 1, 2. \tag{9}$$

In Equation (9), $\omega = \omega^{JS}$ is recommended according to the theoretical analysis of WENO-M by Henrick in [25] where the good properties of ω^{JS} to guarantee the success of the mapped function have been analyzed very carefully. Actually, $\omega = \omega^{JS}$ is commonly used in almost all maping functions [9,27–32] although some other kind of nonlinear weights may also be available.

We can verify that $\left(g^M\right)_s(\omega)$ meets the conditions in Equation (8) as it is a non-decreasing monotone function on $[0, 1]$ with finite slopes and satisfies the following properties.

Lemma 1. *The mapping function $\left(g^M\right)_s(\omega)$ defined by Equation (9) satisfies:*

C1. $0 \leq \left(g^M\right)_s(\omega) \leq 1$, $\left(g^M\right)_s(0) = 0$, $\left(g^M\right)_s(1) = 1$;
C2. $\left(g^M\right)_s(d_s) = d_s$;
C3. $\left(g^M\right)'_s(d_s) = \left(g^M\right)''_s(d_s) = 0$.

Following Henrick's idea, a great many improved mapping functions were successfully proposed [9,27–32]. To clarify our major concern and provide convenience to readers but for brevity in the description, we only state some mapping functions in the following context, and we refer to references for properties similar to Lemma 1 and more details of these mapping functions.

■ WENO-IM(k, A) [27]

$$\left(g^{IM}\right)_s(\omega; k, A) = d_s + \frac{(\omega - d_s)^{k+1} A}{(\omega - d_s)^k A + \omega(1 - \omega)}, \qquad A > 0, k = 2n, n \in \mathbb{N}^+. \tag{10}$$

■ WENO-PMk [29]

$$\left(g^{PM}\right)_s(\omega) = c_1(\omega - d_s)^{k+1}(\omega + c_2) + d_s, \qquad k \geq 2, \tag{11}$$

where c_1, c_2 are constants with specified parameters k and d_s, taking the following forms

$$c_1 = \begin{cases} (-1)^k \dfrac{k+1}{d_s^{k+1}}, & 0 \leq \omega \leq d_s, \\ -\dfrac{k+1}{(1-d_s)^{k+1}}, & d_s < \omega \leq 1, \end{cases} \qquad c_2 = \begin{cases} \dfrac{d_s}{k+1}, & 0 \leq \omega \leq d_s, \\ \dfrac{d_s - (k+2)}{k+1}, & d_s < \omega \leq 1. \end{cases}$$

■ WENO-PPMn [30]

$$\left(g_s^{PPMn}\right)_s(\omega) = \begin{cases} \left(g_{s,L}^{PPMn}\right)_s(\omega), & \omega \in [0, d_s] \\ \left(g_{s,R}^{PPMn}\right)_s(\omega), & \omega \in (d_s, 1], \end{cases} \tag{12}$$

and for $n = 5$,

$$\left(g_{s,L}^{PPM5}\right)_s(\omega) = d_s\left(1 + (a - 1)^5\right), \qquad \left(g_{s,R}^{PPM5}\right)_s(\omega) = d_s + b^4(\omega - d_s)^5. \tag{13}$$

where $a = \omega/d_s, b = 1/(d_s - 1)$.

■ WENO-RM($mn0$) [28]

$$(g^{\text{RM}})_s(\omega) = d_s + \frac{(\omega - d_s)^{n+1}}{a_0 + a_1\omega + \cdots + a_{m+1}\omega^{m+1}}, \quad m \leq n \leq 8, \tag{14}$$

where

$$\begin{cases} a_i = C_{n+1}^i(-d_s)^{n-i}, & i = 0, 1, \cdots, m, \\ a_{m+1} = (1-d_s)^n - \sum_{i=0}^m a_i. \end{cases} \tag{15}$$

Furthermore, $m = 2, n = 6$ is recommended in [28], then

$$(g^{\text{RM}})_s(\omega) = d_s + \frac{(\omega - d_s)^7}{a_0 + a_1\omega + a_2\omega^2 + a_3\omega^3}, \quad \omega \in [0,1] \tag{16}$$

where

$$a_0 = d_s^6, \quad a_1 = -7d_s^5, \quad a_2 = 21d_s^4, \quad a_3 = (1-d_s)^6 - \sum_{i=0}^2 a_i. \tag{17}$$

■ WENO-MAIM1 [31]

$$(g^{\text{MAIM1}})_s(\omega) = d_s + \frac{f^{\text{FIM}} \cdot (\omega - d_s)^{k+1}}{f^{\text{FIM}} \cdot (\omega - d_s)^k + \omega^{\frac{d_s}{m_s\omega + \epsilon_A}}(1-\omega)^{\frac{1-d_s}{m_s(1-\omega)+\epsilon_A}}}, \tag{18}$$

with

$$f^{\text{FIM}} = A\left(\frac{1+(-1)^k}{2} + \frac{1+(-1)^{k+1}}{2} \cdot \text{sgm}(\omega - d_s, \delta, 1, k)\right), \tag{19}$$

and

$$\text{sgm}(x, \delta, B, k) = \begin{cases} \dfrac{x}{|x|}, & |x| \geq \delta, \\ \dfrac{x}{\left(B(\delta^2 - x^2)\right)^{k+3} + |x|}, & |x| < \delta. \end{cases} \tag{20}$$

In Equations (18)–(20), $k \in \mathbb{N}^+$, $A > 0$, $\delta > 0$ with $\delta \to 0$, ϵ_A is a very small positive number to prevent the denominator from becoming zero, and $m_s \in \left[\frac{\alpha_s}{k+1}, M\right)$ with M being a finite positive constant real number and α_s a positive constant that only depends on s in the fifth-order WENO-MAIM1 scheme. In Equation (20), the positive parameter B is a scale transformation factor introduced to adjust the shape of the mapping function and it is set to be 1 in WENO-MAIM1 while to be other values in the following WENO-ACM schemes.

■ WENO-ACM [32]

$$(g^{\text{ACM}})_s(\omega) = \begin{cases} \dfrac{d_s}{2}\text{sgm}(\omega - \text{CFS}_s, \delta_s, B, k) + \dfrac{d_s}{2}, & \omega \leq d_s, \\ \dfrac{1-d_s}{2}\text{sgm}(\omega - \overline{\text{CFS}}_s, \delta_s, B, k) + \dfrac{1+d_s}{2}, & \omega > d_s, \end{cases} \tag{21}$$

where $\text{CFS}_s \in (0, d_s)$, $\overline{\text{CFS}}_s = 1 - \frac{1-d_s}{d_s} \times \text{CFS}_s$ with $\overline{\text{CFS}}_s \in (d_s, 1)$, and $\delta_s < \min\left\{\text{CFS}_s, d_s - \text{CFS}_s, (1-d_s)\left(1 - \frac{\text{CFS}_s}{d_s}\right), \frac{1-d_s}{d_s}\text{CFS}_s\right\}$.

■ MIP-WENO-ACMk [9]

$$(g^{\text{MIP-ACM}k})_s(\omega) = \begin{cases} k_s\omega, & \omega \in [0, \text{CFS}_s), \\ d_s, & \omega \in [\text{CFS}_s, \overline{\text{CFS}}_s], \\ 1 - k_s(1-\omega), & \omega \in (\overline{\text{CFS}}_s, 1], \end{cases} \tag{22}$$

where $\text{CFS}_s \in (0, d_s)$, $\overline{\text{CFS}}_s = 1 - \frac{1-d_s}{d_s} \times \text{CFS}_s$ with $\overline{\text{CFS}}_s \in (d_s, 1)$, and $k_s \in \left[0, \frac{d_s}{\text{CFS}_s}\right]$.

By using the mapping function $\left(g^X\right)_s(\omega)$, where the superscript "X" corresponds to "M," "PM6," or "IM," etc., the nonlinear weights of the associated WENO-X scheme are defined as

$$\omega_s^X = \frac{\alpha_s^X}{\sum_{l=0}^{2} \alpha_l^X}, \alpha_s^X = \left(g^X\right)_s(\omega_s^{JS}), \quad s = 0, 1, 2,$$

where ω_s^{JS} are calculated by Equation (6).

In other studies, it has been analyzed and proved in detail that the WENO-X schemes can retain the optimal order of accuracy in smooth regions even at or near critical points.

3. A General Method to Introduce *Order-Preserving* Mapping for Mapped WENO Schemes

3.1. The OP-Mapped WENO

Before giving Definition 3 below, to maintain coherence and for the readers' convenience, we state the definition of *order-preserving/non-order-preserving* mapping and *OP/non-OP* point proposed in [9].

Definition 1 (*order-preserving/non-order-preserving* mapping). *Suppose that* $\left(g^X\right)_s(\omega)$, $s = 0, \cdots, r-1$ *is a monotone increasing piecewise mapping function of the* $(2r-1)$*th-order mapped WENO-X scheme. If for* $\forall m, n \in \{0, \cdots, r-1\}$*, when* $\omega_m > \omega_n$*, we have*

$$\left(g^X\right)_m(\omega_m) \geq \left(g^X\right)_n(\omega_n). \tag{23}$$

and when $\omega_m = \omega_n$*, we have* $\left(g^X\right)_m(\omega_m) = \left(g^X\right)_n(\omega_n)$*, then we say the set of mapping functions* $\left\{\left(g^X\right)_s(\omega), s = 0, \cdots, r-1\right\}$ *is* **order-preserving (OP)**. *Otherwise, we say the set of mapping functions* $\left\{\left(g^X\right)_s(\omega), s = 0, \cdots, r-1\right\}$ *is* **non-order-preserving (non-OP)**.

Definition 2 (OP/non-OP point). *Let* S^{2r-1} *denote the* $(2r-1)$*-point global stencil centered around* x_j*. Assume that* S^{2r-1} *is subdivided into r-point substencils* $\{S_0, \cdots, S_{r-1}\}$ *and* ω_s *are the nonlinear weights corresponding to the substencils* S_s *with* $s = 0, \cdots, r-1$*, which are used as the independent variables by the mapping function. Suppose that* $\left(g^X\right)_s(\omega), s = 0, \cdots, r-1$ *is the mapping function of the mapped WENO-X scheme; then we say that a* **non-OP** *mapping process occurs at* x_j*, if* $\exists m, n \in \{0, \cdots, r-1\}$*, s.t.*

$$\begin{cases} (\omega_m - \omega_n)\left(\left(g^X\right)_m(\omega_m) - \left(g^X\right)_n(\omega_n)\right) < 0, & \text{if } \omega_m \neq \omega_n, \\ \left(g^X\right)_m(\omega_m) \neq \left(g^X\right)_n(\omega_n), & \text{if } \omega_m = \omega_n. \end{cases} \tag{24}$$

In addition, we say x_j *is a* **non-OP point**. *Otherwise, we say* x_j *is an* **OP point**.

Definition 3 (OP-Mapped WENO). *The family of the mapped WENO schemes with OP mappings is collectively referred to as* **OP-Mapped WENO** *in our study.*

3.2. A General Formula for the Existing Mapping Functions

We rewrite the mapping function of the WENO-X scheme, that is, $\left(g^X\right)_s(\omega), s = 0, 1, \cdots, r-1$, to be a general formula, given as

$$g^X(\omega; m_P, P_{s,1}, \cdots, P_{s,m_P}) = \left(g^X\right)_s(\omega), \tag{25}$$

where m_P is the number of the parameters related with s indicating the substencil, and $P_{s,1}, \cdots, P_{s,m_P}$ are these parameters. Taking the WENO-IM(k, A) scheme as an example, besides the independent variable ω, there are the other three parameters in its mapping function (see Equation (10)), namely, d_s, k and A. It is easy to know that d_s is related to the substencil S_s, and k and A are not. Thus, for the WENO-IM(k, A) scheme, we have $m_P = 1$ and $P_{s,1} = d_s$. We can also determine the value of m_P and the corresponding $P_{s,1}, \cdots, P_{s,m_P}$

of other WENO schemes. Clearly, we have $m_P = 0$ for the WENO-JS scheme and $m_P \geq 1$ for other mapped WENO schemes. In Table 1, taking nine different WENO schemes as examples, we have presented their parameters of m_P and $P_{s,1}, \cdots, P_{s,m_P}$. Let n_X denote the order of the specified critical point, namely, $\omega = d_s$, of the mapping function of the WENO-X scheme, that is, $(g^X)'_s(d_s) = \cdots = (g^X)_s^{(n_X)}(d_s) = 0, (g^X)_s^{(n_X+1)}(d_s) \neq 0$. To simplify the description of Theorem 2 below, we present n_X of the WENO-X scheme in the sixth column of Table 1.

Table 1. The parameters m_P and $P_{s,1}, \cdots, P_{s,m_P}$ for the WENO-JS scheme and some existing mapped WENO schemes whose mapping functions are *non-OP*.

No.	Scheme, WENO-X	m_P	$P_{s,1}, \cdots, P_{s,m_P}$	Parameters	n_X	Ref.
1	WENO-JS	0	None	None	None	See [8]
2	WENO-M	1	$P_{s,1} = d_s$	None	2	See [25]
3	WENO-IM(k, A)	1	$P_{s,1} = d_s$	$k = 2.0, A = 0.1$	k	See [27]
4	WENO-PMk	1	$P_{s,1} = d_s$	$k = 6$	k	See [29]
5	WENO-PPMn	1	$P_{s,1} = d_s$	$n = 5$	4	See [30]
6	WENO-RM(mn0)	1	$P_{s,1} = d_s$	$m = 2, n = 6$	3, 4	See [28]
7	WENO-MAIM1	2	$P_{s,1} = d_s, P_{s,2} = m_s$	$k = 10, A = 1.0e-6, m_s = 0.06$	$k, k+1$	See [31]
8	WENO-ACM	2	$P_{s,1} = d_s, P_{s,2} = \text{CFS}_s$	$A = 20, k = 2, \mu = 1e-6, \text{CFS}_s = d_s/10$	∞	See [32]
9	MIP-WENO-ACMk	3	$P_{s,1} = d_s, P_{s,2} = \text{CFS}_s, P_{s,3} = k_s$	$k_s = 0.0, \text{CFS}_s = d_s/10$	∞	See [9]

Lemma 2. *For the WENO-X scheme shown in Table 1, the mapping function $(g^X)_s(\omega), s = 0, 1, \cdots, r - 1$ is monotonically increasing over $[0, 1]$.*

Proof. See the corresponding references given in the last column of Table 1. □

3.3. The New Mapping Functions

Firstly, we give the **minDist** function by the following definition.

Definition 4 (**minDist** function). *Define the **minDist** function as follows:*

$$\begin{cases} \mathbf{minDist}(x_0, \cdots, x_{r-1}; d_0, \cdots, d_{r-1}; \omega) = x_{k^*}, \\ k^* = \min\left(\text{IndexOf}\left(\min\left\{|\omega - d_0|, |\omega - d_1|, \cdots, |\omega - d_{r-1}|\right\}\right)\right), \end{cases} \quad (26)$$

where $d_s, s = 0, \cdots, r - 1$ is the optimal weight; ω is the nonlinear weight, being the independent variable of the mapping function; and the function IndexOf$(\cdot)$ *returns a set of the subscripts of "·"—that is, if* $\min\left\{|\omega - d_0|, |\omega - d_1|, \cdots, |\omega - d_{r-1}|\right\} = |\omega - d_{m_1}| = |\omega - d_{m_2}| = \cdots = |\omega - d_{m_M}|$, *then*

$$\text{IndexOf}\left(\min\left\{|\omega - d_0|, |\omega - d_1|, \cdots, |\omega - d_{r-1}|\right\}\right) = \{m_1, m_2, \cdots, m_M\}. \quad (27)$$

Let $\mathcal{D} = \{d_0, d_1, \cdots, d_{r-1}\}$ be an array of all the ideal weights of the $(2r - 1)$th-order WENO schemes. We build a new array by sorting the elements of $\mathcal{D}$ in ascending order—that is, $\widetilde{\mathcal{D}} = \{\tilde{d}_0, \tilde{d}_1, \cdots, \tilde{d}_{r-1}\}$. In other words, the arrays $\mathcal{D}$ and $\widetilde{\mathcal{D}}$ have the same elements with different arrangements, and the elements of $\widetilde{\mathcal{D}}$ satisfy

$$0 < \tilde{d}_0 < \tilde{d}_1 < \cdots < \tilde{d}_{r-1} < 1. \quad (28)$$

Definition 5. *Let $\mathcal{G} = \left\{(g^X)_0(\omega), (g^X)_1(\omega), \cdots, (g^X)_{r-1}(\omega)\right\}$ be an array of all the mapping functions of the $(2r - 1)$th-order mapped WENO-X scheme. We define a new array by sorting* the elements of $\mathcal{G}$ in a new order—that is, $\widetilde{\mathcal{G}} = \left\{ \widetilde{(g^X)}_0(\omega), \widetilde{(g^X)}_1(\omega), \cdots, \widetilde{(g^X)}_{r-1}(\omega) \right\}$, where $\widetilde{(g^X)}_s(\omega)$ is the mapping function associated with $\tilde{d}_s$.

Lemma 3. Denote $\tilde{d}_{-1} = 0, \tilde{d}_r = 1$. Let $\mathring{d}_{-1} = \tilde{d}_{-1}, \mathring{d}_0 = \frac{\tilde{d}_0 + \tilde{d}_1}{2}, \cdots, \mathring{d}_{r-2} = \frac{\tilde{d}_{r-2} + \tilde{d}_{r-1}}{2}, \mathring{d}_{r-1} = \tilde{d}_r$. For $\forall i = 0, 1, \cdots, r-1$, if $\omega \in (\mathring{d}_{i-1}, \mathring{d}_i]$, then

$$\min\left(\mathrm{IndexOf}\left(\min\left\{|\omega - \tilde{d}_0|, |\omega - \tilde{d}_1|, \cdots, |\omega - \tilde{d}_{r-1}|\right\}\right)\right) = i.$$

Proof. (1) We first prove the cases of $i = 1, \cdots, r-2$. When $\tilde{d}_i \leq \omega \leq \frac{\tilde{d}_i + \tilde{d}_{i+1}}{2}$, as Equation (28) holds, we get

$$\begin{cases} 0 \leq \omega - \tilde{d}_i \leq \tilde{d}_{i+1} - \omega < \cdots < \tilde{d}_{r-1} - \omega, \\ 0 \leq \omega - \tilde{d}_i < \omega - \tilde{d}_{i-1} < \cdots < \omega - \tilde{d}_0. \end{cases} \quad (29)$$

Similarly, when $\frac{\tilde{d}_{i-1} + \tilde{d}_i}{2} < \omega < \tilde{d}_i$, we get

$$\begin{cases} 0 < \tilde{d}_i - \omega < \omega - \tilde{d}_{i-1} < \cdots < \omega - \tilde{d}_0, \\ 0 < \tilde{d}_i - \omega < \tilde{d}_{i+1} - \omega < \cdots < \tilde{d}_{r-1} - \omega. \end{cases} \quad (30)$$

Then, according to Equations (29) and (30), we obtain

$$\begin{aligned} &\min\left\{|\omega - \tilde{d}_0|, \cdots, |\omega - \tilde{d}_{i-1}|, |\omega - \tilde{d}_i|, |\omega - \tilde{d}_{i+1}|, \cdots, |\omega - \tilde{d}_{r-1}|\right\} \\ &= |\omega - \tilde{d}_i| = |\omega - \tilde{d}_{i+1}|, \end{aligned} \quad (31)$$

where $i = 1, \cdots, r-2$ and the last equality holds if and only if $\omega - \tilde{d}_i = \tilde{d}_{i+1} - \omega$.

(2) For the case of $i = 0$, we know that $\omega \in (\mathring{d}_{-1}, \mathring{d}_0] = \left(0, \frac{\tilde{d}_0 + \tilde{d}_1}{2}\right]$. When $\tilde{d}_0 \leq \omega \leq \frac{\tilde{d}_0 + \tilde{d}_1}{2}$, we have

$$0 \leq \omega - \tilde{d}_0 \leq \tilde{d}_1 - \omega < \cdots < \tilde{d}_{r-1} - \omega. \quad (32)$$

Additionally, when $0 < \omega < \tilde{d}_0$, we have

$$0 < \tilde{d}_0 - \omega < \tilde{d}_1 - \omega < \cdots < \tilde{d}_{r-1} - \omega. \quad (33)$$

Then, according to Equations (32) and (33), we obtain

$$\begin{aligned} &\min\left\{|\omega - \tilde{d}_0|, \cdots, |\omega - \tilde{d}_{i-1}|, |\omega - \tilde{d}_i|, |\omega - \tilde{d}_{i+1}|, \cdots, |\omega - \tilde{d}_{r-1}|\right\} \\ &= |\omega - \tilde{d}_0| = |\omega - \tilde{d}_1|, \end{aligned} \quad (34)$$

where the last equality holds if and only if $\omega - \tilde{d}_0 = \tilde{d}_1 - \omega$.

(3) As the proof of the case of $i = r - 1$ is very similar to that of the case $i = 0$, we do not state it here for simplicity. Additionally, we can get that, if $\omega \in (\mathring{d}_{r-2}, \mathring{d}_{r-1}]$, then

$$\min\left\{|\omega - \tilde{d}_0|, \cdots, |\omega - \tilde{d}_{i-1}|, |\omega - \tilde{d}_i|, |\omega - \tilde{d}_{i+1}|, \cdots, |\omega - \tilde{d}_{r-1}|\right\} = |\omega - \tilde{d}_{r-1}|. \quad (35)$$

(4) Thus, according to Equation (4) and Equations (31), (34), and (35), we obtain

$$\min\left(\text{IndexOf}\left(\min\left\{|\omega-\tilde{d}_0|,\cdots,|\omega-\tilde{d}_{i-1}|,|\omega-\tilde{d}_i|,\right.\right.\right.$$
$$\left.\left.\left.|\omega-\tilde{d}_{i+1}|,\cdots,|\omega-\tilde{d}_{r-1}|\right\}\right)\right)=i.$$

Now, we have finished the proof of Lemma 3. □

For simplicity of description and according to Lemma 3, we introduce intervals Ω_i defined as follows.

$$\Omega_i = \left\{\omega\,|\,\mathbf{minDist}(\tilde{d}_0,\tilde{d}_1,\cdots,\tilde{d}_{r-1};\tilde{d}_0,\tilde{d}_1,\cdots,\tilde{d}_{r-1};\omega)=\tilde{d}_i\right\} = (\mathring{d}_{i-1},\mathring{d}_i], \quad (36)$$

where $i = 0, 1, \cdots, r-1$.

If $\omega \in \Omega = (0,1]$, it is trivial to verify that: (1) $\Omega = \Omega_0 \bigcup \Omega_1 \bigcup \cdots \bigcup \Omega_{r-1}$; (2) for $\forall i, j = 0, 1, \cdots, r-1$ and $i \neq j$, $\Omega_i \bigcap \Omega_j = \varnothing$.

Lemma 4. *Let $a, b \in \{0, 1, \cdots, r-1\}$ and WENO-X be the scheme shown in Table 1. For $\forall a \geq b$ and $\omega_\alpha \in \Omega_a, \omega_\beta \in \Omega_b$, we have the following properties: C1. If $a = b$ and $\omega_\alpha > \omega_\beta$, then $\widetilde{(g^X)}_a(\omega_\alpha) \geq \widetilde{(g^X)}_b(\omega_\beta)$; C2. If $a = b$ and $\omega_\alpha = \omega_\beta$, then $\widetilde{(g^X)}_a(\omega_\alpha) = \widetilde{(g^X)}_b(\omega_\beta)$; C3. If $a > b$, then $\omega_\alpha > \omega_\beta$, $\widetilde{(g^X)}_a(\omega_\alpha) > \widetilde{(g^X)}_b(\omega_\beta)$.*

Proof. (1) We can directly get properties C1 and C2 from Lemma 2. (2) As $a > b$, according to Equations (28) and (36), we know that the interval Ω_a must be on the right side of the interval Ω_b, and $\omega_\alpha \in \Omega_a, \omega_\beta \in \Omega_b$ is given, then we get $\omega_\alpha > \omega_\beta$. Trivially, according to Definition 5, or by intuitively observing the curves of the mapping function $\widetilde{(g^X)}_s(\omega)$ as shown in Figure 1, we can obtain $\widetilde{(g^X)}_a(\omega_\alpha) > \widetilde{(g^X)}_b(\omega_\beta)$. Thus, C3 is proved. □

By employing the **minDist** function, we built a general method to introduce the *OP* criterion into the existing mappings which are *non-OP*. The general method is stated in Algorithm 1. It is worthy to note that Algorithm 1 actually does some sorting of the parameters of $P_{s,1}, \cdots, P_{s,m_P}$ in Equation (25), and this plays an important role in constructing the *OP* mappings from the existing *non-OP* mappings.

Theorem 1. *The set of mapping functions $\left\{(g^{\text{MOP-X}})_s(\omega_s^{\text{JS}}), s = 0, 1, \cdots, r-1\right\}$ obtained through Algorithm 1 is OP.*

Proof. Let $\omega_m^{\text{JS}}, \omega_n^{\text{JS}} \in [0,1]$ and $\forall m, n \in \{0, 1, \cdots, r-1\}$. According to Algorithm 1 and without loss of generality, we can assume that $\omega_m^{\text{JS}} \in \Omega_{k_m^*}, \omega_n^{\text{JS}} \in \Omega_{k_n^*}$, and then we get

$$\begin{cases} (g^{\text{MOP-X}})_m(\omega_m^{\text{JS}}) = g^X\left(\omega_m^{\text{JS}}; m_P, P_{k_m^*,1}, \cdots, P_{k_m^*,m_P}\right), \\ (g^{\text{MOP-X}})_n(\omega_n^{\text{JS}}) = g^X\left(\omega_n^{\text{JS}}; m_P, P_{k_n^*,1}, \cdots, P_{k_n^*,m_P}\right). \end{cases}$$

It is easy to verify that

$$\begin{cases} g^X\left(\omega_m^{\text{JS}}; m_P, P_{k_m^*,1}, \cdots, P_{k_m^*,m_P}\right) = \widetilde{(g^X)}_{k_m^*}(\omega_m^{\text{JS}}), \\ g^X\left(\omega_n^{\text{JS}}; m_P, P_{k_n^*,1}, \cdots, P_{k_n^*,m_P}\right) = \widetilde{(g^X)}_{k_n^*}(\omega_n^{\text{JS}}). \end{cases}$$

Therefore, according to Lemma 4, we can finish the proof trivially. □

We now define the modified weights which are *OP* as follows:

$$\omega_s^{\text{MOP-X}} = \frac{\alpha_s^{\text{MOP-X}}}{\sum_{l=0}^{r-1} \alpha_l^{\text{MOP-X}}}, \quad \alpha_s^{\text{MOP-X}} = \left(g^{\text{MOP-X}}\right)_s(\omega_s^{\text{JS}}), \quad s = 0, \cdots, r-1, \quad (37)$$

where $\left(g^{\text{MOP-X}}\right)_s(\omega_s^{\text{JS}})$ is obtained from Algorithm 1. The associated scheme will be referred to as MOP-WENO-X.

The mapping functions of the WENO-X schemes presented in Table 1 and those of the associated MOP-WENO-X schemes are shown in Figure 1. We can find that, for the mapping functions of the MOP-WENO-X schemes: (1) the monotonicity over the whole domain $(0,1)$ is maintained; (2) the differentiability is reduced and limited to the neighborhood of the optimal weights d_s; (3) the OP property is obtained. We summarize these properties as follows.

Algorithm 1: A general method to construct OP mappings.

---

input : s, index indicating the substencil S_s and $s = 0, 1, \cdots, r-1$
  d_s, optimal weights
  ω_s^{JS}, nonlinear weights computed by the WENO-JS scheme
  m_P, the number of the parameters related with s
  $P_{s,j}$, parameters related with s and $j = 1, \cdots, m_P$

output: $\left\{\left(g^{\text{MOP-X}}\right)_s(\omega_s^{\text{JS}}), s = 0, 1, \cdots, r-1\right\}$, the new set of mapping functions that is OP

1 $\left(g^X\right)_s(\omega), s = 0, 1, \cdots, r-1$ is a monotonically increasing mapping function over $[0,1]$, and the set of mapping functions $\left\{\left(g^X\right)_s(\omega), s = 0, 1, \cdots, r-1\right\}$ is non-OP;
2 // implementation of the "minDist" function in Definition 4
3 **for** $s = 0; s \leq r-1; s++$ **do**
4   // get k^* in Equation (26)
5   set $d^{\min} = |\omega_s^{\text{JS}} - d_0|, k_s^* = 0$;
6   **for** $i = 1; i \leq r-1; i++$ **do**
7   **if** $|\omega_s^{\text{JS}} - d_i| < d^{\min}$ **then**
8       $d^{\min} = |\omega_s^{\text{JS}} - d_i|$,
9       $k_s^* = i$;
10   **end**
11   **end**
12   // remark: the for loop above indicates that $\omega_s^{\text{JS}} \in \Omega_{k_s^*}$
13   // get x_{k^*} in Equation (26)
14   **for** $j = 1; j \leq m_P; j++$ **do**
15     $\overline{P}_{s,j} = P_{k_s^*,j}$;
16   **end**
17 **end**
18 // get $\left(g^{\text{MOP-X}}\right)_s(\omega_s^{\text{JS}})$
19 **for** $s = 0; s \leq r-1; s++$ **do**
20   $\left(g^{\text{MOP-X}}\right)_s(\omega_s^{\text{JS}}) = g^X\left(\omega_s^{\text{JS}}; m_P, \overline{P}_{s,1}, \cdots, \overline{P}_{s,m_P}\right)$.
21 **end**

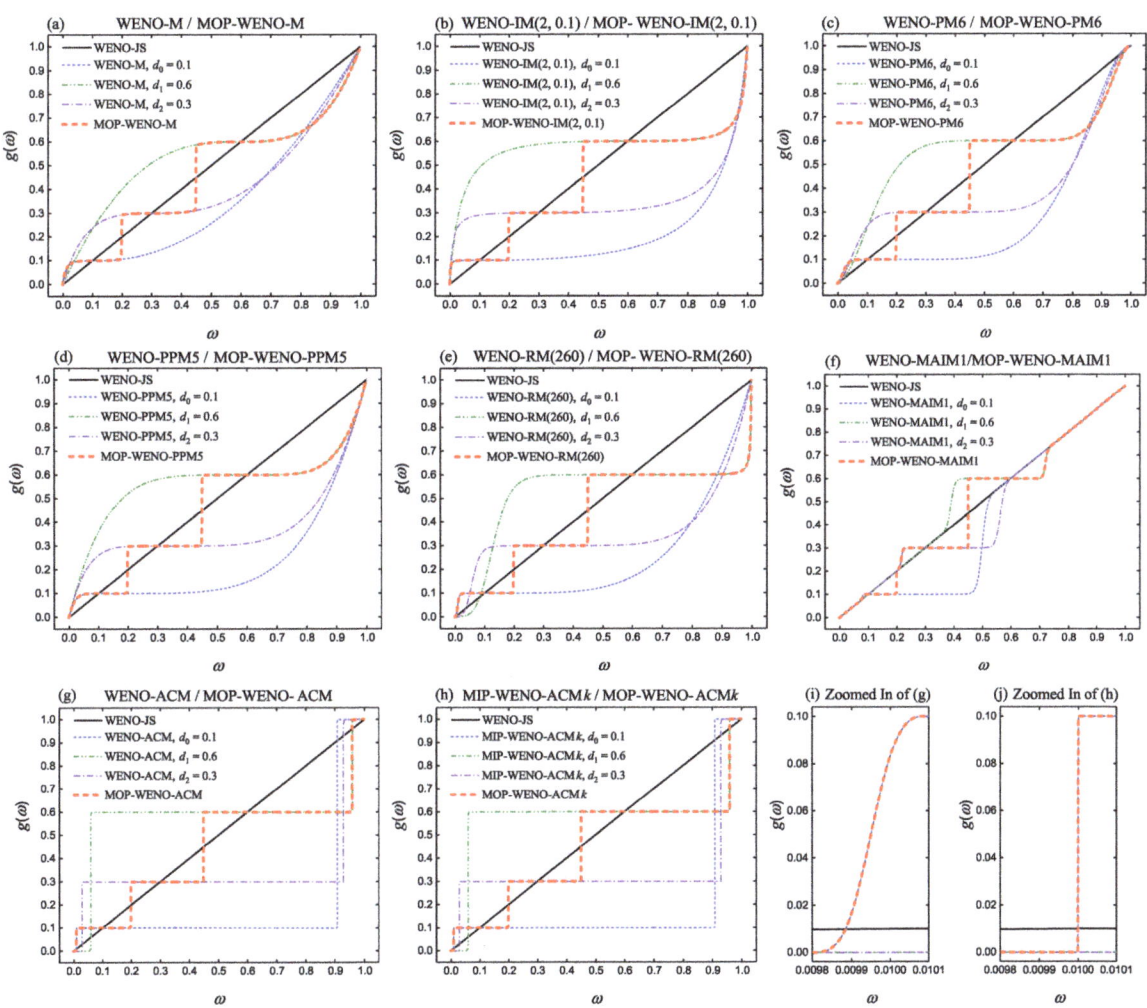

Figure 1. A comparison of the mapping functions for WENO-X (shown in Table 1) and MOP-WENO-X.

Theorem 2. *Let $\overline{\Omega}_i = \{\omega \in \Omega_i \cap \omega \neq \partial\Omega_i\}, i = 0, 1, \cdots, r-1$. The mapping function $\left(g^{\mathrm{MOP-X}}\right)_s(\omega)$ obtained from Althorithm 1 satisfies the following properties:*

C1. For $\forall \omega \in \overline{\Omega}_i, i = 0, 1, \cdots, r-1, \left(g^{\mathrm{MOP-X}}\right)_s'(\omega) \geq 0$;

C2. For $\forall \omega \in \Omega, 0 \leq \left(g^{\mathrm{MOP-X}}\right)_s(\omega) \leq 1$, and $\left(g^{\mathrm{MOP-X}}\right)_s(0) = 0, \left(g^{\mathrm{MOP-X}}\right)_s(1) = 1$;

C3. For $\forall s \in \{0, 1, \cdots, r-1\}, \tilde{d}_s \in \Omega_s$, and $\left(g^{\mathrm{MOP-X}}\right)_s(\tilde{d}_s) = \tilde{d}_s, \left(g^{\mathrm{MOP-X}}\right)_s'(\tilde{d}_s) = \cdots = \left(g^{\mathrm{MOP-X}}\right)_s^{(n_X)}(\tilde{d}_s) = 0$ where n_X is given in Table 1;

C4. $\left(g^{\mathrm{MOP-X}}\right)_s'(0) = \left(g^X\right)_s'(0), \left(g^{\mathrm{MOP-X}}\right)_s'(1) = \left(g^X\right)_s'(1)$;

C5. For $\forall m, n \in \{0, \cdots, r-1\}$, if $\omega_m > \omega_n$, then $\left(g^{\mathrm{MOP-X}}\right)_m(\omega_m) \geq \left(g^{\mathrm{MOP-X}}\right)_n(\omega_n)$, and if $\omega_m = \omega_n$, then $\left(g^{\mathrm{MOP-X}}\right)_m(\omega_m) = \left(g^{\mathrm{MOP-X}}\right)_n(\omega_n)$.

Remark 1. *(1) The properties C1–C3 are designed to recover the optimal convergence rate of accuracy in a smooth region even in the presence of critical points, and the detailed theoretical analysis has been proposed in Section 5 of [25], Section 3.2 of [27], Section 3.1 of [29], etc. (2) The property C4 is designed to decrease the effect from non-smooth stencils, and we refer to Sections 3.1* and 3.2 of [29], Remark 1 of [28], Section 2.2 of [30], and Section 3.3 of [31] for more details. (3) The property C5 is designed to enhance the performance for long-output-time simulations and to remove or reduce post-shock numerical oscillations, and we have analyzed this in [9] systematically and carefully.

3.4. Convergence Properties

According to Theorem 2, we get the convergence properties for the $(2r-1)$th-order MOP-WENO-X schemes as given in Theorem 3. The proof is almost identical to that of the associated WENO-X schemes in the references presented in Table 1.

Theorem 3. *The requirements for the $(2r-1)$th-order MOP-WENO-X schemes to achieve the optimal order of accuracy are identical to that of the associated $(2r-1)$th-order WENO-X schemes.*

For the integrity of this paper and the benefit of the reader, we concisely express the following Corollaries of Theorem 3.

Corollary 1. *If n mapping is used in the $(2r-1)$th-order MOP-WENO-M scheme, then for different values of n_{cp}, the weights $\omega_s^{\text{MOP-M}}$ in the $(2r-1)$th-order MOP-WENO-M scheme satisfy*

$$\omega_s^{\text{MOP-M}} - d_s = O\Big((\Delta x)^{3^n \times (r-1-n_{cp})}\Big), \quad r = 2,3,\cdots,9, \quad n_{cp} = 0,1,\cdots,r-1,$$

and the rate of convergence is

$$r_c = \begin{cases} 2r-1, & \text{if } n_{cp} = 0,\cdots,\left\lfloor \dfrac{3^n-1}{3^n}r-1 \right\rfloor, \\ (3^n+1)(r-1) - 3^n \times n_{cp}, & \text{if } n_{cp} = \left\lfloor \dfrac{3^n-1}{3^n}r-1 \right\rfloor + 1,\cdots,r-1, \end{cases}$$

where $\lfloor x \rfloor$ is a floor function of x.

Proof. The proof is almost identical to that of Lemma 6 in [31]. □

Corollary 2. *When $n_{cp} = 1$, the $(2r-1)$th-order MOP-WENO-IM(k, A) schemes can achieve the optimal order of accuracy if the mapping function $\big(g^{\text{MOP-IM}}\big)_s(\omega)$ is applied to the original weights in the $(2r-1)$th-order WENO-JS schemes with the requirement of $k \geq 2$ (except for the case of $r = 2$).*

Proof. The proof is almost identical to that of Theorem 2 in [27]. □

Corollary 3. *The $(2r-1)$th-order MOP-WENO-PMk schemes can achieve the optimal order of accuracy if the mapping function $\big(g^{\text{MOP-PM}}\big)_s(\omega)$ is applied to the original weights in the $(2r-1)$th-order WENO-JS schemes with specific requirements for k in following different cases: (I) require $k \geq 1$ for $n_{cp} = 0$; (II) require $k \geq 1$ for $n_{cp} = 1$; (III) require $k \geq 3$ for $n_{cp} = 2$.*

Proof. The proof is almost identical to that of Proposition 1 in [29]. □

Corollary 4. *The $(2r-1)$th-order MOP-WENO-RM(mn0) schemes can recover the optimal order of accuracy if the mapping function $\big(g^{\text{MOP-RM}}\big)_s(\omega)$ is applied to the original weights in the $(2r-1)$th-order WENO-JS schemes with requirement of $n \geq \frac{1+n_{cp}}{r-1-n_{cp}}$ for different values of n_{cp} with $1 \leq n_{cp} < r-1$.*

Proof. The proof is almost identical to that of Theorem 3 in [28]. □

Corollary 5. Let $\lceil x \rceil$ be a ceiling function of x. For $n_{cp} < r - 1$, the $(2r - 1)$th-order MOP-WENO-MAIM1 schemes can achieve the optimal order of accuracy if the mapping function $(g^{MOP-MAIM1})_s(\omega)$ is applied to the original weights in the $(2r - 1)$th-order WENO-JS schemes with requirement of $k \geq k^{MAIM}$, where

$$k^{MAIM} = \left\lceil \frac{r}{r - 1 - n_{cp}} - 2 \right\rceil + \frac{1 + (-1)^{\left\lceil \frac{r}{r-1-n_{cp}} - 2 \right\rceil}}{2}.$$

Proof. The proof is almost identical to that of Theorem 2 in [31]. □

Corollary 6. For $n_{cp} < r - 1$, the $(2r - 1)$th-order MOP-WENO-ACM schemes can achieve the optimal order of accuracy if the mapping function $(g^{MOP-ACM})_s(\omega)$ is applied to the original weights in the $(2r - 1)$th-order WENO-JS schemes.

Proof. The proof is almost identical to that of Theorem 2 in [32]. □

Corollary 7. When $CFS_s \ll \tilde{d}_0$, for $n_{cp} < r - 1$, the $(2r - 1)$th-order MOP-WENO-ACMk schemes can achieve the optimal order of accuracy if the mapping function $(g^{MOP-ACMk})_s(\omega)$ is applied to the original weights in the $(2r - 1)$th-order WENO-JS schemes.

Proof. The proof is almost identical to that of Theorem 2 in [9]. □

4. Numerical Results

In this section, we compare the numerical performances of the MOP-WENO-X schemes with the associated existing mapped WENO-X schemes shown in Table 1, and the classic WENO-JS scheme. To further demonstrate the superiority of the MOP-WENO-X schemes, some comparisons with other WENO type reconstructions, e.g., WENO-Z [26] (in Sections 4.1 and 4.2) and the central WENO schemes of WENO-NW6 [36], WENO-CU6 [37], and WENO-θ6 [38] (in Section 4.3), have also been performed. As the performances of the WENO-ACM scheme and the MOP-WENO-ACM scheme are almost identical to those of the MIP-WENO-ACMk scheme and the MOP-WENO-ACMk scheme, respectively, we do not present the solutions of the WENO-ACM scheme and the MOP-WENO-ACM scheme below for simplicity. It should be noted that although we mainly provide the solutions of the fifth-order WENO methods (WENO5) in present study, the methodology proposed in this paper can be successfully extended to higher order WENO methods, such as WENO-7 or WENO-9, and because of the space limitations, we do not show their solutions here.

Typical one-dimensional linear advection equation and two-dimensional Euler equations, with different initial conditions, are used to test the considered schemes. The presentation of these numerical tests in this section starts with the accuracy test of one-dimensional linear advection equation with four different initial conditions, followed by the long-output-time simulations of it with two different initial conditions, including discontinuities, and finishes with two-dimensional simulations on the shock-vortex interaction and the 2D Riemann problem. In all calculations below, ϵ is taken to be 10^{-40} for all schemes following the recommendations in [25,27].

In the following numerical tests, the ODEs resulting from the semi-discretized PDEs are marched in time using the following explicit, third-order, strong stability preserving (SSP) Runge–Kutta method [5,39,40]:

$$\vec{U}^* = \vec{U}^n + \Delta t \mathcal{L}(\vec{U}^n),$$
$$\vec{U}^{**} = \frac{3}{4}\vec{U}^n + \frac{1}{4}\vec{U}^* + \frac{1}{4}\Delta t \mathcal{L}(\vec{U}^*),$$
$$\vec{U}^{n+1} = \frac{1}{3}\vec{U}^n + \frac{2}{3}\vec{U}^{**} + \frac{2}{3}\Delta t \mathcal{L}(\vec{U}^{**}),$$

where $\vec{U}^*, \vec{U}^{**}$ are the intermediate stages, $\vec{U}^n$ is the value of $\vec{U}$ at time level $t^n = n\Delta t$, and Δt is the time step satisfying some proper CFL condition. The spatial operator $\mathcal{L}$ is defined as in Equation (3), and the WENO reconstructions will be applied to obtain it.

4.1. Accuracy Test

In this subsection, we solve the following one-dimensional linear advection equation:

$$\frac{\partial u}{\partial t} + \frac{\partial u}{\partial x} = 0, \quad -1 \leq x \leq 1, \tag{38}$$

with different initial conditions to test the accuracy of the considered WENO schemes. In all accuracy tests, the L_1, L_2, L_∞ norms of the error are given as

$$L_1 = h \cdot \sum_j \left| u_j^{\text{exact}} - (u_h)_j \right|,$$

$$L_2 = \sqrt{h \cdot \sum_j (u_j^{\text{exact}} - (u_h)_j)^2},$$

$$L_\infty = \max_j \left| u_j^{\text{exact}} - (u_h)_j \right|,$$

where $h = \Delta x$ is the uniform spatial step size, $(u_h)_j$ is the numerical solution, and u_j^{exact} is the exact solution.

Example 1. *We calculate Equation (38) with the periodic boundary condition using the following initial condition [27]:*

$$u(x, 0) = \sin(\pi x). \tag{39}$$

It is trivial to verify that although the initial condition in Equation (39) has two first-order critical points, their first and third derivatives vanish simultaneously. It is known that the rate of the temporal convergence is $O(\Delta t^3)$ for the third-order Runge–Kutta method [5,39,40] and the CFL number is defined by CFL $= \frac{|\alpha|\Delta t}{\Delta x}$ leading to $\Delta t = \text{CFL} \cdot \frac{\Delta x}{|\alpha|}$ where $|\alpha| = 1$ here. Therefore, note that we consider only the fifth-order methods here, and to ensure that the error for the overall scheme is a measure of the spatial convergence only, we set the CFL number to be $(\Delta x)^{2/3}$. The calculation was run until a time of $t = 2.0$.

In Table 2, we show the L_1, L_2, L_∞ errors and corresponding convergence orders of various considered WENO schemes. Unsurprisingly, the MOP-WENO-X schemes and the associated WENO-X schemes, along with the WENO-Z scheme, provide more accurate results than the WENO-JS scheme do in general. Naturally and as expected, all the considered schemes have gained the fifth-order convergence rate of accuracy. It can be found that the results of the MOP-WENO-X schemes are identical to those of the associated WENO-X schemes for all grid numbers except $N = 10$. As discussed in [9], the cause of the accuracy loss for the computing cases of all MOP-WENO-X schemes with $N = 10$ is that the mapping functions of the MOP-WENO-X schemes have narrower optimal weight intervals (standing for the intervals about $\omega = d_s$ over which the mapping process attempts to use the corresponding optimal weights; see [31,32]) than the associated WENO-X schemes.

Figure 2 shows the overall L_∞ convergence behavior of various considered schemes. We can observe that: (1) the solutions of all schemes converge at fifth-order, as evidenced by the slope of the lines; (2) the MOP-WENO-X schemes and their associated WENO-X schemes, along with the WENO-Z scheme, are significantly more accurate than the classic WENO-JS scheme; (3) the errors and convergence orders of the MOP-WENO-X schemes are almost identical to those of their associated WENO-X schemes.

We use this example to discuss the computational cost of the MOP-WENO-X scheme compared with its associated WENO-X scheme and the classic WENO-JS scheme. In Figure 3, we drew the graphs for the CPU time versus the computing errors (we only present the results of the L_∞-norm error here just for the sake of brevity in the presentation, hereinafter the same). From Figure 3, we can easily see that: (1) generally speaking, the MOP-WENO-X schemes have better efficiency than the WENO-JS scheme; (2) for all MOP-WENO-X schemes except the case of "X = M," they perform almost identically to their associated WENO-X schemes; (3) for the MOP-WENO-M scheme, it has a slightly lower efficiency than its associated WENO-M scheme and it has significantly higher efficiency than the WENO-JS scheme.

Table 2. Convergence properties of considered schemes on solving $u_t + u_x = 0$ with initial condition $u(x,0) = \sin(\pi x)$. To be continued.

Scheme	N	L_1 Error	L_1 Order	L_2 Error	L_2 Order	L_∞ Error	L_∞ Order
WENO-JS	10	6.18328×10^{-2}	-	4.72306×10^{-2}	-	4.87580×10^{-2}	-
	20	2.96529×10^{-3}	4.3821	2.42673×10^{-3}	4.2826	2.57899×10^{-3}	4.2408
	40	9.27609×10^{-5}	4.9985	7.64332×10^{-5}	4.9887	9.05453×10^{-5}	4.8320
	80	2.89265×10^{-6}	5.0031	2.33581×10^{-6}	5.0322	2.90709×10^{-6}	4.9610
	160	9.03392×10^{-8}	5.0009	7.19259×10^{-8}	5.0213	8.85753×10^{-8}	5.0365
	320	2.82330×10^{-9}	4.9999	2.23105×10^{-9}	5.0107	2.72458×10^{-9}	5.0228
WENO-Z	10	1.64485×10^{-2}	-	1.27535×10^{-2}	-	1.18974×10^{-2}	-
	20	5.04450×10^{-4}	5.0271	3.98253×10^{-4}	5.0011	3.94040×10^{-4}	4.9162
	40	1.59132×10^{-5}	4.9864	1.25050×10^{-5}	4.9931	1.24948×10^{-5}	4.9789
	80	4.98858×10^{-7}	4.9955	3.91834×10^{-7}	4.9961	3.91804×10^{-7}	4.9951
	160	1.56020×10^{-8}	4.9988	1.22541×10^{-8}	4.9989	1.22538×10^{-8}	4.9988
	320	4.88356×10^{-10}	4.9977	3.83568×10^{-10}	4.9976	3.83541×10^{-10}	4.9977
WENO-M	10	2.01781×10^{-2}	-	1.55809×10^{-2}	-	1.47767×10^{-2}	-
	20	5.18291×10^{-4}	5.2829	4.06148×10^{-4}	5.2616	3.94913×10^{-4}	5.2256
	40	1.59422×10^{-5}	5.0228	1.25236×10^{-5}	5.0193	1.24993×10^{-5}	4.9816
	80	4.98914×10^{-7}	4.9979	3.91875×10^{-7}	4.9981	3.91808×10^{-7}	4.9956
	160	1.56021×10^{-8}	4.9990	1.22541×10^{-8}	4.9991	1.22538×10^{-8}	4.9988
	320	4.88356×10^{-10}	4.9977	3.83568×10^{-10}	4.9976	3.83541×10^{-10}	4.9977
MOP-WENO-M	10	3.64427×10^{-2}	-	2.95270×10^{-2}	-	2.81876×10^{-2}	-
	20	5.18291×10^{-4}	6.1357	4.06148×10^{-4}	6.1839	3.94913×10^{-4}	6.1574
	40	1.59422×10^{-5}	5.0228	1.25236×10^{-5}	5.0193	1.24993×10^{-5}	4.9816
	80	4.98914×10^{-7}	4.9979	3.91875×10^{-7}	4.9981	3.91808×10^{-7}	4.9956
	160	1.56021×10^{-8}	4.9990	1.22541×10^{-8}	4.9991	1.22538×10^{-8}	4.9988
	320	4.88356×10^{-10}	4.9977	3.83568×10^{-10}	4.9976	3.83541×10^{-10}	4.9977
WENO-IM(2, 0.1)	10	1.58051×10^{-2}	-	1.23553×10^{-2}	-	1.19178×10^{-2}	-
	20	5.04401×10^{-4}	4.9697	3.96236×10^{-4}	4.9626	3.94458×10^{-4}	4.9171
	40	1.59160×10^{-5}	4.9860	1.25033×10^{-5}	4.9860	1.24963×10^{-5}	4.9803
	80	4.98863×10^{-7}	4.9957	3.91836×10^{-7}	4.9959	3.91797×10^{-7}	4.9953
	160	1.56020×10^{-8}	4.9988	1.22541×10^{-8}	4.9989	1.22538×10^{-8}	4.9988
	320	4.88355×10^{-10}	4.9977	3.83568×10^{-10}	4.9976	3.83547×10^{-10}	4.9977
MOP-WENO-IM(2, 0.1)	10	3.35513×10^{-2}	-	2.75968×10^{-2}	-	2.71898×10^{-2}	-
	20	5.04401×10^{-4}	6.0557	3.96236×10^{-4}	6.1220	3.94458×10^{-4}	6.1071

Table 2. *Cont.*

Scheme	N	L_1 Error	L_1 Order	L_2 Error	L_2 Order	L_∞ Error	L_∞ Order
	40	1.59160×10^{-5}	4.9860	1.25033×10^{-5}	4.9860	1.24963×10^{-5}	4.9803
	80	4.98863×10^{-7}	4.9957	3.91836×10^{-7}	4.9959	3.91797×10^{-7}	4.9953
	160	1.56020×10^{-8}	4.9988	1.22541×10^{-8}	4.9989	1.22538×10^{-8}	4.9988
	320	4.88355×10^{-10}	4.9977	3.83568×10^{-10}	4.9976	3.83547×10^{-10}	4.9977
WENO-PM6	10	1.74869×10^{-2}	-	1.35606×10^{-2}	-	1.27577×10^{-2}	-
	20	5.02923×10^{-4}	5.1198	3.95215×10^{-4}	5.1006	3.94515×10^{-4}	5.0151
	40	1.59130×10^{-5}	4.9821	1.25010×10^{-5}	4.9825	1.24960×10^{-5}	4.9805
	80	4.98858×10^{-7}	4.9954	3.91831×10^{-7}	4.9957	3.91795×10^{-7}	4.9952
	160	1.56020×10^{-8}	4.9988	1.22541×10^{-8}	4.9989	1.22538×10^{-8}	4.9988
	320	4.88355×10^{-10}	4.9977	3.83568×10^{-10}	4.9976	3.83543×10^{-10}	4.9977
MOP-WENO-PM6	10	3.54584×10^{-2}	-	2.88246×10^{-2}	-	2.76902×10^{-2}	-
	20	5.02923×10^{-4}	6.1396	3.95215×10^{-4}	6.1885	3.94515×10^{-4}	6.1332
	40	1.59130×10^{-5}	4.9821	1.25010×10^{-5}	4.9825	1.24960×10^{-5}	4.9805
	80	4.98858×10^{-7}	4.9954	3.91831×10^{-7}	4.9957	3.91795×10^{-7}	4.9952
	160	1.56020×10^{-8}	4.9988	1.22541×10^{-8}	4.9989	1.22538×10^{-8}	4.9988
	320	4.88355×10^{-10}	4.9977	3.83568×10^{-10}	4.9976	3.83543×10^{-10}	4.9977
WENO-PPM5	10	1.73978×10^{-2}	-	1.34998×10^{-2}	-	1.27018×10^{-2}	-
	20	5.03464×10^{-4}	5.1109	3.95644×10^{-4}	5.0926	3.94865×10^{-4}	5.0075
	40	1.59131×10^{-5}	4.9836	1.25011×10^{-5}	4.9841	1.24961×10^{-5}	4.9818
	80	4.98858×10^{-7}	4.9954	3.91831×10^{-7}	4.9957	3.91795×10^{-7}	4.9952
	160	1.56020×10^{-8}	4.9988	1.22541×10^{-8}	4.9989	1.22538×10^{-8}	4.9988
	320	4.88356×10^{-10}	4.9977	3.83568×10^{-10}	4.9976	3.83528×10^{-10}	4.9978
MOP-WENO-PPM5	10	3.49872×10^{-2}	-	2.85173×10^{-2}	-	2.75955×10^{-2}	-
	20	5.03464×10^{-4}	6.1188	3.95644×10^{-4}	6.1715	3.94865×10^{-4}	6.1269
	40	1.59131×10^{-5}	4.9836	1.25011×10^{-5}	4.9841	1.24961×10^{-5}	4.9818
	80	4.98858×10^{-7}	4.9954	3.91831×10^{-7}	4.9957	3.91795×10^{-7}	4.9952
	160	1.56020×10^{-8}	4.9988	1.22541×10^{-8}	4.9989	1.22538×10^{-8}	4.9988
	320	4.88356×10^{-10}	4.9977	3.83568×10^{-10}	4.9976	3.83528×10^{-10}	4.9978
WENO-RM(260)	10	1.52661×10^{-2}	-	1.19792×10^{-2}	-	1.17698×10^{-2}	-
	20	5.02845×10^{-4}	4.9241	3.95138×10^{-4}	4.9220	3.94406×10^{-4}	4.8993
	40	1.59130×10^{-5}	4.9818	1.25010×10^{-5}	4.9822	1.24960×10^{-5}	4.9801
	80	4.98858×10^{-7}	4.9954	3.91831×10^{-7}	4.9957	3.91795×10^{-7}	4.9952
	160	1.56020×10^{-8}	4.9988	1.22541×10^{-8}	4.9989	1.22538×10^{-8}	4.9988
	320	4.88355×10^{-10}	4.9977	3.83568×10^{-10}	4.9976	3.83543×10^{-10}	4.9977
MOP-WENO-RM(260)	10	3.29243×10^{-2}	-	2.73131×10^{-2}	-	2.73015×10^{-2}	-
	20	5.02845×10^{-4}	6.0329	3.95138×10^{-4}	6.1111	3.94406×10^{-4}	6.1132
	40	1.59130×10^{-5}	4.9818	1.25010×10^{-5}	4.9822	1.24960×10^{-5}	4.9801
	80	4.98858×10^{-7}	4.9954	3.91831×10^{-7}	4.9957	3.91795×10^{-7}	4.9952
	160	1.56020×10^{-8}	4.9988	1.22541×10^{-8}	4.9989	1.22538×10^{-8}	4.9988
	320	4.88355×10^{-10}	4.9977	3.83568×10^{-10}	4.9976	3.83543×10^{-10}	4.9977
WENO-MAIM1	10	6.13264×10^{-2}	-	4.81375×10^{-2}	-	4.86913×10^{-2}	-
	20	5.08205×10^{-4}	6.9150	4.26155×10^{-4}	6.8196	5.03701×10^{-4}	6.5950
	40	1.59130×10^{-5}	4.9971	1.25010×10^{-5}	5.0913	1.24960×10^{-5}	5.3330
	80	4.98858×10^{-7}	4.9954	3.91831×10^{-7}	4.9957	3.91795×10^{-7}	4.9952
	160	1.56020×10^{-8}	4.9988	1.22541×10^{-8}	4.9989	1.22538×10^{-8}	4.9988
	320	4.88355×10^{-10}	4.9977	3.83568×10^{-10}	4.9976	3.83543×10^{-10}	4.9977
MOP-WENO-MAIM1	10	6.63923×10^{-2}	-	5.17462×10^{-2}	-	5.19799×10^{-2}	-
	20	5.08205×10^{-4}	7.0295	4.26155×10^{-4}	6.9239	5.03701×10^{-4}	6.6892
	40	1.59130×10^{-5}	4.9971	1.25010×10^{-5}	5.0913	1.24960×10^{-5}	5.3330
	80	4.98858×10^{-7}	4.9954	3.91831×10^{-7}	4.9957	3.91795×10^{-7}	4.9952
	160	1.56020×10^{-8}	4.9988	1.22541×10^{-8}	4.9989	1.22538×10^{-8}	4.9988
	320	4.88355×10^{-10}	4.9977	3.83568×10^{-10}	4.9976	3.83543×10^{-10}	4.9977

Table 2. *Cont.*

Scheme	N	L_1 Error	L_1 Order	L_2 Error	L_2 Order	L_∞ Error	L_∞ Order
MIP-WENO-ACMk	10	1.52184×10^{-2}	-	1.19442×10^{-2}	-	1.17569×10^{-2}	-
	20	5.02844×10^{-4}	4.9196	3.95138×10^{-4}	4.9178	3.94406×10^{-4}	4.8977
	40	1.59130×10^{-5}	4.9818	1.25010×10^{-5}	4.9822	1.24960×10^{-5}	4.9801
	80	4.98858×10^{-7}	4.9954	3.91831×10^{-7}	4.9957	3.91795×10^{-7}	4.9952
	160	1.56020×10^{-8}	4.9988	1.22541×10^{-8}	4.9989	1.22538×10^{-8}	4.9988
	320	4.88355×10^{-10}	4.9977	3.83568×10^{-10}	4.9976	3.83543×10^{-10}	4.9977
MOP-WENO-ACMk	10	3.29609×10^{-2}	-	2.72363×10^{-2}	-	2.70295×10^{-2}	-
	20	5.02844×10^{-4}	6.0345	3.95138×10^{-4}	6.1070	3.94406×10^{-4}	6.0987
	40	1.59130×10^{-5}	4.9818	1.25010×10^{-5}	4.9822	1.24960×10^{-5}	4.9801
	80	4.98858×10^{-7}	4.9954	3.91831×10^{-7}	4.9957	3.91795×10^{-7}	4.9952
	160	1.56020×10^{-8}	4.9988	1.22541×10^{-8}	4.9989	1.22538×10^{-8}	4.9988
	320	4.88355×10^{-10}	4.9977	3.83568×10^{-10}	4.9976	3.83543×10^{-10}	4.9977

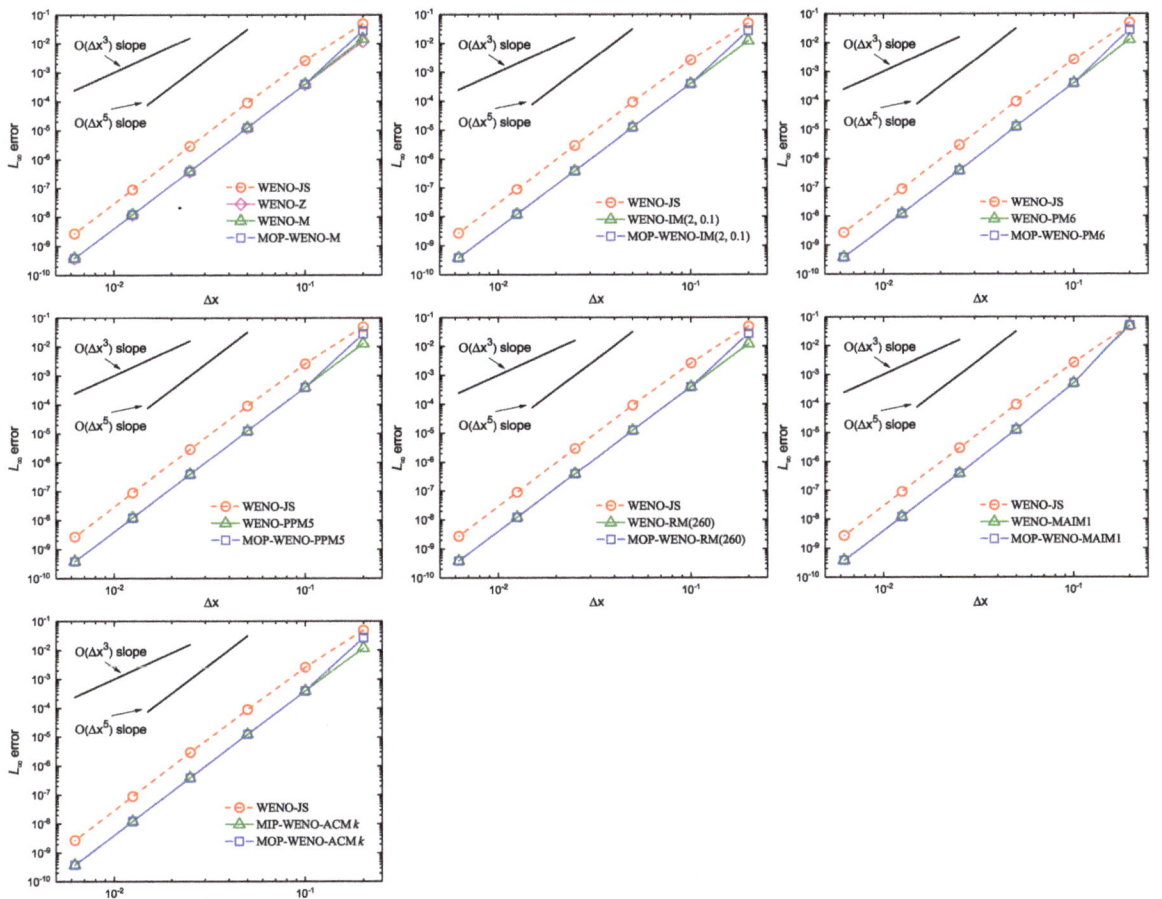

Figure 2. L_∞-norm error plots for various WENO schemes for Example 1.

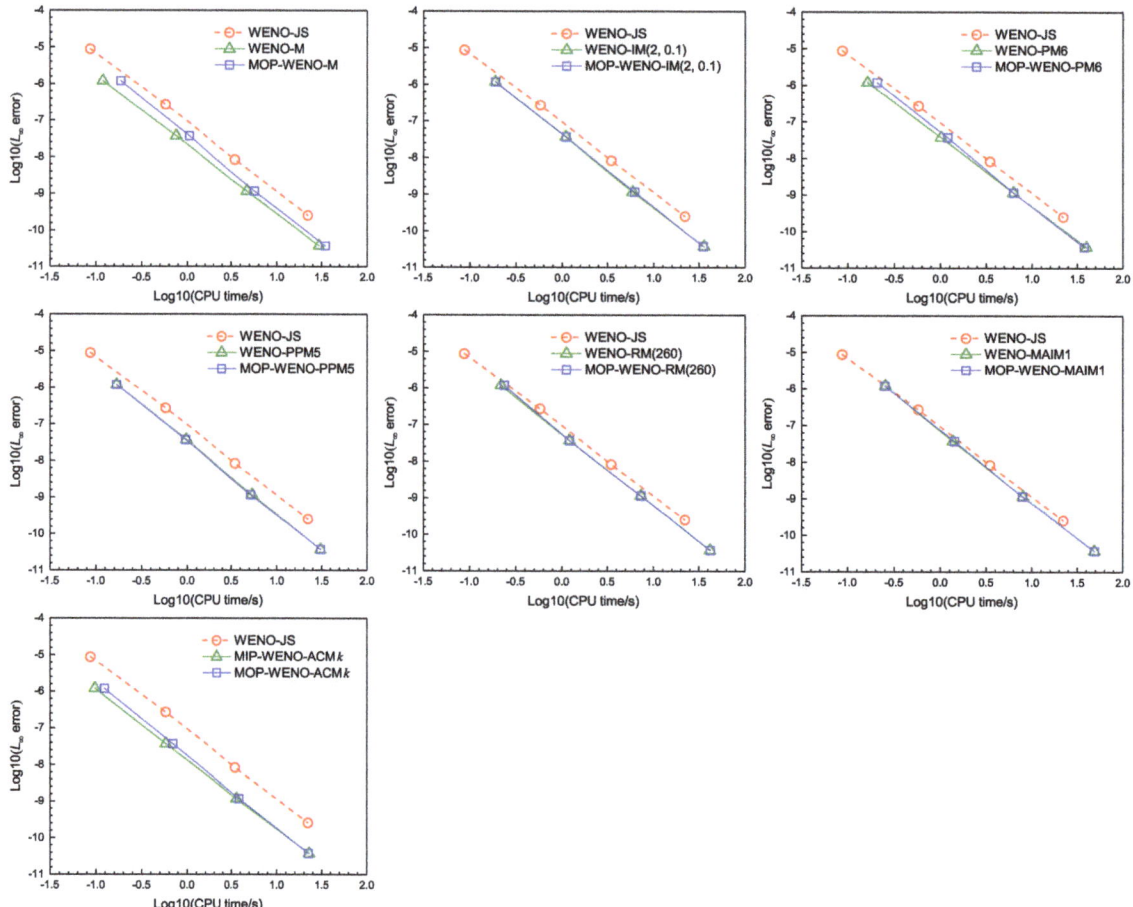

Figure 3. Comparison of various WENO schemes for Example 1 in CPU time and L_∞-norm computing errors.

Example 2. *We calculate Equation (38) with the periodic boundary condition using the following initial condition [25]:*

$$u(x,0) = \sin\left(\pi x - \frac{\sin(\pi x)}{\pi}\right). \tag{40}$$

This particular initial condition has two first-order critical points, which both have a non-vanishing third derivative. Again, the CFL number was set to be $(\Delta x)^{2/3}$ and the calculation was run until a time of $t = 2.0$.

Table 3 compares the L_1, L_2, L_∞ errors and corresponding convergence orders obtained from the considered schemes. It is evident that the WENO-X schemes and the associated MOP-WENO-X schemes can achieve the optimal convergence orders, and this verifies the properties $C1 \sim C3$ of Theorem 2. Unsurprisingly, the WENO-JS scheme gives less accurate results than the other schemes, and its L_∞ convergence order decreases by almost 2 orders leading to the noticeable drops of the L_1 and L_2 convergence orders. It is noteworthy that when the grid number is too small, such as $N \leq 40$, in terms of accuracy, the MOP-WENO-X schemes provide less accurate results than those of the associated WENO-X schemes. As mentioned in Example 1, the cause of this kind of accuracy loss is that the mapping functions of the MOP-WENO-X schemes have narrower optimal weight intervals than the associated WENO-X schemes, and this issue can surely be addressed by increasing the grid number. Therefore, as expected, the MOP-WENO-X schemes show equally accurate numerical solutions like those of the associated WENO-X schemes when the grid number $N \geq 80$.

Figure 4 shows the overall L_∞ convergence behavior of various considered schemes. We can observe that: (1) the solutions of all MOP-WENO-X schemes and their associated WENO-X schemes, and of the WENO-Z scheme, converge at fifth-order, as evidenced by the slope of the lines, especially for larger (slightly) grid numbers; (2) for the classic WENO-JS scheme, its solution converges at third-order, as evidenced by its slope of the line; (3) naturally, the MOP-WENO-X schemes and their associated WENO-X schemes, and the WENO-Z scheme, are significantly more accurate than the classic WENO-JS scheme; (4) the errors and convergence orders of the MOP-WENO-X schemes are very close to those of their associated WENO-X schemes.

We also use this example to discuss the computational cost of the MOP-WENO-X scheme compared with its associated WENO-X scheme and the classic WENO-JS scheme. In Figure 5, we drew the graphs for the CPU time versus the L_∞-norm computing errors. From Figure 5, we can easily see that: (1) as expected, the WENO-JS scheme has the lowest efficiency; (2) again, for all MOP-WENO-X schemes except the case of "X = M," they perform almost identically to their associated WENO-X schemes; (3) for the MOP-WENO-M scheme, despite the fact that it has slightly less efficiency than its associated WENO-M scheme, it has significantly superior efficiency to the WENO-JS scheme.

Table 3. Convergence properties of considered schemes on solving $u_t + u_x = 0$ with initial condition $u(x,0) = \sin(\pi x - \sin(\pi x)/\pi)$. To be continued.

Scheme	N	L_1 Error	L_1 Order	L_2 Error	L_2 Order	L_∞ Error	L_∞ Order
WENO-JS	10	1.24488×10^{-1}	-	1.09463×10^{-1}	-	1.24471×10^{-1}	-
	20	1.01260×10^{-2}	3.6199	8.72198×10^{-3}	3.6496	1.43499×10^{-2}	3.1167
	40	7.22169×10^{-4}	3.8096	6.76133×10^{-4}	3.6893	1.09663×10^{-3}	3.7099
	80	3.42286×10^{-5}	4.3991	3.63761×10^{-5}	4.2162	9.02485×10^{-5}	3.6030
	160	1.58510×10^{-6}	4.4326	2.29598×10^{-6}	3.9858	8.24022×10^{-6}	3.4531
	320	7.95517×10^{-8}	4.3165	1.68304×10^{-7}	3.7700	8.31702×10^{-7}	3.3085
WENO-Z	10	5.85966×10^{-2}	-	4.83441×10^{-2}	-	5.14928×10^{-2}	-
	20	3.21455×10^{-3}	4.1881	2.72340×10^{-3}	4.1499	3.67979×10^{-3}	3.8067
	40	1.35382×10^{-4}	4.5695	1.35344×10^{-4}	4.3307	2.31013×10^{-4}	3.9936
	80	4.67008×10^{-6}	4.8574	4.50404×10^{-6}	4.9093	6.79475×10^{-6}	5.0874
	160	1.50985×10^{-7}	4.9510	1.42363×10^{-7}	4.9836	2.14556×10^{-7}	4.9850
	320	4.76201×10^{-9}	4.9867	4.45798×10^{-9}	4.9970	6.71078×10^{-9}	4.9987
WENO-M	10	7.53259×10^{-2}	-	6.39017×10^{-2}	-	7.49250×10^{-2}	-
	20	3.70838×10^{-3}	4.3443	3.36224×10^{-3}	4.2484	5.43666×10^{-3}	3.7847
	40	1.45082×10^{-4}	4.6758	1.39007×10^{-4}	4.5962	2.18799×10^{-4}	4.6350
	80	4.80253×10^{-6}	4.9169	4.52646×10^{-6}	4.9406	6.81451×10^{-6}	5.0049
	160	1.52120×10^{-7}	4.9805	1.42463×10^{-7}	4.9897	2.14545×10^{-7}	4.9893
	320	4.77083×10^{-9}	4.9948	4.45822×10^{-9}	4.9980	6.71080×10^{-9}	4.9987
MOP-WENO-M	10	9.41832×10^{-2}	-	8.03446×10^{-2}	-	9.78919×10^{-2}	-
	20	6.59540×10^{-3}	3.8359	6.37937×10^{-3}	3.6547	8.97094×10^{-3}	3.4479
	40	2.60456×10^{-4}	4.6623	2.50868×10^{-4}	4.6684	4.10480×10^{-4}	4.4499
	80	4.80253×10^{-6}	5.7611	4.52646×10^{-6}	5.7924	6.81451×10^{-6}	5.9126
	160	1.52120×10^{-7}	4.9805	1.42463×10^{-7}	4.9897	2.14545×10^{-7}	4.9893
	320	4.77083×10^{-9}	4.9948	4.45822×10^{-9}	4.9980	6.71080×10^{-9}	4.9987

Table 3. *Cont.*

Scheme	N	L_1 Error	L_1 Order	L_2 Error	L_2 Order	L_∞ Error	L_∞ Order
WENO-IM(2, 0.1)	10	8.38131×10^{-2}	-	6.71285×10^{-2}	-	7.62798×10^{-2}	-
	20	4.30725×10^{-3}	4.2823	3.93700×10^{-3}	4.0918	5.84039×10^{-3}	3.7072
	40	1.51327×10^{-4}	4.8310	1.41737×10^{-4}	4.7958	2.10531×10^{-4}	4.7940
	80	4.85592×10^{-6}	4.9618	4.53602×10^{-6}	4.9656	6.82606×10^{-6}	4.9468
	160	1.52659×10^{-7}	4.9914	1.42479×10^{-7}	4.9926	2.14534×10^{-7}	4.9918
	320	4.77654×10^{-9}	4.9982	4.45805×10^{-9}	4.9982	6.71079×10^{-9}	4.9986
MOP-WENO-IM(2, 0.1)	10	8.49795×10^{-2}	-	7.29388×10^{-2}	-	9.47429×10^{-2}	-
	20	7.01287×10^{-3}	3.5990	6.80019×10^{-3}	3.4230	9.96943×10^{-3}	3.2484
	40	2.59767×10^{-4}	4.7547	2.51121×10^{-4}	4.7591	4.01785×10^{-4}	4.6330
	80	4.85592×10^{-6}	5.7413	4.53602×10^{-6}	5.7908	6.82606×10^{-6}	5.8792
	160	1.52659×10^{-7}	4.9914	1.42479×10^{-7}	4.9926	2.14534×10^{-7}	4.9918
	320	4.77654×10^{-9}	4.9982	4.45805×10^{-9}	4.9982	6.71079×10^{-9}	4.9986
WENO-PM6	10	9.51313×10^{-2}	-	7.83600×10^{-2}	-	9.32356×10^{-2}	-
	20	4.82173×10^{-3}	4.3023	4.29510×10^{-3}	4.1894	5.91037×10^{-3}	3.9796
	40	1.55428×10^{-4}	4.9552	1.43841×10^{-4}	4.9001	2.09540×10^{-4}	4.8180
	80	4.87327×10^{-6}	4.9952	4.54036×10^{-6}	4.9855	6.83270×10^{-6}	4.9386
	160	1.52750×10^{-7}	4.9956	1.42488×10^{-7}	4.9939	2.14532×10^{-7}	4.9932
	320	4.77729×10^{-9}	4.9988	4.45807×10^{-9}	4.9983	6.71079×10^{-9}	4.9986
MOP-WENO-PM6	10	1.00298×10^{-1}	-	8.49034×10^{-2}	-	9.88357×10^{-2}	-
	20	5.84504×10^{-3}	4.1009	5.80703×10^{-3}	3.8699	9.01779×10^{-3}	3.4542
	40	2.51725×10^{-4}	4.5373	2.40678×10^{-4}	4.5926	3.66822×10^{-4}	4.6196
	80	4.87327×10^{-6}	5.6908	4.54036×10^{-6}	5.7282	6.83270×10^{-6}	5.7465
	160	1.52750×10^{-7}	4.9956	1.42488×10^{-7}	4.9939	2.14532×10^{-7}	4.9932
	320	4.77729×10^{-9}	4.9988	4.45807×10^{-9}	4.9983	6.71079×10^{-9}	4.9986
WENO-PPM5	10	9.22982×10^{-2}	-	7.46925×10^{-2}	-	8.46229×10^{-2}	-
	20	4.68376×10^{-3}	4.3006	4.18882×10^{-3}	4.1563	5.92748×10^{-3}	3.8356
	40	1.55745×10^{-4}	4.9104	1.44018×10^{-4}	4.8622	2.09420×10^{-4}	4.8229
	80	4.88795×10^{-6}	4.9938	4.54528×10^{-6}	4.9857	6.83617×10^{-6}	4.9371
	160	1.52852×10^{-7}	4.9990	1.42506×10^{-7}	4.9953	2.14527×10^{-7}	4.9940
	320	4.77759×10^{-9}	4.9997	4.45812×10^{-9}	4.9984	6.71080×10^{-9}	4.9985
MOP-WENO-PPM5	10	9.50369×10^{-2}	-	8.08190×10^{-2}	-	9.65522×10^{-2}	-
	20	6.27179×10^{-3}	3.9215	6.11267×10^{-3}	3.7248	8.98120×10^{-3}	3.4263
	40	2.52600×10^{-4}	4.6340	2.41656×10^{-4}	4.6608	3.69338×10^{-4}	4.6039
	80	4.88795×10^{-6}	5.6915	4.54528×10^{-6}	5.7324	6.83617×10^{-6}	5.7556
	160	1.52852×10^{-7}	4.9990	1.42506×10^{-7}	4.9953	2.14527×10^{-7}	4.9940
	320	4.77759×10^{-9}	4.9997	4.45812×10^{-9}	4.9984	6.71080×10^{-9}	4.9985
WENO-RM(260)	10	8.24328×10^{-2}	-	6.64590×10^{-2}	-	7.64206×10^{-2}	-
	20	4.37642×10^{-3}	4.2354	4.00547×10^{-3}	4.0524	5.88375×10^{-3}	3.6992
	40	1.52200×10^{-4}	4.8457	1.42162×10^{-4}	4.8164	2.09889×10^{-4}	4.8090
	80	4.86434×10^{-6}	4.9676	4.53769×10^{-6}	4.9694	6.83016×10^{-6}	4.9416
	160	1.52735×10^{-7}	4.9931	1.42486×10^{-7}	4.9931	2.14533×10^{-7}	4.9926
	320	4.77728×10^{-9}	4.9987	4.45807×10^{-9}	4.9983	6.71079×10^{-9}	4.9986
MOP-WENO-RM(260)	10	8.96509×10^{-2}	-	7.51169×10^{-2}	-	9.20962×10^{-2}	-
	20	6.87612×10^{-3}	3.7047	6.65488×10^{-3}	3.4967	9.75043×10^{-3}	3.2396
	40	2.59418×10^{-4}	4.7282	2.51194×10^{-4}	4.7275	4.03065×10^{-4}	4.5964
	80	4.86434×10^{-6}	5.7369	4.53769×10^{-6}	5.7907	6.83016×10^{-6}	5.8829
	160	1.52735×10^{-7}	4.9931	1.42486×10^{-7}	4.9931	2.14533×10^{-7}	4.9926
	320	4.77728×10^{-9}	4.9987	4.45807×10^{-9}	4.9983	6.71079×10^{-9}	4.9986

Table 3. Cont.

Scheme	N	L_1 Error	L_1 Order	L_2 Error	L_2 Order	L_∞ Error	L_∞ Order
WENO-MAIM1	10	1.24659×10^{-1}	-	1.14152×10^{-1}	-	1.40438×10^{-1}	-
	20	8.07923×10^{-3}	3.9476	7.08117×10^{-3}	4.0108	1.03772×10^{-2}	3.7584
	40	3.32483×10^{-4}	4.6029	3.36264×10^{-4}	4.3963	6.62891×10^{-4}	3.9685
	80	1.01162×10^{-5}	5.0385	1.49724×10^{-5}	4.4892	4.48554×10^{-5}	3.8854
	160	1.52910×10^{-7}	6.0478	1.42515×10^{-7}	6.7150	2.14522×10^{-7}	7.7080
	320	4.77728×10^{-9}	5.0003	4.45807×10^{-9}	4.9986	6.71079×10^{-9}	4.9985
MOP-WENO-MAIM1	10	1.27999×10^{-1}	-	1.12692×10^{-1}	-	1.31113×10^{-1}	-
	20	7.62753×10^{-3}	4.0688	6.93240×10^{-3}	4.0229	1.27480×10^{-2}	3.3625
	40	3.37132×10^{-4}	4.4998	3.36497×10^{-4}	4.3647	6.40953×10^{-4}	4.3139
	80	1.01162×10^{-5}	5.0586	1.49724×10^{-5}	4.4902	4.48554×10^{-5}	3.8369
	160	1.52910×10^{-7}	6.0478	1.42515×10^{-7}	6.7150	2.14522×10^{-7}	7.7080
	320	4.77728×10^{-9}	5.0003	4.45807×10^{-9}	4.9986	6.71079×10^{-9}	4.9985
MIP-WENO-ACMk	10	8.75629×10^{-2}	-	6.98131×10^{-2}	-	7.91292×10^{-2}	-
	20	4.39527×10^{-3}	4.3163	4.02909×10^{-3}	4.1150	5.89045×10^{-3}	3.7478
	40	1.52219×10^{-4}	4.8517	1.42172×10^{-4}	4.8247	2.09893×10^{-4}	4.8107
	80	4.86436×10^{-6}	4.9678	4.53770×10^{-6}	4.9695	6.83017×10^{-6}	4.9416
	160	1.52735×10^{-7}	4.9931	1.42486×10^{-7}	4.9931	2.14533×10^{-7}	4.9926
	320	4.77728×10^{-9}	4.9987	4.45807×10^{-9}	4.9983	6.71079×10^{-9}	4.9986
MOP-WENO-ACMk	10	9.08634×10^{-2}	-	7.58160×10^{-2}	-	9.29135×10^{-2}	-
	20	7.09246×10^{-3}	3.6793	6.88532×10^{-3}	3.4609	1.01479×10^{-2}	3.1947
	40	2.59429×10^{-4}	4.7729	2.51208×10^{-4}	4.7766	4.03069×10^{-4}	4.6540
	80	4.86436×10^{-6}	5.7369	4.53770×10^{-6}	5.7908	6.83017×10^{-6}	5.8830
	160	1.52735×10^{-7}	4.9931	1.42486×10^{-7}	4.9931	2.14533×10^{-7}	4.9926
	320	4.77728×10^{-9}	4.9987	4.45807×10^{-9}	4.9983	6.71079×10^{-9}	4.9986

Example 3. *We calculate Equation (38) using the following initial condition [29]:*

$$u(x,0) = \sin^9(\pi x), \tag{41}$$

with the periodic boundary condition. It is trivial to verify that this initial condition has high-order critical points. We also set the CFL number to be $(\Delta x)^{2/3}$.

We use the L_1- and L_∞-norm of numerical errors to measure the dissipations of the schemes. It is easy to check that the exact solution is $u(x,t) = \sin^9(\pi(x-t))$. Moreover, we consider the increased errors (in percentage) compared to the MIP-WENO-ACMk scheme that gives solutions with highly low dissipations. For the L_1- and L_∞-norms of numerical errors of the scheme "Y," their associated increased errors at output time t are defined by

$$\chi_1 = \frac{L_1^Y(t) - L_1^{MIP-WENO-ACMk}(t)}{L_1^{MIP-WENO-ACMk}(t)} \times 100\%,$$

$$\chi_\infty = \frac{L_\infty^Y(t) - L_\infty^{MIP-WENO-ACMk}(t)}{L_\infty^{MIP-WENO-ACMk}(t)} \times 100\%,$$

where $L_1^{MIP-WENO-ACMk}(t)$ and $L_\infty^{MIP-WENO-ACMk}(t)$ are the L_1- and L_∞-norms of numerical errors of the MIP-WENO-ACMk scheme.

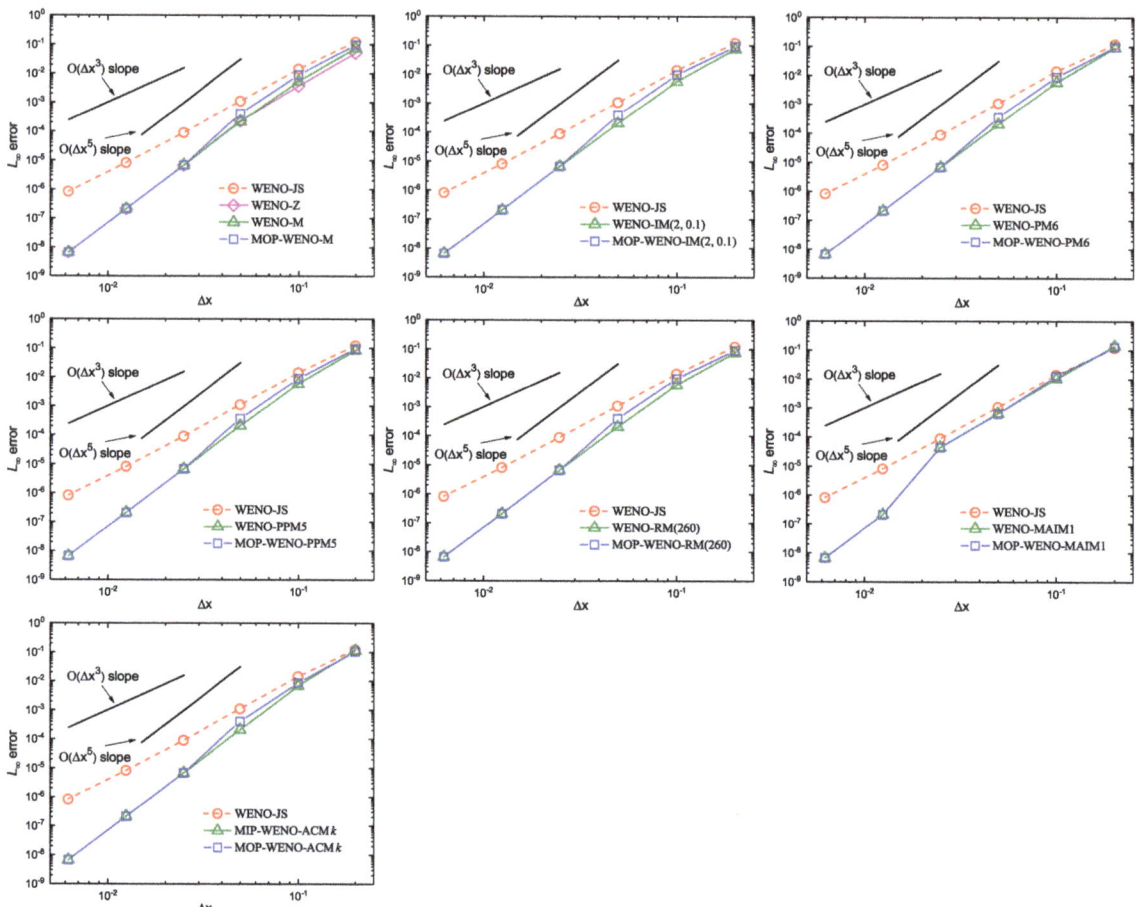

Figure 4. L_∞-norm error plots for various WENO schemes for Example 2.

Table 4 shows the L_1- and L_∞-norm numerical errors and their increased errors by using a uniform grid cell of $N = 200$ at different output times of $t = 10, 100, 200, 500, 1000$. From Table 4, we can observe that: (1) the WENO-JS scheme has the largest increased errors for no matter short or long output times; (2) for short output times, such as $t \leq 100$, the solutions computed by the WENO-M scheme are closer to those of the MIP-WENO-ACMk scheme, leading to smaller increased errors than the associated MOP-WENO-M scheme; (3) however, when the output time is larger, such as $t \geq 200$, the solutions computed by the MOP-WENO-M scheme, whose increased errors do not get larger but evidently decreased, are closer to those of the MIP-WENO-ACMk scheme than the associated WENO-M scheme, whose errors increases dramatically, leading to significantly larger increased errors; (4) the performance of the WENO-Z scheme is very similar to that of the WENO-M scheme; (5) although the errors of the MOP-WENO-X schemes except the MOP-WENO-M scheme are not as small as those of the associated WENO-X schemes, these errors can be maintained considerable levels leading to acceptable increases in errors that are much lower than those of the WENO-JS and WENO-M schemes.

Actually, as mentioned in Examples 1 and 2, the cause of the slight accuracy loss discussed above is that the mapping function of the MOP-WENO-X scheme has narrower optimal weight intervals than the associated WENO-X schemes, and one can easily overcome this drawback by increasing the grid number. To demonstrate this, we calculate this problem using the same schemes at the same output times with a larger grid number of $N = 800$. The results are shown in Table 5, and we can see that: (1) the errors of the MOP-WENO-X schemes get closer to those of the MIP-WENO-ACMk scheme when the grid number increases from $N = 200$ to $N = 800$, resulting in the significant decrease of the increased errors, and in different words, the errors of the MOP-WENO-X schemes and the MIP-WENO-ACMk scheme are so close that one can ignore their differences; (2) although the errors of the WENO-JS, WENO-M and WENO-Z schemes get smaller when the grid number increases from $N = 200$ to $N = 800$, their increased errors become very large; (3) naturally, the increased errors of the MOP-WENO-X schemes are far smaller than those of the WENO-JS, WENO-M and WENO-Z schemes. Actually, it is an important advantage of the MOP-WENO-X schemes that can maintain comparably high resolution for long output times. In the next subsection we have further discussion of this.

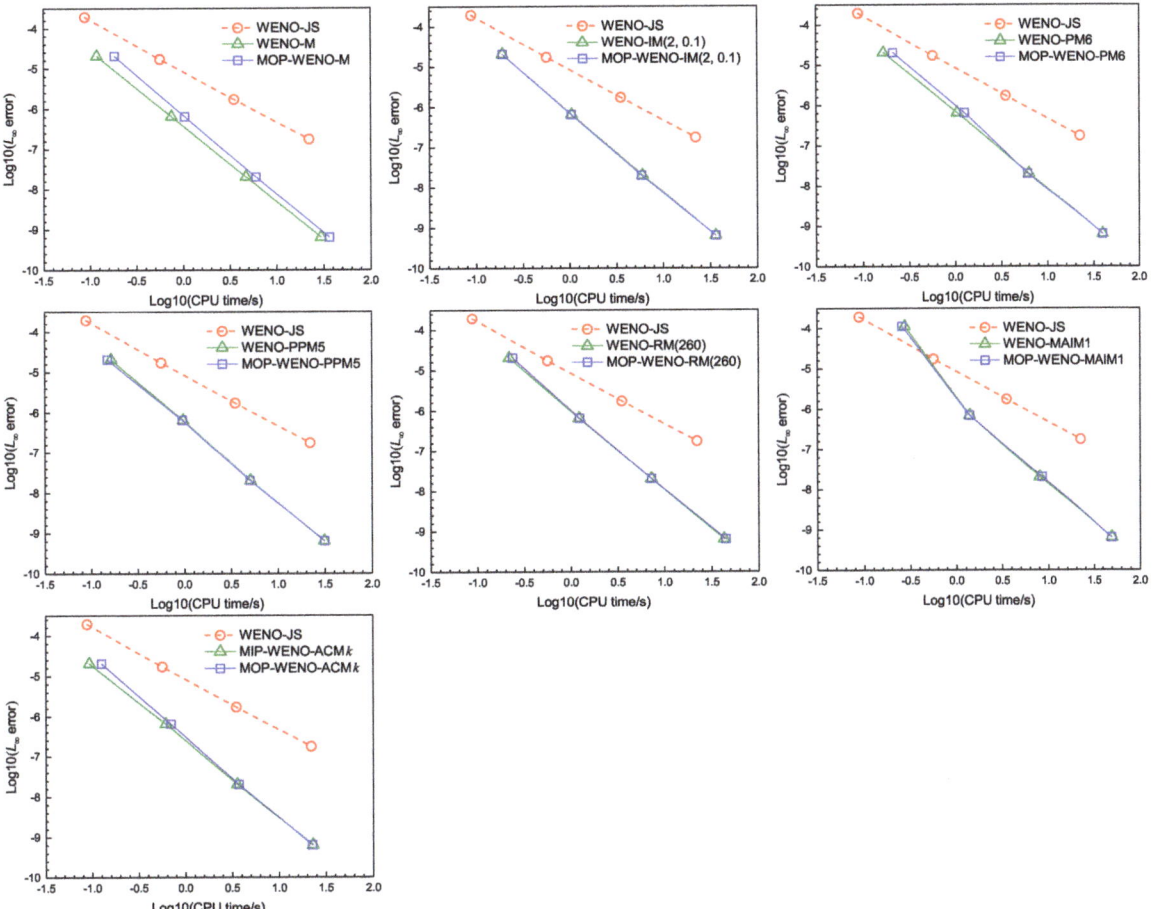

Figure 5. Comparison of various WENO schemes for Example 2 in CPU time and L_∞-norm computing errors.

Table 4. Performances of various considered schemes on solving $u_t + u_x = 0$ with $u(x,0) = \sin^9(\pi x)$, $N = 200$.

	MIP-WENO-ACMk				MOP-WENO-ACMk			
Time, t	L_1 error	χ_1	L_∞ error	χ_∞	L_1 error	χ_1	L_∞ error	χ_∞
10	8.42873×10^{-5}	–	1.38205×10^{-4}	–	1.55900×10^{-4}	85%	5.22964×10^{-4}	278%
100	8.35747×10^{-4}	–	1.36404×10^{-3}	–	2.72470×10^{-3}	226%	9.83147×10^{-3}	621%
200	1.65557×10^{-3}	–	2.68955×10^{-3}	–	4.11740×10^{-3}	149%	6.66166×10^{-3}	148%
500	3.95849×10^{-3}	–	6.45564×10^{-3}	–	8.34435×10^{-3}	111%	1.83215×10^{-2}	184%
1000	7.24723×10^{-3}	–	1.21593×10^{-2}	–	1.54830×10^{-2}	114%	3.16523×10^{-2}	160%
	WENO–JS				WENO–Z			
Time, t	L_1 error	χ_1	L_∞ error	χ_∞	L_1 error	χ_1	L_∞ error	χ_∞
10	3.86931×10^{-4}	359%	5.36940×10^{-4}	289%	9.25912×10^{-5}	10%	1.38334×10^{-4}	0%
100	5.42288×10^{-3}	549%	1.20056×10^{-2}	780%	1.45856×10^{-3}	75%	3.76895×10^{-3}	176%
200	2.35657×10^{-2}	1323%	6.47820×10^{-2}	2309%	8.32696×10^{-3}	403%	3.37176×10^{-2}	1154%
500	1.55650×10^{-1}	3832%	2.57663×10^{-1}	3891%	8.95980×10^{-2}	2163%	1.94577×10^{-1}	2914%
1000	2.91359×10^{-1}	3920%	4.44664×10^{-1}	3557%	1.42377×10^{-1}	1865%	2.80558×10^{-1}	2207%
	WENO–M				MOP–WENO–M			
Time, t	L_1 error	χ_1	L_∞ error	χ_∞	L_1 error	χ_1	L_∞ error	χ_∞
10	8.90890×10^{-5}	6%	1.38348×10^{-4}	0%	1.56466×10^{-4}	86%	5.08956×10^{-4}	268%
100	1.29154×10^{-3}	55%	3.32665×10^{-3}	144%	2.88442×10^{-3}	245%	1.01393×10^{-2}	643%
200	5.74021×10^{-3}	247%	2.37125×10^{-2}	782%	5.11795×10^{-3}	209%	1.02172×10^{-2}	280%
500	4.89290×10^{-2}	1136%	1.78294×10^{-1}	2662%	9.09352×10^{-3}	130%	1.98022×10^{-2}	207%
1000	1.34933×10^{-1}	1762%	3.17199×10^{-1}	2509%	1.75990×10^{-2}	143%	4.01776×10^{-2}	230%
	WENO-IM(2, 0.1)				MOP-WENO-IM(2, 0.1)			
Time, t	L_1 error	χ_1	L_∞ error	χ_∞	L_1 error	χ_1	L_∞ error	χ_∞
10	8.46989×10^{-5}	0%	1.38220×10^{-4}	0%	1.55777×10^{-4}	85%	5.08361×10^{-4}	268%
100	8.39425×10^{-4}	0%	1.36420×10^{-3}	0%	2.74109×10^{-3}	228%	9.88287×10^{-3}	625%
200	1.67834×10^{-3}	1%	2.68977×10^{-3}	0%	4.16210×10^{-3}	151%	6.81406×10^{-3}	153%
500	4.17514×10^{-3}	5%	8.13666×10^{-3}	12%	8.37898×10^{-3}	112%	1.84998×10^{-2}	187%
1000	6.45231×10^{-3}	0%	1.21388×10^{-2}	0%	1.25166×10^{-2}	73%	2.02754×10^{-2}	67%
	WENO–PM6				MOP–WENO–PM6			
Time, t	L_1 error	χ_1	L_∞ error	χ_∞	L_1 error	χ_1	L_∞ error	χ_∞
10	8.40259×10^{-5}	0%	1.38205×10^{-4}	0%	1.53937×10^{-4}	83%	4.92116×10^{-4}	256%
100	8.30374×10^{-4}	−1%	1.36410×10^{-3}	0%	2.70283×10^{-3}	223%	9.52154×10^{-3}	598%
200	1.63963×10^{-3}	−1%	2.68938×10^{-3}	0%	4.07454×10^{-3}	146%	6.49923×10^{-3}	142%
500	3.88864×10^{-3}	−2%	6.45650×10^{-3}	0%	8.46326×10^{-3}	114%	1.83171×10^{-2}	184%
1000	7.17606×10^{-3}	−1%	1.21637×10^{-2}	0%	1.54196×10^{-2}	113%	3.15065×10^{-2}	159%
	WENO–PPM5				MOP–WENO–PPM5			
Time, t	L_1 error	χ_1	L_∞ error	χ_∞	L_1 error	χ_1	L_∞ error	χ_∞
10	8.40198×10^{-5}	0%	1.38206×10^{-4}	0%	1.53322×10^{-4}	82%	4.97691×10^{-4}	260%
100	8.30119×10^{-4}	−1%	1.36411×10^{-3}	0%	2.70476×10^{-3}	224%	9.71919×10^{-3}	613%
200	1.63931×10^{-3}	−1%	2.68939×10^{-3}	0%	4.17894×10^{-3}	152%	6.89990×10^{-3}	157%
500	3.89396×10^{-3}	−2%	6.45658×10^{-3}	0%	8.34997×10^{-3}	111%	1.83470×10^{-2}	184%
1000	7.20573×10^{-3}	−1%	1.21629×10^{-2}	0%	1.21149×10^{-2}	67%	1.87607×10^{-2}	54%
	WENO–RM(260)				MOP–WENO–RM(260)			
Time, t	L_1 error	χ_1	L_∞ error	χ_∞	L_1 error	χ_1	L_∞ error	χ_∞
10	8.43348×10^{-5}	0%	1.38206×10^{-4}	0%	1.55787×10^{-4}	85%	5.05390×10^{-4}	266%
100	8.35534×10^{-4}	0%	1.36404×10^{-3}	0%	2.72147×10^{-3}	226%	9.74612×10^{-3}	615%
200	1.65314×10^{-3}	0%	2.68956×10^{-3}	0%	4.13179×10^{-3}	150%	6.71615×10^{-3}	150%
500	3.94006×10^{-3}	0%	6.45544×10^{-3}	0%	8.32505×10^{-3}	110%	1.83262×10^{-2}	184%
1000	7.25689×10^{-3}	0%	1.21576×10^{-2}	0%	1.57577×10^{-2}	117%	3.30552×10^{-2}	172%
	WENO–MAIM1				MOP–WENO–MAIM1			
Time, t	L_1 error	χ_1	L_∞ error	χ_∞	L_1 error	χ_1	L_∞ error	χ_∞
10	8.24623×10^{-5}	−2%	1.38215×10^{-4}	0%	9.97376×10^{-5}	18%	1.38172×10^{-4}	0%
100	8.03920×10^{-4}	−4%	1.36392×10^{-3}	0%	8.16839×10^{-4}	−2%	1.36470×10^{-3}	0%
200	1.58626×10^{-3}	−4%	2.68849×10^{-3}	0%	1.60912×10^{-3}	−3%	2.68832×10^{-3}	0%
500	3.77900×10^{-3}	−5%	6.46356×10^{-3}	0%	6.83393×10^{-3}	73%	1.63188×10^{-2}	153%
1000	7.04287×10^{-3}	−3%	1.21473×10^{-2}	0%	1.24817×10^{-2}	72%	2.22178×10^{-2}	83%

Table 5. Performance of various considered schemes on solving $u_t + u_x = 0$ with $u(x,0) = \sin^9(\pi x)$, $N = 800$.

Time, t	MIP-WENO-ACMk					MOP-WENO-ACMk			
	L_1 error	χ_1	L_∞ error	χ_∞		L_1 error	χ_1	L_∞ error	χ_∞
10	8.28794×10^{-8}	-	1.36172×10^{-7}	-		8.47930×10^{-8}	2%	1.36172×10^{-7}	0%
100	8.28891×10^{-7}	-	1.36206×10^{-6}	-		9.73202×10^{-7}	17%	1.79160×10^{-6}	32%
200	1.65782×10^{-6}	-	2.72415×10^{-6}	-		1.78369×10^{-6}	8%	2.72415×10^{-6}	0%
500	4.14451×10^{-6}	-	6.81018×10^{-6}	-		4.84739×10^{-6}	17%	8.79296×10^{-6}	29%
1000	8.28868×10^{-6}	-	1.36194×10^{-5}	-		8.61232×10^{-6}	4%	1.36194×10^{-5}	0%

Time, t	WENO-JS					WENO-Z			
	L_1 error	χ_1	L_∞ error	χ_∞		L_1 error	χ_1	L_∞ error	χ_∞
10	4.23531×10^{-7}	411%	6.95290×10^{-7}	411%		8.28830×10^{-8}	0%	1.36173×10^{-7}	0%
100	4.74028×10^{-6}	472%	1.09481×10^{-5}	704%		8.28938×10^{-7}	0%	1.36207×10^{-6}	0%
200	7.29285×10^{-5}	4299%	9.51604×10^{-4}	34832%		2.10734×10^{-6}	27%	9.02795×10^{-6}	231%
500	3.11698×10^{-2}	751974%	8.63989×10^{-2}	1268573%		9.91182×10^{-4}	23816%	1.65219×10^{-2}	242506%
1000	1.01278×10^{-1}	1221783%	2.13485×10^{-1}	1567407%		2.82670×10^{-3}	34003%	1.85472×10^{-2}	136082%

Time, t	WENO-M					MOP-WENO-M			
	L_1 error	χ_1	L_∞ error	χ_∞		L_1 error	χ_1	L_∞ error	χ_∞
10	8.28912×10^{-8}	0%	1.36173×10^{-7}	0%		8.48762×10^{-8}	2%	1.36173×10^{-7}	0%
100	8.29015×10^{-7}	0%	1.36207×10^{-6}	0%		9.93577×10^{-7}	20%	2.03738×10^{-6}	50%
200	2.27991×10^{-6}	38%	1.22731×10^{-5}	351%		1.81123×10^{-6}	9%	2.72417×10^{-6}	0%
500	1.41413×10^{-3}	34021%	1.90785×10^{-2}	280047%		4.68314×10^{-6}	13%	6.81022×10^{-6}	0%
1000	1.83325×10^{-2}	221075%	1.38215×10^{-1}	1014739%		8.53126×10^{-6}	3%	1.36195×10^{-5}	0%

Time, t	WENO-IM(2, 0.1)					MOP-WENO-IM(2, 0.1)			
	L_1 error	χ_1	L_∞ error	χ_∞		L_1 error	χ_1	L_∞ error	χ_∞
10	8.28803×10^{-8}	0%	1.36172×10^{-7}	0%		8.48292×10^{-8}	2%	1.36172×10^{-7}	0%
100	8.28891×10^{-7}	0%	1.36206×10^{-6}	0%		9.80868×10^{-7}	18%	1.87953×10^{-6}	38%
200	1.65781×10^{-6}	0%	2.72415×10^{-6}	0%		1.79137×10^{-6}	8%	2.72415×10^{-6}	0%
500	4.14443×10^{-6}	0%	6.81019×10^{-6}	0%		4.88306×10^{-6}	18%	9.14624×10^{-6}	34%
1000	8.28840×10^{-6}	0%	1.36194×10^{-5}	0%		8.63424×10^{-6}	4%	1.36194×10^{-5}	0%

Time, t	WENO-PM6					MOP-WENO-PM6			
	L_1 error	χ_1	L_∞ error	χ_∞		L_1 error	χ_1	L_∞ error	χ_∞
10	8.28795×10^{-8}	0%	1.36172×10^{-7}	0%		8.47719×10^{-8}	2%	1.36172×10^{-7}	0%
100	8.28892×10^{-7}	0%	1.36206×10^{-6}	0%		9.71688×10^{-7}	17%	1.78452×10^{-6}	31%
200	1.65782×10^{-6}	0%	2.72415×10^{-6}	0%		1.78163×10^{-6}	7%	2.72415×10^{-6}	0%
500	4.14452×10^{-6}	0%	6.81018×10^{-6}	0%		4.93547×10^{-6}	19%	1.08735×10^{-5}	60%
1000	8.84565×10^{-6}	7%	1.38461×10^{-5}	2%		8.65269×10^{-6}	4%	1.36194×10^{-5}	0%

Time, t	WENO-PPM5					MOP-WENO-PPM5			
	L_1 error	χ_1	L_∞ error	χ_∞		L_1 error	χ_1	L_∞ error	χ_∞
10	8.28794×10^{-8}	0%	1.36172×10^{-7}	0%		8.47367×10^{-8}	2%	1.36172×10^{-7}	0%
100	8.28890×10^{-7}	0%	1.36206×10^{-6}	0%		1.04103×10^{-6}	26%	1.78285×10^{-6}	31%
200	1.65781×10^{-6}	0%	2.72415×10^{-6}	0%		1.83725×10^{-6}	11%	2.72415×10^{-6}	0%
500	4.14448×10^{-6}	0%	6.81018×10^{-6}	0%		4.30721×10^{-6}	4%	6.81018×10^{-6}	0%
1000	8.28862×10^{-6}	0%	1.36194×10^{-5}	0%		8.27506×10^{-6}	0%	1.36194×10^{-5}	0%

Time, t	WENO-RM(260)					MOP-WENO-RM(260)			
	L_1 error	χ_1	L_∞ error	χ_∞		L_1 error	χ_1	L_∞ error	χ_∞
10	8.28794×10^{-8}	0%	1.36172×10^{-7}	0%		8.48225×10^{-8}	2%	1.36172×10^{-7}	0%
100	8.28889×10^{-7}	0%	1.36206×10^{-6}	0%		9.56819×10^{-7}	15%	1.58572×10^{-6}	16%
200	1.65781×10^{-6}	0%	2.72415×10^{-6}	0%		1.77008×10^{-6}	7%	2.72415×10^{-6}	0%
500	4.14448×10^{-6}	0%	6.81018×10^{-6}	0%		4.72311×10^{-6}	14%	6.81018×10^{-6}	0%
1000	8.28860×10^{-6}	0%	1.36194×10^{-5}	0%		8.55573×10^{-6}	3%	1.36194×10^{-5}	0%

Time, t	WENO-MAIM1					MOP-WENO-MAIM1			
	L_1 error	χ_1	L_∞ error	χ_∞		L_1 error	χ_1	L_∞ error	χ_∞
10	8.28796×10^{-8}	0%	1.36172×10^{-7}	0%		8.28791×10^{-8}	0%	1.36172×10^{-7}	0%
100	8.28893×10^{-7}	0%	1.36206×10^{-6}	0%		8.28894×10^{-7}	0%	1.36206×10^{-6}	0%
200	1.65782×10^{-6}	0%	2.72415×10^{-6}	0%		1.65783×10^{-6}	0%	2.72415×10^{-6}	0%
500	4.14450×10^{-6}	0%	6.81018×10^{-6}	0%		4.14454×10^{-6}	0%	6.81018×10^{-6}	0%
1000	8.28865×10^{-6}	0%	1.36194×10^{-5}	0%		8.28830×10^{-6}	0%	1.36194×10^{-5}	0%

In Figures 6 and 7, we plot the solutions computed by various schemes at output time $t = 1000$ with the grid numbers of $N = 200$ and $N = 800$, respectively. For $N = 200$, Figure 6 shows that: (1) the MOP-WENO-M scheme provides results with far higher resolution than the associated WENO-M scheme and the WENO-Z scheme, which give results with slightly better resolution than the worst one computed by the WENO-JS scheme; (2) the results of the MOP-WENO-MAIM1 scheme are very close to those of its associated WENO-MAIM1 scheme; (3) the results of the other MOP-WENO-X schemes show far better resolutions than the WENO-M, WENO-Z, and WENO-JS schemes, although they give results with very slightly lower resolutions than their associated WENO-X schemes because of the narrower optimal weight intervals. Actually, we can amend this minor issue by using a larger grid number. Consequently, for $N = 800$, it can be seen from Figure 7 that: (1) all the MOP-WENO-X schemes produce results very close to those of their associated mapped WENO-X schemes with extremely high resolutions except the case of X = M; (2) the MOP-WENO-M scheme also produces results with very high resolution, whereas the resolutions of the results from the WENO-M, WENO-Z, and WENO-JS schemes have far lower resolutions.

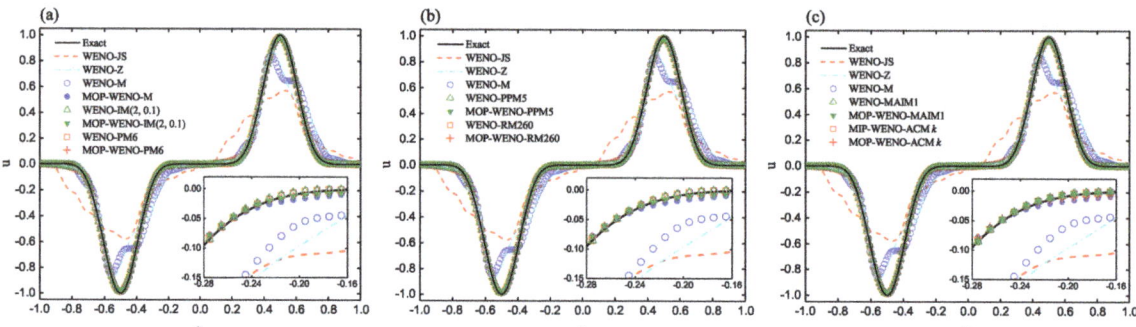

Figure 6. Performances of various WENO schemes for Example 3 at output time $t = 1000$ with a uniform mesh size of $N = 200$.

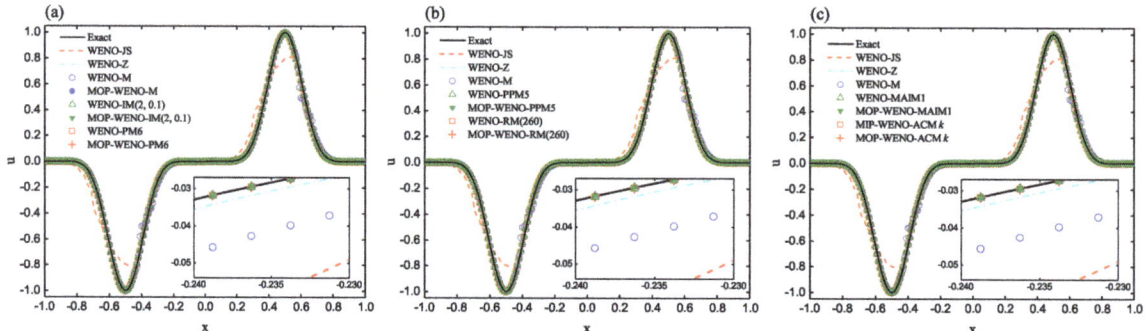

Figure 7. Performances of various WENO schemes for Example 3 at output time $t = 1000$ with a uniform mesh size of $N = 800$.

Example 4. *We calculate Equation (38) using the following initial condition [8]:*

$$u(x,0) = \begin{cases} \frac{1}{6}\left[G(x, \beta, z - \hat{\delta}) + 4G(x, \beta, z) + G(x, \beta, z + \hat{\delta})\right], & x \in [-0.8, -0.6], \\ 1, & x \in [-0.4, -0.2], \\ 1 - |10(x - 0.1)|, & x \in [0.0, 0.2], \\ \frac{1}{6}\left[F(x, \alpha, a - \hat{\delta}) + 4F(x, \alpha, a) + F(x, \alpha, a + \hat{\delta})\right], & x \in [0.4, 0.6], \\ 0, & \text{otherwise,} \end{cases} \quad (42)$$

where $G(x, \beta, z) = e^{-\beta(x-z)^2}, F(x, \alpha, a) = \sqrt{\max\left(1 - \alpha^2(x - a)^2, 0\right)}$, *and the constants are* $z = -0.7, \hat{\delta} = 0.005, \beta = \frac{\log 2}{36\hat{\delta}^2}, a = 0.5,$ *and* $\alpha = 10$. *The periodic boundary condition is used. Although the CFL number can be chosen from a wide range of values—for example, CFL = 0.6 usually works well—we set CFL = 0.1 here to keep the consistent with the literatures [27,29,31,32] having strong relevance to the present study and to make thorough comparisons with the results of these literature. For brevity in the presentation, we call* this *linear problem* SLP as it is presented by Shu et al. in [8]. It is known that this problem consists of a Gaussian, a square wave, a sharp triangle, and a semi-ellipse.

In Tables 6 and 7, we present the L_1, L_2, L_∞ errors and the corresponding convergence rates of accuracy with $t = 2$ and $t = 2000$, respectively. For the case of $t = 2$, it can be seen that: (1) the L_1 and L_2 orders of all considered schemes are approximately 1.0 and about 0.35 to 0.5, respectively; (2) negative values of the L_∞ orders of all considered schemes are generated; (3) in terms of accuracy, the MOP-WENO-X schemes produce less accurate results than the associated WENO-X schemes. For the case of $t = 2000$, it can be seen that: (1) the L_1, L_2 orders of the WENO-JS, WENO-M, and WENO-Z schemes decrease to very small values and even become negative; (2) however, the L_1 and L_2 orders of all the MOP-WENO-X schemes, and the associated mapped WENO-X schemes without WENO-M, are clearly larger than 1.0 and around 0.5 to 0.9, respectively; (3) the L_∞ orders of all WENO-X schemes are very small, and some of them are even negative (e.g., the WENO-JS, WENO-PPM5 and MIP-WENO-ACMk schemes), and those of the MOP-WENO-X schemes are all positive, although they are also very small; (4) in terms of accuracy, on the whole, the MOP-WENO-X schemes produce accurate and comparable results to the associated WENO-X schemes, except the WENO-M scheme. However, if we take a closer look, we can find that the resolution of the results computed by the WENO-M scheme is significantly lower than that of the MOP-WENO-M scheme, and the other mapped WENO-X schemes generate spurious oscillations, but the associated MOP-WENO-X schemes do not. Detailed tests are conducted and the solutions are presented carefully to demonstrate this in the following subsection.

Table 6. Convergence properties of various considered schemes on solving $u_t + u_x = 0$ with initial condition Equation (42), $t = 2$. To be continued.

Scheme	N	L_1 Error	L_1 Order	L_2 Error	L_2 Order	L_∞ Error	L_∞ Order
WENO-JS	200	6.30497×10^{-2}	-	1.08621×10^{-1}	-	4.09733×10^{-1}	-
	400	2.81654×10^{-2}	1.2103	7.71111×10^{-2}	0.4943	4.19594×10^{-1}	−0.0343
	800	1.41364×10^{-2}	0.9945	5.69922×10^{-2}	0.4362	4.28463×10^{-1}	−0.0302
WENO-Z	200	4.98422×10^{-2}	-	9.59452×10^{-2}	-	3.92478×10^{-1}	-
	400	2.37836×10^{-2}	1.0674	6.98647×10^{-2}	0.4576	4.03601×10^{-1}	−0.0403
	800	1.19851×10^{-2}	0.9887	5.14607×10^{-2}	0.4411	4.13262×10^{-1}	−0.0341
WENO-M	200	4.77201×10^{-2}	-	9.53073×10^{-2}	-	3.94243×10^{-1}	-
	400	2.23407×10^{-2}	1.0949	6.91333×10^{-2}	0.4632	4.05856×10^{-1}	−0.0419
	800	1.11758×10^{-2}	0.9993	5.09232×10^{-2}	0.4411	4.16937×10^{-1}	−0.0389
MOP-WENO-M	200	5.72690×10^{-2}	-	1.00827×10^{-1}	-	4.14785×10^{-1}	-
	400	2.72999×10^{-2}	1.0689	7.33765×10^{-2}	0.4585	4.45144×10^{-1}	-0.1019
	800	1.42908×10^{-2}	0.9338	5.57886×10^{-2}	0.3953	4.64024×10^{-1}	−0.0599
WENO-IM(2, 0.1)	200	4.40293×10^{-2}	-	9.19118×10^{-2}	-	3.86789×10^{-1}	-
	400	2.02331×10^{-2}	1.1217	6.68479×10^{-2}	0.4594	3.98769×10^{-1}	−0.0441
	800	1.01805×10^{-2}	0.9909	4.95333×10^{-2}	0.4325	4.09515×10^{-1}	−0.0383
MOP-WENO-IM(2, 0.1)	200	6.09985×10^{-2}	-	1.03438×10^{-1}	-	4.35238×10^{-1}	-
	400	2.86731×10^{-2}	1.0891	7.56598×10^{-2}	0.4512	4.62098×10^{-1}	−0.0864
	800	1.45601×10^{-2}	0.9777	5.61842×10^{-2}	0.4294	4.64674×10^{-1}	−0.0080
WENO-PM6	200	4.66681×10^{-2}	-	9.45566×10^{-2}	-	3.96866×10^{-1}	-
	400	2.13883×10^{-2}	1.1256	6.82948×10^{-2}	0.4694	4.06118×10^{-1}	−0.0332
	800	1.06477×10^{-2}	1.0063	5.03724×10^{-2}	0.4391	4.15277×10^{-1}	−0.0322
MOP-WENO-PM6	200	5.45129×10^{-2}	-	9.95654×10^{-2}	-	4.02785×10^{-1}	-
	400	2.61755×10^{-2}	1.0584	7.16656×10^{-2}	0.4744	4.26334×10^{-1}	−0.0820
	800	1.38981×10^{-2}	0.9133	5.44733×10^{-2}	0.3957	4.63134×10^{-1}	−0.1194

Table 6. *Cont.*

Scheme	N	L_1 Error	L_1 Order	L_2 Error	L_2 Order	L_∞ Error	L_∞ Order
WENO-PPM5	200	4.54081×10^{-2}	-	9.33165×10^{-2}	-	3.91076×10^{-1}	-
	400	2.07948×10^{-2}	1.1267	6.76172×10^{-2}	0.4647	4.02214×10^{-1}	−0.0405
	800	1.04018×10^{-2}	0.9994	4.99580×10^{-2}	0.4367	4.12113×10^{-1}	−0.0351
MOP-WENO-PPM5	200	5.51553×10^{-2}	-	9.94592×10^{-2}	-	4.04763×10^{-1}	-
	400	2.65464×10^{-2}	1.0550	7.19973×10^{-2}	0.4662	4.32887×10^{-1}	−0.0969
	800	1.41381×10^{-2}	0.9089	5.52704×10^{-2}	0.3814	4.68577×10^{-1}	−0.1143
WENO-RM(260)	200	4.63072×10^{-2}	-	9.40674×10^{-2}	-	3.96762×10^{-1}	-
	400	2.13545×10^{-2}	1.1167	6.81954×10^{-2}	0.4640	4.08044×10^{-1}	−0.0405
	800	1.06392×10^{-2}	1.0052	5.03289×10^{-2}	0.4383	4.16722×10^{-1}	−0.0304
MOP-WENO-RM(260)	200	5.54343×10^{-2}	-	9.93009×10^{-2}	-	4.04041×10^{-1}	-
	400	2.71415×10^{-2}	1.0303	7.22823×10^{-2}	0.4582	4.38358×10^{-1}	−0.1176
	800	1.45563×10^{-2}	0.8989	5.66845×10^{-2}	0.3507	4.70380×10^{-1}	−0.1017
WENO-MAIM1	200	5.71142×10^{-2}	-	1.03257×10^{-1}	-	4.15051×10^{-1}	-
	400	2.48065×10^{-2}	1.2031	7.29236×10^{-2}	0.5018	4.23185×10^{-1}	−0.0280
	800	1.21078×10^{-2}	1.0348	5.32803×10^{-2}	0.4528	4.28710×10^{-1}	−0.0187
MOP-WENO-MAIM1	200	5.98640×10^{-2}	-	1.05066×10^{-1}	-	4.12365×10^{-1}	-
	400	2.64819×10^{-2}	1.1767	7.38102×10^{-2}	0.5094	4.26841×10^{-1}	−0.0498
	800	1.33647×10^{-2}	0.9866	5.44089×10^{-2}	0.4400	4.38310×10^{-1}	−0.0383
MIP-WENO-ACMk	200	4.45059×10^{-2}	-	9.24356×10^{-2}	-	3.92505×10^{-1}	-
	400	2.03633×10^{-2}	1.1280	6.69718×10^{-2}	0.4649	4.03456×10^{-1}	−0.0397
	800	1.02139×10^{-2}	0.9954	4.95672×10^{-2}	0.4342	4.13217×10^{-1}	−0.0345
MOP-WENO-ACMk	200	5.56533×10^{-2}	-	9.94223×10^{-2}	-	4.03765×10^{-1}	-
	400	2.79028×10^{-2}	0.9961	7.33101×10^{-2}	0.4396	4.48412×10^{-1}	−0.1513
	800	1.43891×10^{-2}	0.9554	5.51602×10^{-2}	0.4104	4.67036×10^{-1}	−0.0587

Table 7. Convergence properties of various considered schemes on solving $u_t + u_x = 0$ with initial condition Equation (42), $t = 2000$.

Scheme	N	L_1 Error	L_1 Order	L_2 Error	L_2 Order	L_∞ Error	L_∞ Order
WENO-JS	200	6.12899×10^{-1}	-	5.08726×10^{-1}	-	7.99265×10^{-1}	-
	400	5.99215×10^{-1}	0.0326	5.01160×10^{-1}	0.0216	8.20493×10^{-1}	−0.0378
	800	5.50158×10^{-1}	0.1232	4.67585×10^{-1}	0.1000	8.14650×10^{-1}	0.0103
WENO-Z	200	3.86995×10^{-1}	-	3.42335×10^{-1}	-	6.85835×10^{-1}	-
	400	2.02287×10^{-1}	0.9359	2.18125×10^{-1}	0.6503	5.18993×10^{-1}	0.4021
	800	1.66703×10^{-1}	0.2791	1.94240×10^{-1}	0.1673	5.04564×10^{-1}	0.0407
WENO-M	200	3.81597×10^{-1}	-	3.59205×10^{-1}	-	6.89414×10^{-1}	-
	400	3.25323×10^{-1}	0.2302	3.12970×10^{-1}	0.1988	6.75473×10^{-1}	0.0295
	800	3.48528×10^{-1}	−0.0994	3.24373×10^{-1}	−0.0516	6.25645×10^{-1}	0.1106
MOP-WENO-M	200	3.85134×10^{-1}	-	3.48164×10^{-1}	-	7.41230×10^{-1}	-
	400	1.74987×10^{-1}	1.1381	1.86418×10^{-1}	0.9012	5.04987×10^{-1}	0.5537
	800	6.40251×10^{-2}	1.4505	1.07629×10^{-1}	0.7925	4.81305×10^{-1}	0.0693
WENO-IM(2, 0.1)	200	2.17411×10^{-1}	-	2.30000×10^{-1}	-	5.69864×10^{-1}	-
	400	1.12590×10^{-1}	0.9493	1.64458×10^{-1}	0.4839	4.82180×10^{-1}	0.2410
	800	5.18367×10^{-2}	1.1190	9.98968×10^{-2}	0.7192	4.73102×10^{-1}	0.02784
MOP-WENO-IM(2, 0.1)	200	3.83289×10^{-1}	-	3.47817×10^{-1}	-	7.25185×10^{-1}	-
	400	1.67452×10^{-1}	1.1947	1.76550×10^{-1}	0.9783	5.24538×10^{-1}	0.4673
	800	6.44253×10^{-2}	1.3780	1.05858×10^{-1}	0.7379	5.19333×10^{-1}	0.0144
WENO-PM6	200	2.17323×10^{-1}	-	2.28655×10^{-1}	-	5.63042×10^{-1}	-
	400	1.05197×10^{-1}	1.0467	1.47518×10^{-1}	0.6323	5.04977×10^{-1}	0.1570
	80	4.47030×10^{-2}	1.2347	9.34250×10^{-2}	0.6590	4.71368×10^{-1}	0.0994

Table 7. Cont.

Scheme	N	L_1 Error	L_1 Order	L_2 Error	L_2 Order	L_∞ Error	L_∞ Order
MOP-WENO-PM6	200	4.51487×10^{-1}	-	4.01683×10^{-1}	-	7.71539×10^{-1}	-
	400	1.75875×10^{-1}	1.3601	1.83478×10^{-1}	1.1305	5.06314×10^{-1}	0.6077
	800	6.32990×10^{-2}	1.4743	1.04688×10^{-1}	0.8095	4.76091×10^{-1}	0.0888
WENO-PPM5	200	2.17174×10^{-1}	-	2.29008×10^{-1}	-	5.65575×10^{-1}	-
	400	1.03201×10^{-1}	1.0734	1.46610×10^{-1}	0.6434	5.06463×10^{-1}	0.1593
	800	4.81637×10^{-2}	1.0994	9.47748×10^{-2}	0.6294	5.14402×10^{-1}	−0.0224
MOP-WENO-PPM5	200	3.86292×10^{-1}	-	3.49072×10^{-1}	-	7.36405×10^{-1}	-
	400	1.75232×10^{-1}	1.1404	1.88491×10^{-1}	0.8890	5.14732×10^{-1}	0.5167
	800	6.36336×10^{-2}	1.4614	1.06801×10^{-1}	0.8196	4.98424×10^{-1}	0.0464
WENO-RM(260)	200	2.17363×10^{-1}	-	2.28662×10^{-1}	-	5.62933×10^{-1}	-
	400	1.04347×10^{-1}	1.0587	1.47093×10^{-1}	0.6365	4.98644×10^{-1}	0.1750
	800	4.45176×10^{-2}	1.2289	9.33066×10^{-2}	0.6567	4.71450×10^{-1}	0.0809
MOP-WENO-RM(260)	200	4.56942×10^{-1}	-	4.06524×10^{-1}	-	7.71747×10^{-1}	-
	400	2.25420×10^{-1}	1.0194	2.25814×10^{-1}	0.8482	5.12018×10^{-1}	0.5919
	800	8.02414×10^{-2}	1.4902	1.18512×10^{-1}	0.9301	4.90610×10^{-1}	0.0616
WENO-MAIM1	200	2.18238×10^{-1}	-	2.29151×10^{-1}	-	5.63682×10^{-1}	-
	400	1.09902×10^{-1}	0.9897	1.51024×10^{-1}	0.6015	4.94657×10^{-1}	0.1885
	800	4.41601×10^{-2}	1.3154	9.35506×10^{-2}	0.6910	4.72393×10^{-1}	0.0664
MOP-WENO-MAIM1	200	2.39900×10^{-1}	-	2.47191×10^{-1}	-	6.06985×10^{-1}	-
	400	1.41890×10^{-1}	0.7577	1.71855×10^{-1}	0.5244	5.61908×10^{-1}	0.1113
	800	5.43475×10^{-2}	1.3845	1.02170×10^{-1}	0.7502	5.10242×10^{-1}	0.1392
MIP-WENO-ACMk	200	2.21312×10^{-1}	-	2.28433×10^{-1}	-	5.36234×10^{-1}	-
	400	1.06583×10^{-1}	1.0541	1.46401×10^{-1}	0.6418	5.03925×10^{-1}	0.0897
	800	4.76305×10^{-2}	1.1620	9.40930×10^{-2}	0.6378	5.15924×10^{-1}	−0.0339
MOP-WENO-ACMk	200	3.83033×10^{-1}	-	3.46814×10^{-1}	-	7.18464×10^{-1}	-
	400	1.77114×10^{-1}	1.1128	1.87369×10^{-1}	0.8883	5.05980×10^{-1}	0.5058
	800	6.70535×10^{-2}	1.4013	1.09368×10^{-1}	0.7767	4.80890×10^{-1}	0.0734

4.2. 1D Linear Advection Problems with Long Output Times

The objective of this subsection is to demonstrate the advantage of the MOP-WENO-X schemes on long-output-time simulations that can obtain high resolution and meanwhile do not generate spurious oscillations.

The one-dimensional linear advection problem Equation (38) is solved with the periodic boundary condition by taking the following two initial conditions.

Case 1. (SLP) The initial condition is given by Equation (42).

Case 2. (BiCWP) The initial condition is given by

$$u(x,0) = \begin{cases} 0, & x \in [-1.0, -0.8] \cup (-0.2, 0.2) \cup (0.8, 1.0], \\ 0.5, & x \in (-0.6, -0.4] \cup (0.2, 0.4] \cup (0.6, 0.8], \\ 1, & x \in (-0.8, -0.6] \cup (-0.4, -0.2] \cup (0.4, 0.6]. \end{cases} \qquad (43)$$

Case 1 and Case 2 were carefully simulated in [9]. Case 1 is called SLP as mentioned earlier in this paper. Case 2 consists of several constant states separated by sharp discontinuities at $x = \pm 0.8, \pm 0.6, \pm 0.4, \pm 0.2$ and it was called BiCWP for brevity in the presentation as the profile of the exact solution for this *Problem* looks like the *Breach in City Wall*.

In Figures 8–11, we show the comparison of considered schemes for SLP and BiCWP, respectively, by taking $t = 2000$ and $N = 800$. It can be seen that: (1) all the MOP-WENO-X schemes produce results with considerable resolutions which are significantly higher than those of the WENO-JS, WENO-M and WENO-Z schemes, and what is more, they all do not generate spurious oscillations, while most of their associated WENO-X schemes do, when solving both SLP and BiCWP; (2) it should be reminded that the WENO-IM(2, 0.1) scheme appears not to generate spurious oscillations and it gives better resolution than the MOP-WENO-IM(2, 0.1) scheme in most of the region when solving SLP on present computing condition, however, from Figure 8b, one can observe that the MOP-WENO-IM(2, 0.1) scheme gives a better resolution of the Gaussian than the WENO-IM(2, 0.1) scheme, and if taking a closer look, one can see that the WENO-IM(2, 0.1) scheme generates a very slight spurious oscillation near $x = -0.435$ as shown in Figure 8c; (3) it is very evident as shown in Figure 10 that, when solving BiCWP, the WENO-IM(2, 0.1) scheme generates the spurious oscillations.

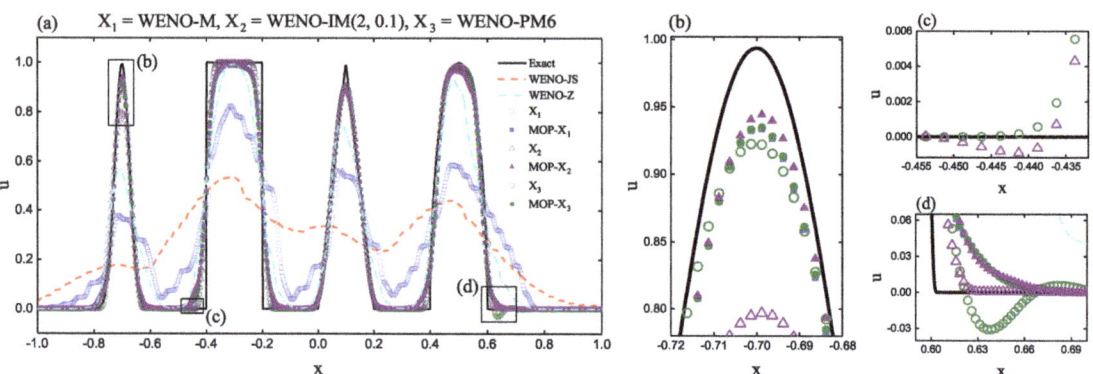

Figure 8. Performance of the WENO-JS, WENO-M, MOP-WENO-M, WENO-IM(2, 0.1), MOP-WENO-IM(2, 0.1), WENO-PM6 and MOP-WENO-PM6 schemes for the SLP at output time $t = 2000$ with a uniform mesh size of $N = 800$.

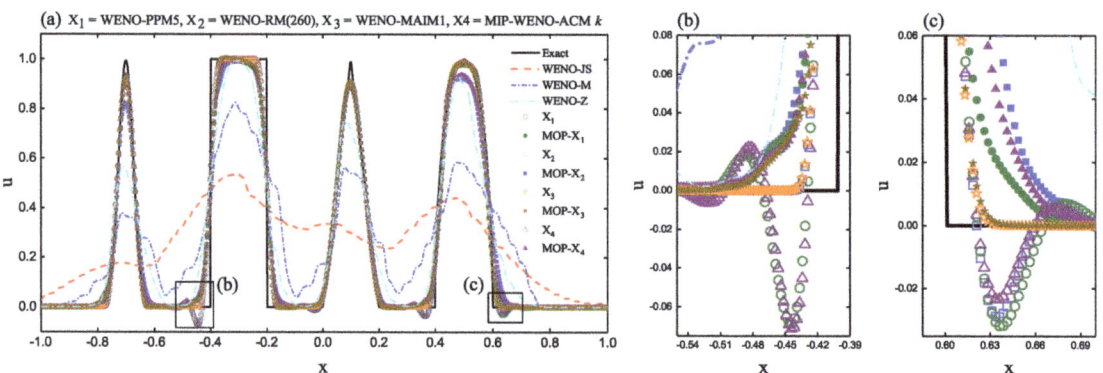

Figure 9. Performance of the WENO-JS, WENO-M, WENO-PPM5, MOP-WENO-PPM5, WENO-RM260, MOP-WENO-RM260, WENO-MAIM1, MOP-WNEO-MAIM1, MIP-WENO-ACMk and MOP-WENO-ACMk schemes for the SLP at output time $t = 2000$ with a uniform mesh size of $N = 800$.

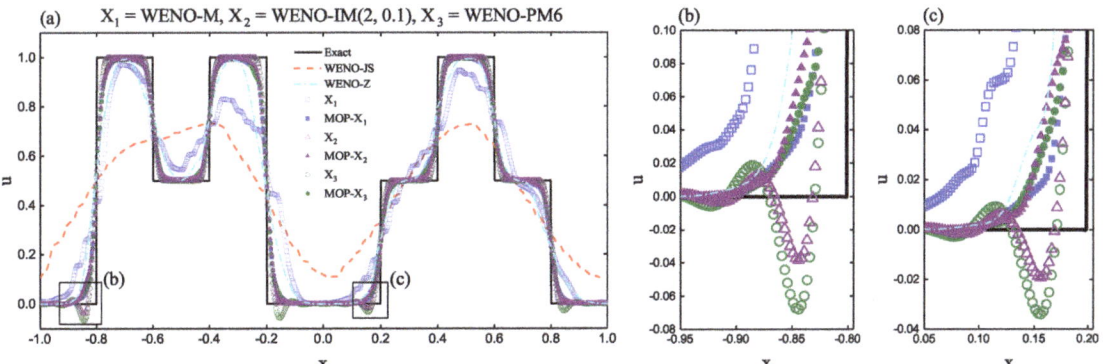

Figure 10. Performance of the WENO-JS, WENO-M, MOP-WENO-M, WENO-IM(2, 0.1), MOP-WENO-IM(2, 0.1), WENO-PM6 and MOP-WENO-PM6 schemes for the BiCWP at output time $t = 2000$ with a uniform mesh size of $N = 800$.

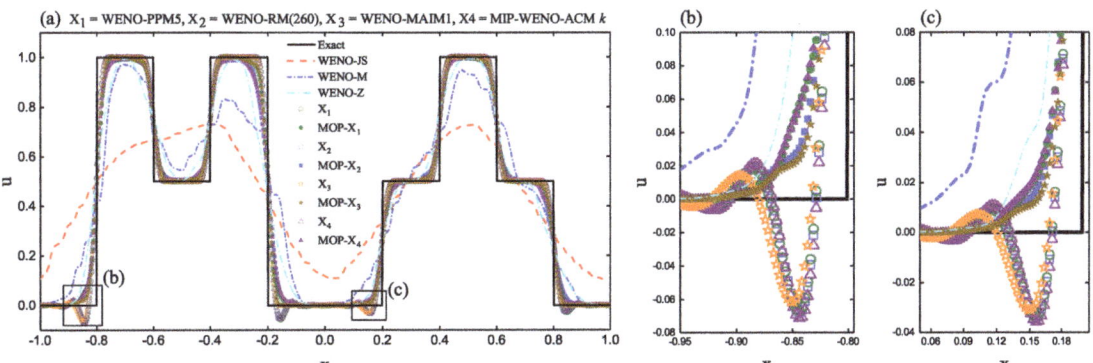

Figure 11. Performance of the WENO-JS, WENO-M, WENO-PPM5, MOP-WENO-PPM5, WENO-RM260, MOP-WENO-RM260, WENO-MAIM1, MOP-WNEO-MAIM1, MIP-WENO-ACMk and MOP-WENO-ACMk schemes for the BiCWP at output time $t = 2000$ with a uniform mesh size of $N = 800$.

In Figures 12–15, we show the comparison of considered schemes for SLP and BiCWP respectively, by taking $t = 200$ and $N = 3200$. From these solutions computed with larger grid numbers and a reduced but still long output time, it can be seen that: (1) firstly, the WENO-IM(2, 0.1) scheme generates spurious oscillations but the MOP-WENO-IM(2, 0.1) scheme does not while provides an improved resolution when solving SLP; (2) although the resolutions of the results computed by the WENO-JS, WENO-M and WENO-Z schemes are significantly improved for both SLP and BiCWP, the MOP-WENO-X schemes still evidently provide much better resolutions; (3) the spurious oscillations generated by the WENO-X schemes appear to be more evident and more intense as the grid number increases, while the associated MOP-WENO-X schemes can still avoid spurious oscillations but obtain higher resolutions, when solving both SLP and BiCWP.

For the further interpretation, without loss of generality, in Figure 16, we present the *non-OP points* of the numerical solutions of SLP computed by the WENO-M and MOP-WENO-M schemes with $N = 800, t = 2000$, and the *non-OP points* of the numerical solutions of BiCWP computed by the WENO-PM6 and MOP-WENO-PM6 schemes with $N = 3200, t = 200$. We can find that there are a great many *non-OP points* in the solutions of the WENO-M and WENO-PM6 schemes while the numbers of the *non-OP points* in the solutions of the MOP-WENO-M and MOP-WENO-PM6 schemes are zero. Actually, there are many *non-OP points* for all considered mapped WENO-X schemes. Furthermore, as expected, there are no *non-OP points* for the associated MOP-WENO-X schemes and the

WENO-JS scheme for all computing cases here. We do not show the results of the *non-OP points* for all computing cases here just for the simplicity of illustration.

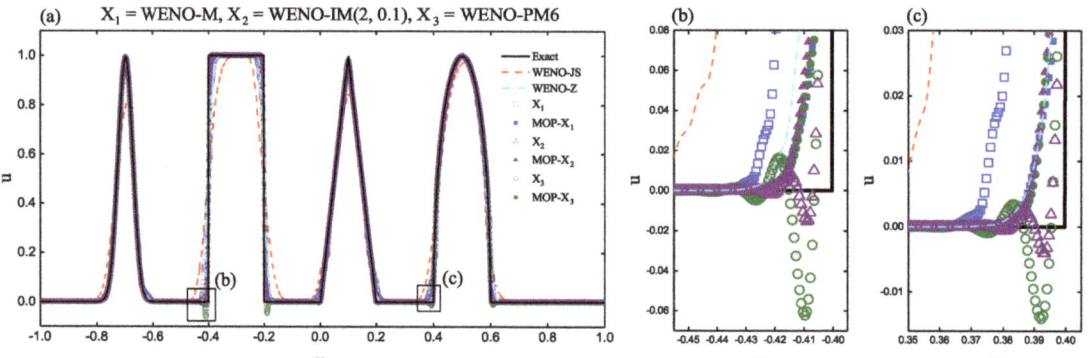

Figure 12. Performance of the WENO-JS, WENO-M, MOP-WENO-M, WENO-IM(2, 0.1), MOP-WENO-IM(2, 0.1), WENO-PM6 and MOP-WENO-PM6 schemes for the SLP at output time $t = 200$ with a uniform mesh size of $N = 3200$.

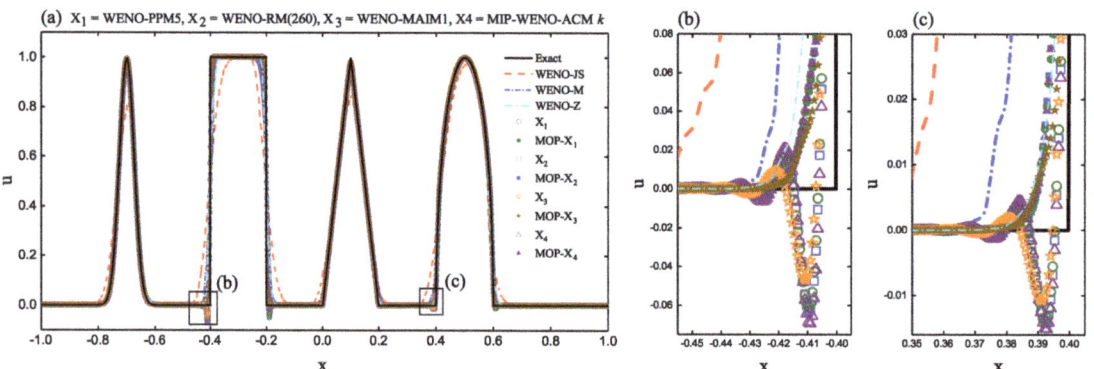

Figure 13. Performance of the WENO-JS, WENO-M, WENO-PPM5, MOP-WENO-PPM5, WENO-RM260, MOP-WENO-RM260, WENO-MAIM1, MOP-WNEO-MAIM1, MIP-WENO-ACMk and MOP-WENO-ACMk schemes for the SLP at output time $t = 200$ with a uniform mesh size of $N = 3200$.

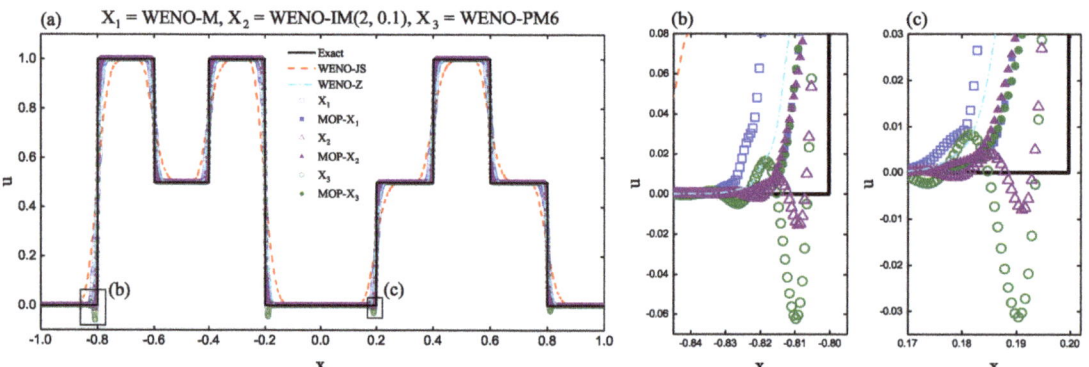

Figure 14. Performance of the WENO-JS, WENO-M, MOP-WENO-M, WENO-IM(2, 0.1), MOP-WENO-IM(2, 0.1), WENO-PM6 and MOP-WENO-PM6 schemes for the BiCWP at output time $t = 200$ with a uniform mesh size of $N = 3200$.

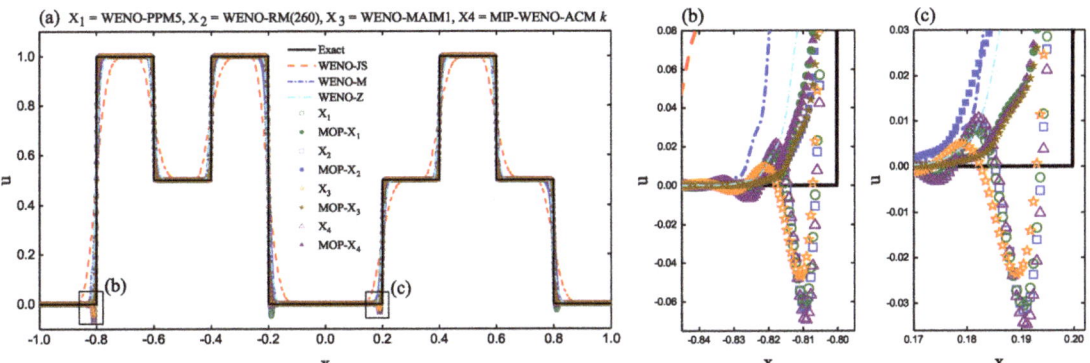

Figure 15. Performance of the WENO-JS, WENO-M, WENO-PPM5, MOP-WENO-PPM5, WENO-RM260, MOP-WENO-RM260, WENO-MAIM1, MOP-WNEO-MAIM1, MIP-WENO-ACMk and MOP-WENO-ACMk schemes for the BiCWP at output time $t = 200$ with a uniform mesh size of $N = 3200$.

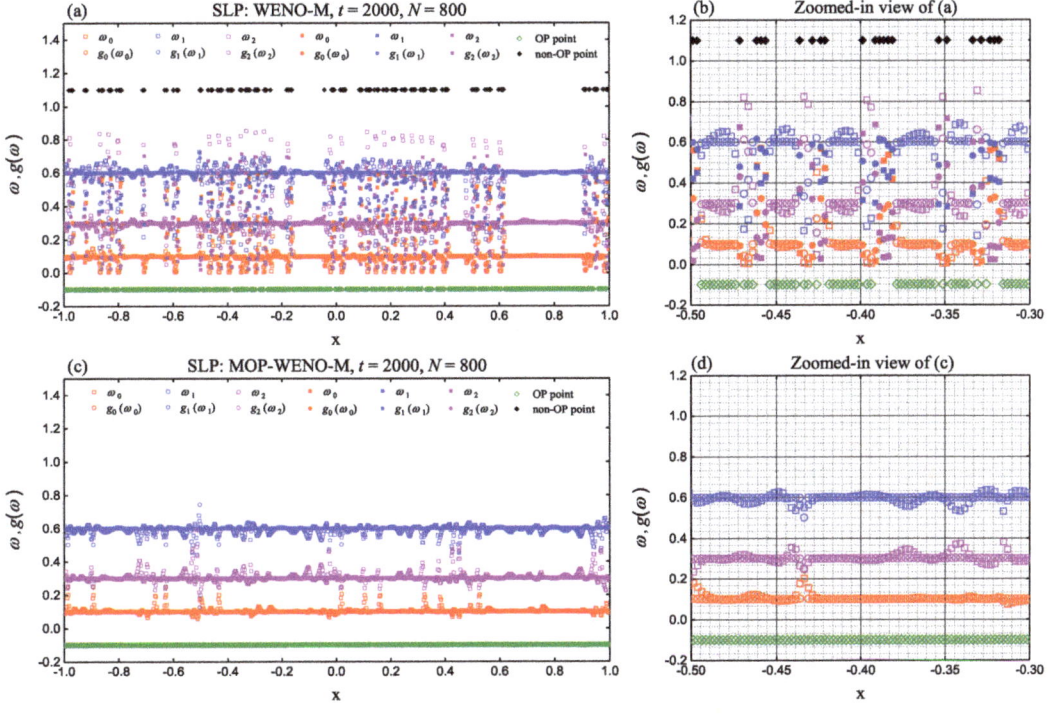

Figure 16. *Cont.*

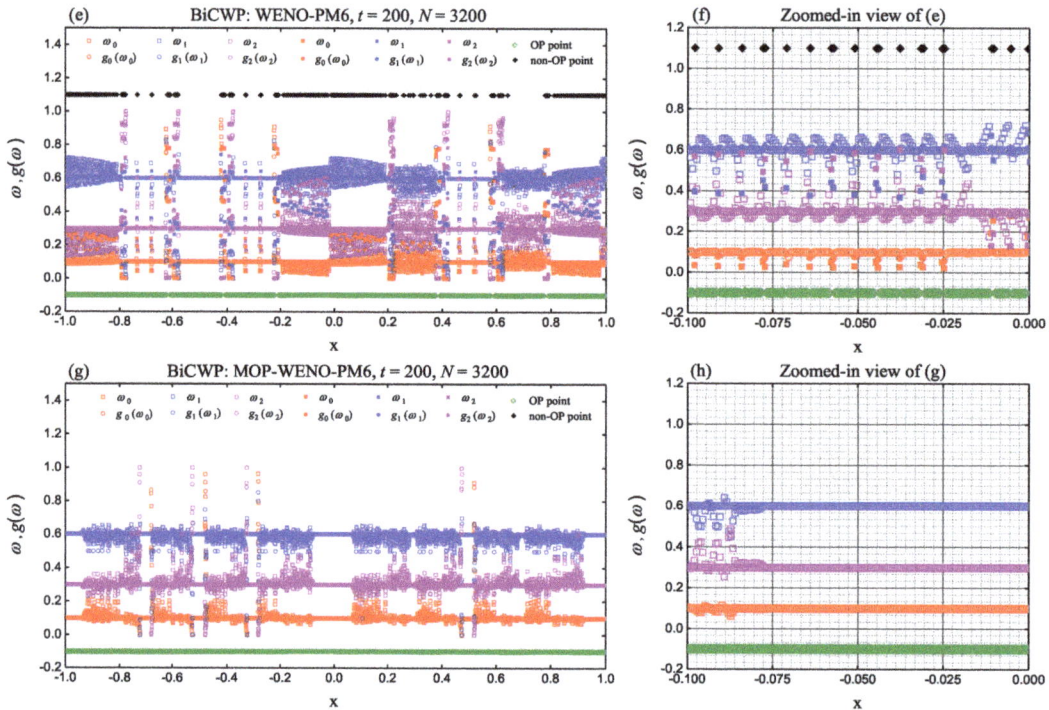

Figure 16. The *non-OP points* in the numerical solutions of SLP computed by the WENO-M and MOP-WENO-M schemes with $N = 800, t = 2000$, and the *non-OP points* in the numerical solutions of BiCWP computed by the WENO-PM6 and MOP-WENO-PM6 schemes with $N = 3200, t = 200$.

In summary, the solutions in this subsection could be regarded as numerical verifications of properties $C4, C5$ of Theorem 2. In other words, it could be indicated that the general method to introduce the *OP* mapping can help to gain the advantage of achieving high resolutions and in the meantime preventing spurious oscillations when solving problems with discontinuities for long output times. Additionally, this is the most important point we want to report in this paper.

4.3. Comparison with Central WENO Schemes

In this subsection, we compare the performances of the MOP-WENO-X schemes with the quite recent approach, called central WENO (CWENO) schemes. For simplicity, only the cases of the WENO-NW6 [36], WENO-CU6 [37] and WENO-θ6 [38] schemes are taken into account in the following discussion.

We firstly consider the following example.

Example 5. *We compute*

$$\begin{cases} u_t + u_x = 0, & x \in (-1, 1), \\ u(x, 0) = \max(-\sin(\pi x), 0), \end{cases} \quad (44)$$

with periodic boundary conditions.

We calculate this problem by the fifth-order MOP-WENO-X schemes proposed in the present work and the sixth-order central schemes of WENO-NW6, WENO-CU6 and WENO-θ6 schemes. The output time is taken to be $t = 2.4$ and the cell number is $N = 200$. The solutions are plotted in Figure 17. It clearly shows that the sixth-order central WENO schemes of WENO-NW6 and WENO-CU6 perform worse than the fifth-order MOP-WENO-

X schemes. It was reported by Jung et al. [38] that this loss of resolution is an important issue since there are many problems whose solution often exhibits the same behavior as this example. Therefore, we claim that the MOP-WENO-X schemes are more favorable than the central WENO schemes of WENO-NW6 and WENO-CU6 for this kind of problems. In addition, the the central WENO scheme of WENO-θ6 performs as well as the MOP-WENO-X schemes in this test. Unfortunately, it performs worse and gives significantly lower resolution than the MOP-WENO-X schemes on solving problems with discontinuities for long output times. We now discuss this in detail.

We calculate the problems of SLP and BiCWP (see Section 4.2) by using the sixth-order central schemes of WENO-NW6, WENO-CU6, and WENO-θ6 schemes. The computing conditions of $t = 200$ and $N = 3200$ are used here. In Figures 18 and 19, the results for SLP and BiCWP are shown. From these figures, we can see that the sixth-order central WENO schemes of WENO-NW6 and WENO-θ6 provide significantly lower resolutions than the fifth-order MOP-WENO-X schemes. The WENO-CU6 scheme appears to obtain the resolution equivalent to, or even better than those of the MOP-WENO-X schemes. However, it generates spurious oscillations, and the MOP-WENO-X schemes do not.

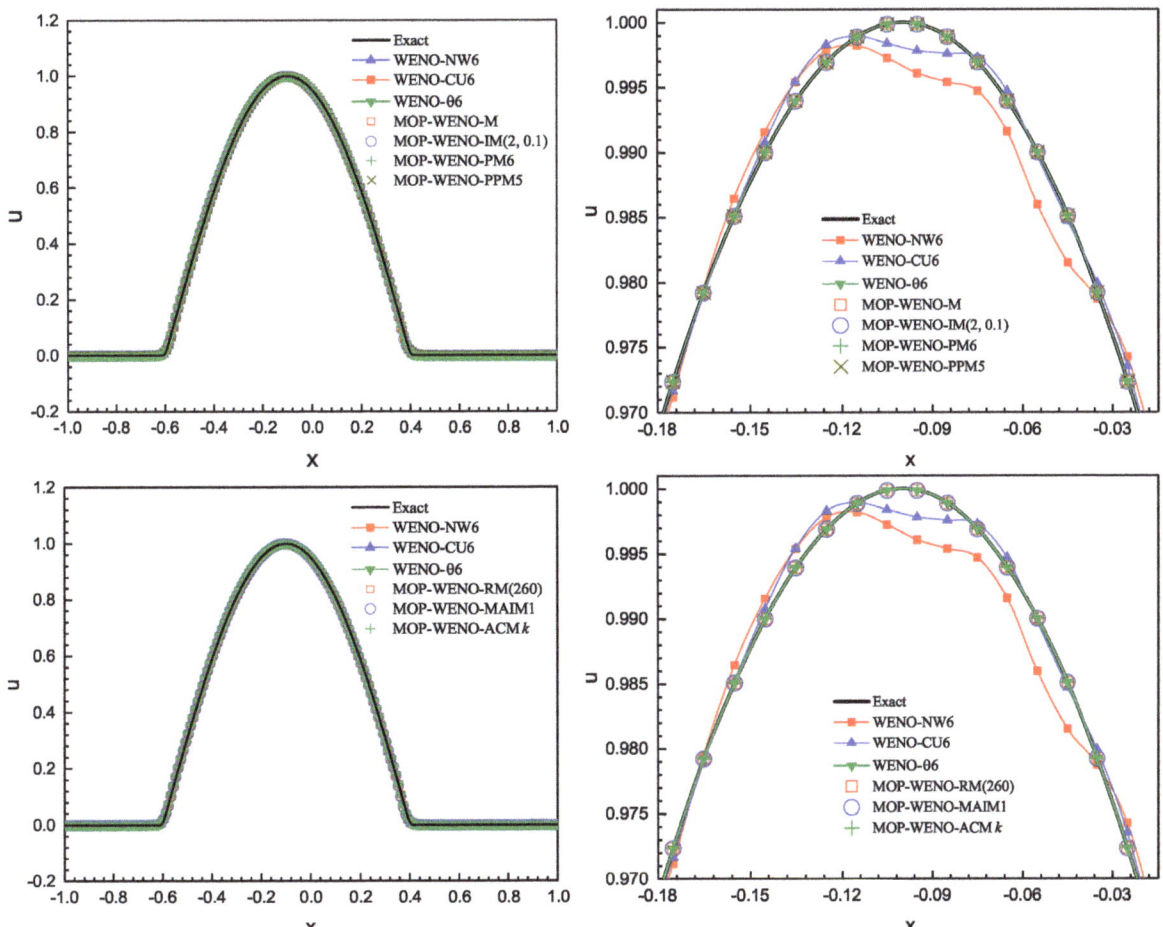

Figure 17. (**Left**): Numerical solutions of Equation (44) at time $t = 2.4$ obtained from different WENO schemes. (**Right**): Zoom near the critical region.

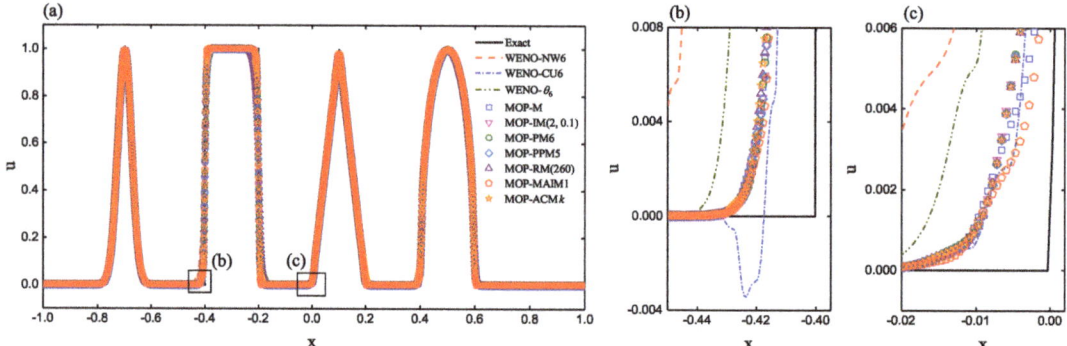

Figure 18. Performance of the WENO-NW6, WENO-CU6, WENO-θ6 and MOP-WENO-X schemes for the SLP at output time $t = 200$ with a uniform mesh size of $N = 3200$.

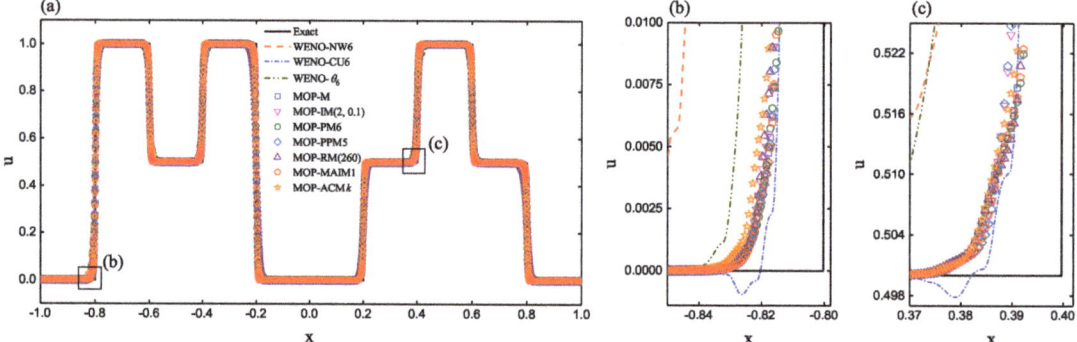

Figure 19. Performance of the WENO-NW6, WENO-CU6, WENO-θ6 and MOP-WENO-X schemes for the BiCWP at output time $t = 200$ with a uniform mesh size of $N = 3200$.

4.4. Euler System in Two Dimension

In this subsection, we focus on the numerical simulations of the shock-vortex interaction problem [41,42] and the 2D Riemann problem [43–45]. They are governed by the two-dimensional Euler system of gas dynamics, taking the following strong conservation form of mass, momentum and energy

$$
\begin{aligned}
\frac{\partial \rho}{\partial t} + \frac{\partial(\rho u)}{\partial x} + \frac{\partial(\rho v)}{\partial y} &= 0, \\
\frac{\partial(\rho u)}{\partial t} + \frac{\partial(\rho u^2 + p)}{\partial x} + \frac{\partial(\rho uv)}{\partial y} &= 0, \\
\frac{\partial(\rho v)}{\partial t} + \frac{\partial(\rho vu)}{\partial x} + \frac{\partial(\rho v^2 + p)}{\partial y} &= 0, \\
\frac{\partial E}{\partial t} + \frac{\partial(uE + up)}{\partial x} + \frac{\partial(vE + vp)}{\partial y} &= 0,
\end{aligned}
\tag{45}
$$

where ρ, u, v, p, and E are the density components of velocity in the x and y coordinate directions, pressure, and total energy, respectively. The following equation of state for an ideal polytropic gas is used to close the two-dimensional Euler system Equation (45)

$$p = (\gamma - 1)\left(E - \frac{1}{2}\rho(u^2 + v^2)\right),$$

where γ is the ratio of specific heat, and we set $\gamma = 1.4$ in this paper. In the computations below, the CFL number is taken to be 0.5. All the considered WENO schemes are applied dimension-by-dimension to solve the two-dimensional Euler system and the local characteristic decomposition [8] is used. In [46], Zhang et al. investigated two commonly used classes of finite volume WENO schemes in two-dimensional Cartesian meshes, and we employ the one denoted as class A in this subsection.

Example 6. (Shock-vortex interaction) *We consider the shock-vortex interaction problem used in [41,42]. It consists of the interaction of a left moving shock wave with a right moving vortex. The computational domain is initialized by*

$$(\rho, u, v, p)(x, y, 0) = \begin{cases} \mathbf{U}_L, & x < 0.5, \\ \mathbf{U}_R, & x \geq 0.5, \end{cases}$$

where $\mathbf{U}_L = (\rho_L, u_L, v_L, p_L) = (1, \sqrt{\gamma}, 0, 1)$, and $\mathbf{U}_R = (\rho_R, u_R, v_R, p_R)$ taking the form

$$p_R = 1.3, \rho_R = \rho_L \left(\frac{\gamma - 1 + (\gamma + 1)p_R}{\gamma + 1 + (\gamma - 1)p_R} \right)$$

$$u_R = u_L \left(\frac{1 - p_R}{\sqrt{\gamma - 1 + p_R(\gamma + 1)}} \right), v_R = 0.$$

The vortex $\delta \mathbf{U} = (\delta\rho, \delta u, \delta v, \delta p)$, defined by the following perturbations, is superimposed onto the left state $\mathbf{U}_L$,

$$\delta\rho = \frac{\rho_L^2}{(\gamma-1)p_L}\delta T, \delta u = \epsilon \frac{y - y_c}{r_c} e^{\alpha(1-r^2)}, \delta vs. = -\epsilon \frac{x - x_c}{r_c} e^{\alpha(1-r^2)}, \delta p = \frac{\gamma \rho_L^2}{(\gamma-1)\rho_L}\delta T,$$

where $\epsilon = 0.3, r_c = 0.05, \alpha = 0.204, x_c = 0.25, y_c = 0.5, r = \sqrt{((x-x_c)^2 + (y-y_c)^2)/r_c^2}$, $\delta T = -(\gamma-1)\epsilon^2 e^{2\alpha(1-r^2)}/(4\alpha\gamma)$. The transmissive boundary condition is used on all boundaries. A uniform mesh size of 800×800 is used and the output time is set to be $t = 0.35$.

We calculate this problem using all the considered mapped WENO-X schemes in Table 1 and their associated MOP-WENO-X schemes. For the sake of brevity though, we only present the solutions of the WENO-M, WENO-IM(2, 0.1), WENO-PPM5, WENO-MAIM1 schemes and their associated MOP-WENO-X schemes in Figures 20 and 21, where the first rows give the final structures of the shock and vortex in density profile of the existing mapped WENO-X schemes, the second rows give those of the associated MOP-WENO-X schemes, and the third rows give the cross-sectional slices of density plot along the plane $y = 0.65$ where $x \in [0.70, 0.76]$. We find that all the considered schemes perform well in capturing the main structure of the shock and vortex after the interaction. It can be seen that there are clear post-shock oscillations in the solutions of the WENO-M, WENO-IM(2, 0.1), and WENO-PPM5 schemes. However, in the solutions of the MOP-WENO-M, MOP-WENO-IM(2, 0.1), and MOP-WENO-PPM5 schemes, the post-shock oscillations are either gone or significantly reduced. The post-shock oscillations of the WENO-MAIM1 scheme are very slight and even hard to be noticed. Actually, it seems difficult to distinguish the solutions of the WENO-MAIM1 scheme from that of the MOP-WENO-MAIM1 scheme only according to the structure of the shock and vortex in the density profile. Nevertheless, when taking a closer look from the cross-sectional slices of the density profile along the plane $y = 0.65$ at the bottom right picture of Figure 21 where the reference solution is obtained using the WENO-JS scheme with a uniform mesh size of 1600×1600, we can see that the post-shock oscillation of the WENO-MAIM1 scheme is very remarkable while it is imperceptible for the MOP-WENO-MAIM1 scheme. Additionally, from the third rows of Figures 20 and 21, we find that the WENO-IM(2, 0.1) and WENO-PPM5 schemes generate the post-shock oscillations with much bigger amplitudes than that of the WENO-MAIM1 scheme. The WENO-M scheme also generates clear post-shock oscillations with the amplitudes slightly smaller than that of the WENO-IM(2, 0.1) and WENO-PPM5 schemes. Evidently, the solutions of the MOP-WENO-M, MOP-WENO-IM(2, 0.1) and

MOP-WENO-PPM5 schemes almost generate no post-shock oscillations or only generate some imperceptible numerical oscillations and their solutions are very close to the reference solution, and this should be an advantage of the mapped WENO schemes whose mapping functions are *OP*.

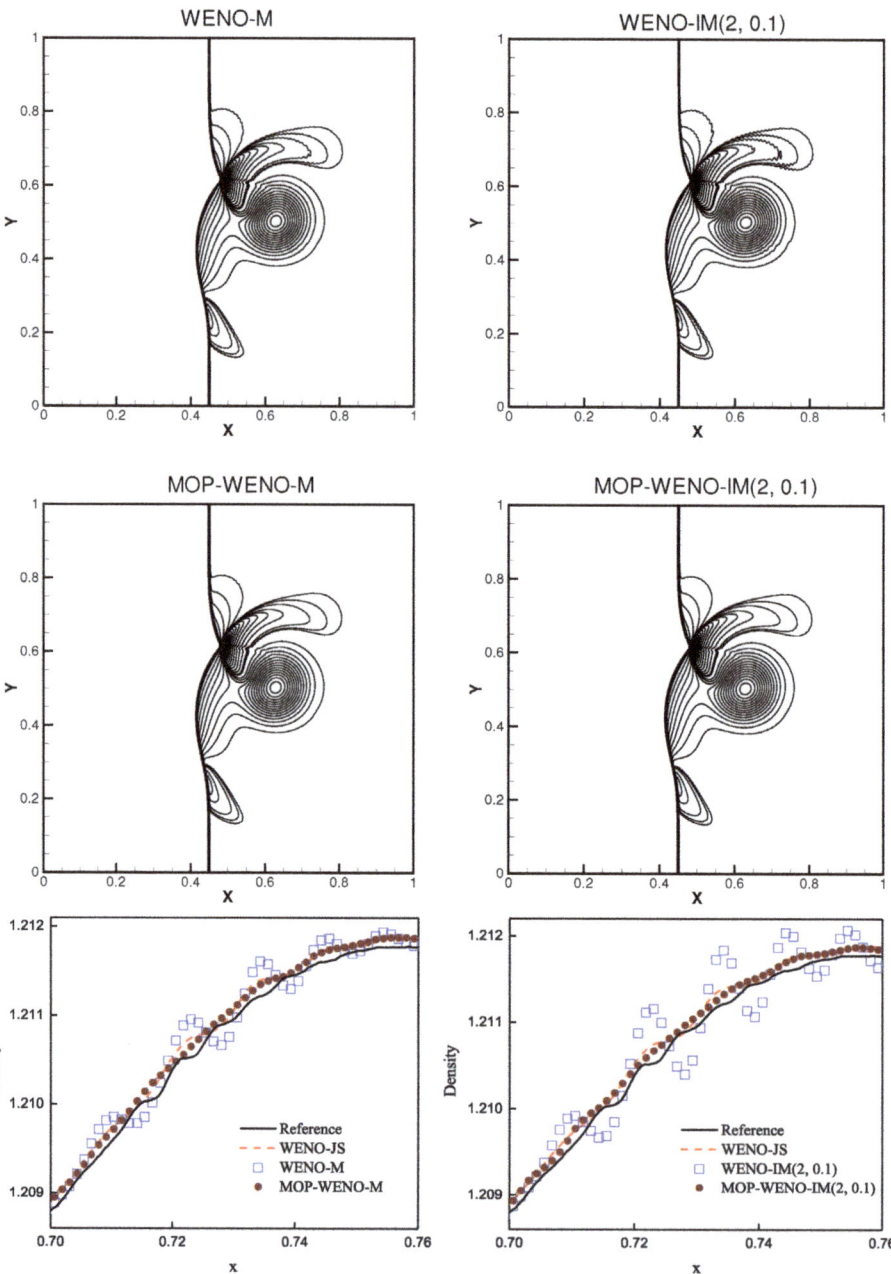

Figure 20. Density plots for the Shock-vortex interaction using 30 contour lines with range from 0.9 to 1.4 (the first two rows) and the cross-sectional slices of density plot along the plane $y = 0.65$ where $x \in [0.70, 0.76]$ (the third row), computed using the WENO-M and MOP-WENO-M (**left column**), WENO-IM(2, 0.1), and MOP-WENO-IM(2, 0.1) (**right column**) schemes.

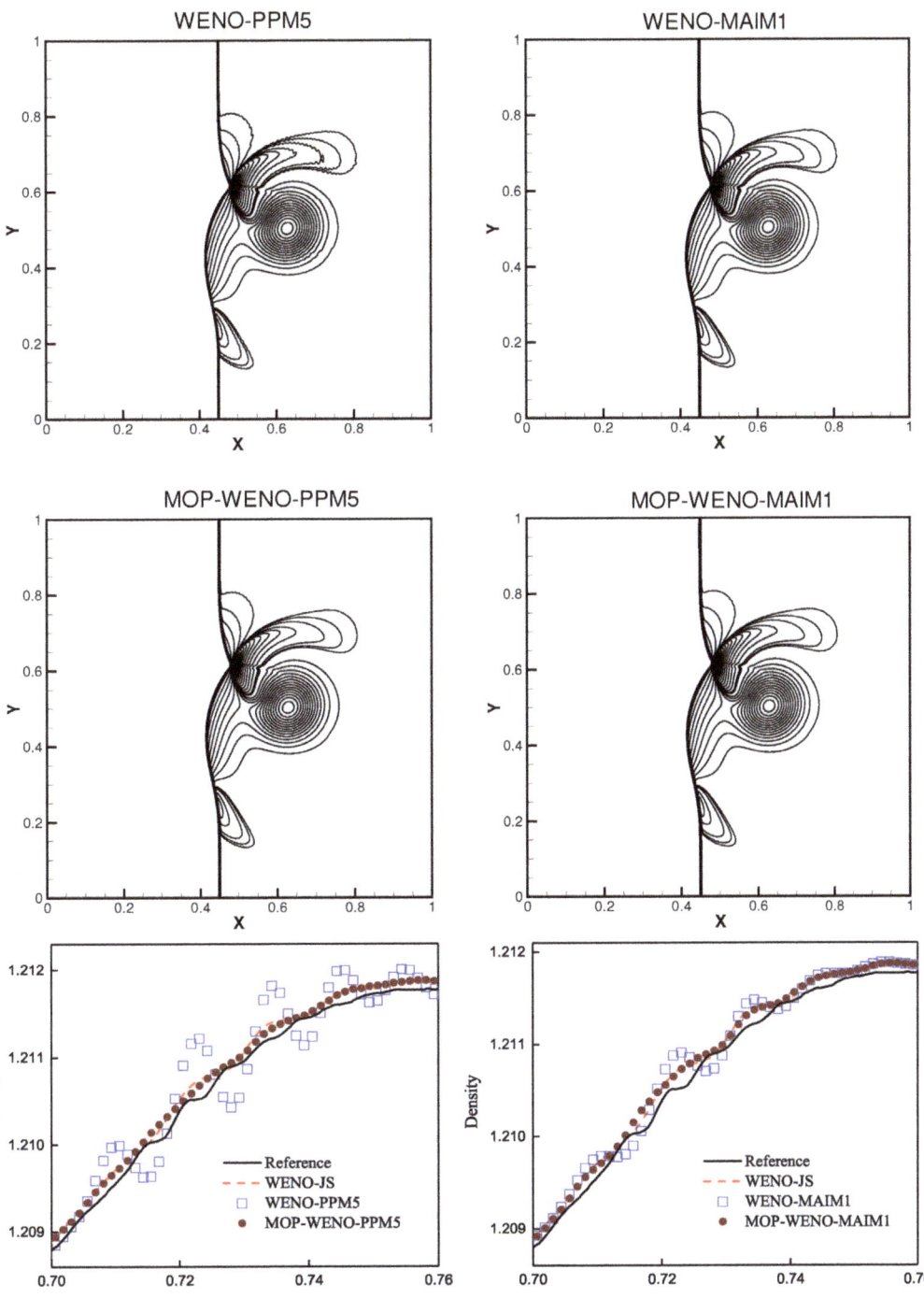

Figure 21. Density plots for the Shock-vortex interaction using 30 contour lines with range from 0.9 to 1.4 (the first two rows) and the cross-sectional slices of density plot along the plane $y = 0.65$ where $x \in [0.70, 0.76]$ (the third row), computed using the WENO-PPM5 and MOP-WENO-PPM5 (**left column**), WENO-MAIM1 and MOP-WENO-MAIM1 (**right column**) schemes.

Example 7. (2D Riemann problem) *It is very favorable to test the high-resolution numerical methods [30,45,47] using the series of 2D Riemann problems [43,44]. In [45], Lax et al. classified a total of 19 genuinely different Configurations for 2D Riemann problem and calculated all the numerical solutions. Configuration 4 is chosen here for the test, and the computational domain is initialized by*

$$(\rho, u, v, p)(x, y, 0) = \begin{cases} (1.1, 0.0, 0.0, 1.1), & 0.5 \leq x \leq 1.0, 0.5 \leq y \leq 1.0, \\ (0.5065, 0.8939, 0.0, 0.35), & 0.0 \leq x \leq 0.5, 0.5 \leq y \leq 1.0, \\ (1.1, 0.8939, 0.8939, 1.1), & 0.0 \leq x \leq 0.5, 0.0 \leq y \leq 0.5, \\ (0.5065, 0.0, 0.8939, 0.35), & 0.5 \leq x \leq 1.0, 0.0 \leq y \leq 0.5. \end{cases}$$

The transmission boundary condition is used on all boundaries, and the numerical solutions are calculated on a uniform mesh size of 800×800. The computations proceed to $t = 0.25$.

Similarly, although we calculate this problem using all the considered mapped WENO-X schemes in Table 1 and their associated MOP-WENO-X schemes, we only present the solutions of the WENO-M, WENO-PM6, WENO-RM260 and MIP-WENO-ACMk schemes and their associated MOP-WENO-X schemes here for the sake of brevity. We have shown the numerical results of density obtained by using these schemes in Figures 22 and 23, where the first rows give the structures of the 2D Riemann problem in density profile of the existing mapped WENO-X schemes, the second rows give those of the associated MOP-WENO-X schemes, and the third rows give the cross-sectional slices of density plot along the plane $y = 0.5$ where $x \in [0.65, 0.692]$. We can see that all schemes can capture the main structure of the solution. However, we can also observe that there are obvious post-shock oscillations (as marked by the pink boxes), which are unfavorable for the fidelity of the results, in the solutions of the WENO-M, WENO-PM6, WENO-RM(260) and MIP-WENO-ACMk schemes. These post-shock oscillations can be seen more clearly from the cross-sectional slices of density profile as presented in the third rows of Figures 22 and 23, where the reference solution is obtained by using the WENO-JS scheme with a uniform mesh size of 3000×3000. Noticeably, there are either almost no or imperceptible post-shock oscillations in the solutions of the MOP-WENO-M, MOP-WENO-PM6, MOP-WENO-RM(RM260) and MOP-WENO-ACMk schemes. Again, we believe that this should be an advantage of the mapped WENO schemes whose mapping functions are *OP*.

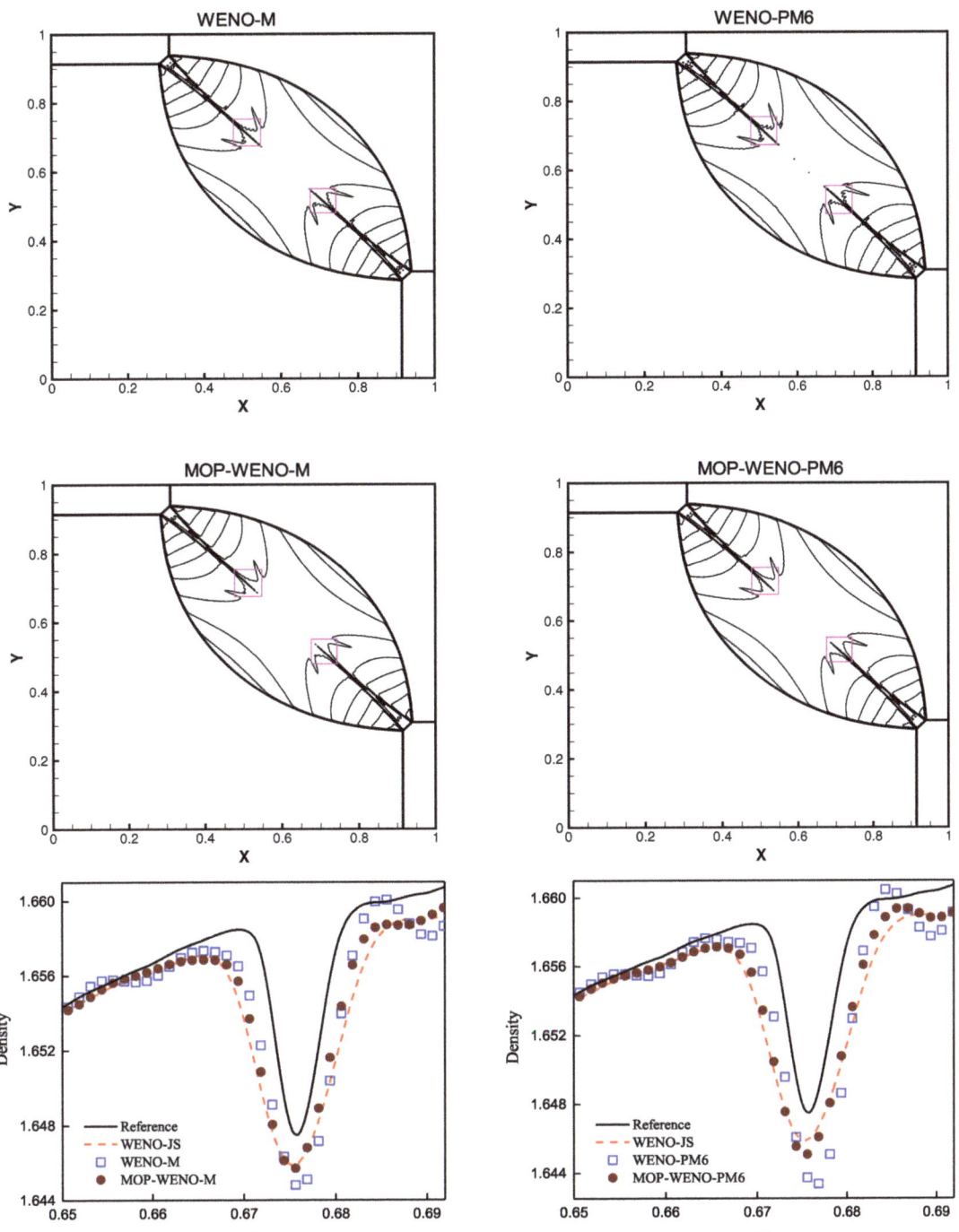

Figure 22. Density plots for the 2D Riemann problem using 30 contour lines with range from 0.5 to 1.9 (the first two rows) and the cross-sectional slices of density plot along the plane $y = 0.5$ where $x \in [0.65, 0.692]$ (the third row), computed using the WENO-M and MOP-WENO-M (**left column**), WENO-PM6 and MOP-WENO-PM6 (**right column**) schemes.

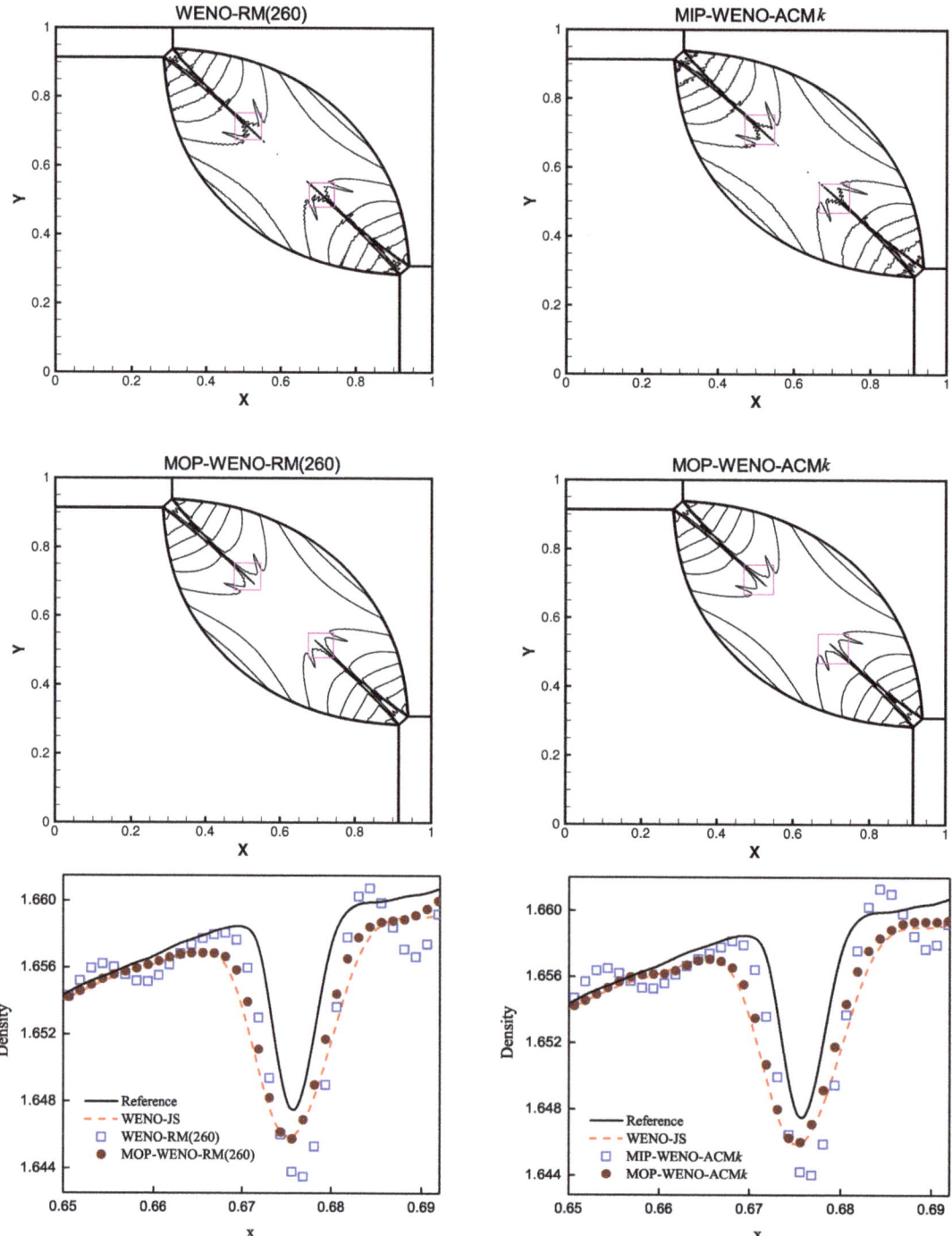

Figure 23. Density plots for the 2D Riemann problem using 30 contour lines with range from 0.5 to 1.9 (the first two rows) and the cross-sectional slices of density plot along the plane $y = 0.5$ where $x \in [0.65, 0.692]$ (the third row), computed using the WENO-RM(260) and MOP-WENO-RM(260) (**left column**), MIP-WENO-ACMk and MOP-WENO-ACMk (**right column**) schemes.

5. Conclusions

The concept of *OP-Mapped WENO* schemes standing for the family of the mapped WENO schemes with *order-preserving (OP)* mappings, as well as a general way to build one group of this kind of schemes, has been proposed in this paper. Specifically, we extended the *OP* mapping introduced in [9] to various existing mapped WENO schemes in references by providing a general formula of their mapping functions. A systematic analysis has been performed to prove that the improved mapped WENO scheme based on the existing mapped WENO-X scheme, denoted as MOP-WENO-X, generates numerical solutions with the same convergence rates of accuracy in smooth regions as the associated WENO-X scheme. Furthermore, numerical experiments were run to show that the MOP-WENO-X schemes have the same advantage as the mapped WENO scheme proposed in [9] in calculating the one-dimensional linear advection problems including discontinuities with long output times. The mapping functions of the MOP-WENO-X schemes are *OP* and hence able to attain high resolutions and avoid spurious oscillations meanwhile. Moreover, numerical results with the 2D Euler system problems were presented to show that the MOP-WENO-X schemes perform well in simulating the two-dimensional problems with strong shock waves to capture the main flow structures and remove or significantly reduce the post-shock oscillations.

Author Contributions: All authors have equally contributed to this paper. They have read and approved the final version of the manuscript. All authors have read and agreed to the published version of the manuscript.

Funding: This research received no external funding.

Conflicts of Interest: The authors declare no conflict of interest.

References

1. Harten, A.; Engquist, B.; Osher, S.; Chakravarthy, S.R. Uniformly high order accurate essentially non-oscillatory schemes III. *J. Comput. Phys.* **1987**, *71*, 231–303. [CrossRef]
2. Harten, A.; Osher, S. Uniformly high order accurate essentially non-oscillatory schemes I. *SIAM J. Numer. Anal.* **1987**, *24*, 279–309. [CrossRef]
3. Harten, A.; Osher, S.; Engquist, B.; Chakravarthy, S.R. Some results on uniformly high order accurate essentially non-oscillatory schemes. *Appl. Numer. Math.* **1986**, *2*, 347–377. [CrossRef]
4. Harten, A. ENO schemes with subcell resolution. *J. Comput. Phys.* **1989**, *83*, 148–184. [CrossRef]
5. Shu, C.W.; Osher, S. Efficient implementation of essentially non-oscillatory shock-capturing schemes. *J. Comput. Phys.* **1988**, *77*, 439–471. [CrossRef]
6. Shu, C.W.; Osher, S. Efficient implementation of essentially non-oscillatory shock-capturing schemes II. *J. Comput. Phys.* **1989**, *83*, 32–78. [CrossRef]
7. Liu, X.D.; Osher, S.; Chan, T. Weighted essentially non-oscillatory schemes. *J. Comput. Phys.* **1994**, *115*, 200–212. [CrossRef]
8. Jiang, G.S.; Shu, C.W. Efficient implementation of weighted ENO schemes. *J. Comput. Phys.* **1996**, *126*, 202–228. [CrossRef]
9. Li, R.; Zhong, W. A new mapped WENO scheme using order-preserving mapping. *Commun. Comput. Phys.* **2021**, under review.
10. Dumbser, M.; Zanotti, O.; Hidalgo, A.; Balsara, D.S. ADER-WENO finite volume schemes with space-time adaptive mesh refinement. *J. Comput. Phys.* **2013**, *248*, 257–286. [CrossRef]
11. Boscheri, W.; Dumbser, M. An Efficient Quadrature-Free Formulation for High Order Arbitrary-Lagrangian–Eulerian ADER-WENO Finite Volume Schemes on Unstructured Meshes. *J. Sci. Comput.* **2016**, *66*, 240–274. [CrossRef]
12. Boscheri, W.; Semplice, M.; Dumbser, M. Central WENO Subcell Finite Volume Limiters for ADER Discontinuous Galerkin Schemes on Fixed and Moving Unstructured Meshes. *Commun. Comput. Phys.* **2019**, *25*, 311–346. [CrossRef]
13. Boscheri, W.; Dumbser, M. High order accurate direct Arbitrary-Lagrangian–Eulerian ADER-WENO finite volume schemes on moving curvilinear unstructured meshes. *Comput. Fluids* **2016**, *136*, 48–66. [CrossRef]
14. Tsoutsanis, P.; Titarev, V.A.; Drikakis, D. WENO schemes on arbitrary mixed-element unstructured meshes in three space dimensions. *J. Comput. Phys.* **2011**, *230*, 1585–1601. [CrossRef]
15. Titarev, V.A.; Tsoutsanis, P.; Drikakis, D. WENO Schemes for Mixed-Element Unstructured Meshes. *Commun. Comput. Phys.* **2010**, *8*, 585–609. [CrossRef]
16. Titarev, V.A.; Toro, E.F. ADER: Arbitrary High Order Godunov Approach. *J. Sci. Comput.* **2002**, *17*, 609–618. [CrossRef]
17. Titarev, V.A.; Toro, E.F. Finite-volume WENO schemes for three-dimensional conservation laws. *J. Comput. Phys.* **2004**, *201*, 238–260. [CrossRef]

18. Titarev, V.A.; Toro, E.F. ADER schemes for three-dimensional non-linear hyperbolic systems. *J. Comput. Phys.* **2005**, *204*, 715–736. [CrossRef]
19. Titarev, V.A.; Toro, E.F. WENO schemes based on upwind and centred TVD fluxes. *Comput. Fluids* **2005**, *34*, 705–720. [CrossRef]
20. Semplice, M.; Coco, A.; Russo, G.; Titarev, V.A.; Toro, E.F. Adaptive Mesh Refinement for Hyperbolic Systems Based on Third-Order Compact WENO Reconstruction. *J. Sci. Comput.* **2016**, *66*, 692–724. [CrossRef]
21. Cravero, I.; Puppo, G.; Semplice, M.; Visconti, G. Cool WENO schemes. *Comput. Fluids* **2018**, *169*, 71–86. [CrossRef]
22. Puppo, G.; Semplice, M. Well-Balanced High Order 1D Schemes on Non-uniform Grids and Entropy Residuals. *J. Sci. Comput.* **2016**, *66*, 1052–1076. [CrossRef]
23. Levy, D.; Puppo, G.; Russo, G. Compact central WENO schemes for multidimensional conservation laws. *SIAM J. Sci. Comput.* **2000**, *22*, 656–672. [CrossRef]
24. Levy, D.; Puppo, G.; Russo, G. A fourth-order central WENO scheme for multidimensional hyperbolic systems of conservation laws. *SIAM J. Sci. Comput.* **2002**, *24*, 480–506. [CrossRef]
25. Henrick, A.K.; Aslam, T.D.; Powers, J.M. Mapped weighted essentially non-oscillatory schemes: Achieving optimal order near critical points. *J. Comput. Phys.* **2005**, *207*, 542–567. [CrossRef]
26. Borges, R.; Carmona, M.; Costa, B.; Don, W.S. An improved weighted essentially non-oscillatory scheme for hyperbolic conservation laws. *J. Comput. Phys.* **2008**, *227*, 3191–3211. [CrossRef]
27. Feng, H.; Huang, C.; Wang, R. An improved mapped weighted essentially non-oscillatory scheme. *Appl. Math. Comput.* **2014**, *232*, 453–468. [CrossRef]
28. Wang, R.; Feng, H.; Huang, C. A New Mapped Weighted Essentially Non-oscillatory Method Using Rational Function. *J. Sci. Comput.* **2016**, *67*, 540–580. [CrossRef]
29. Feng, H.; Hu, F.; Wang, R. A new mapped weighted essentially non-oscillatory scheme. *J. Sci. Comput.* **2012**, *51*, 449–473. [CrossRef]
30. Li, Q.; Liu, P.; Zhang, H. Piecewise Polynomial Mapping Method and Corresponding WENO Scheme with Improved Resolution. *Commun. Comput. Phys.* **2015**, *18*, 1417–1444. [CrossRef]
31. Li, R.; Zhong, W. A modified adaptive improved mapped WENO method. *Commun. Comput. Phys.* **2021**, accepted.
32. Li, R.; Zhong, W. An efficient mapped WENO scheme using approximate constant mapping. *Numer. Math. Theor. Meth. Appl.* **2021**, accepted.
33. Zhang, S.; Shu, C.W. A new smoothness indicator for the WENO schemes and its effect on the convergence to steady state solutions. *J. Sci. Comput.* **2007**, *31*, 273–305. [CrossRef]
34. Wang, R.; Feng, H. Observations on the fifth-order WENO method with non-uniform meshes. *Appl. Math. Comput.* **2008**, *196*, 433–447. [CrossRef]
35. Hu, G.; Li, R.; Tang, T. A robust WENO type finite volume solver for steady Euler equations on unstructured grids. *Commun. Comput. Phys.* **2011**, *9*, 627–648. [CrossRef]
36. Yamaleev, N.K.; Carpenter, M.H. A systematic methodology for constructing high-order energy stable WENO schemes. *J. Comput. Phys.* **2009**, *228*, 4248–4272. [CrossRef]
37. Fu, L.; Hu, X.; Adams, N.A. A family of high-order targeted ENO schemes for compressible-fluid simulations. *J. Comput. Phys.* **2016**, *305*, 333–359. [CrossRef]
38. Jung, C.Y.; Nguyen, T.B. A new adaptive weighted essentially non-oscillatory WENO-θ scheme for hyperbolic conservation laws. *J. Comput. Appl. Math.* **2018**, *328*, 314–339. [CrossRef]
39. Gottlieb, S.; Shu, C.W. Total variation diminishing Runge–Kutta schemes. *Math. Comput.* **1998**, *67*, 73–85. [CrossRef]
40. Gottlieb, S.; Shu, C.W.; Tadmor, E. Strong stability-preserving high-order time discretization methods. *SIAM Rev.* **2001**, *43*, 89–112. [CrossRef]
41. Chatterjee, A. Shock wave deformation in shock-vortex interactions. *Shock Waves* **1999**, *9*, 95–105. [CrossRef]
42. Ren, Y.X.; Liu, M.; Zhang, H. A characteristic-wise hybrid compact-WENO scheme for solving hyperbolic conservation laws. *J. Comput. Phys.* **2003**, *192*, 365–386. [CrossRef]
43. Schulz-Rinne, C.W.; Collins, J.P.; Glaz, H.M. Numerical solution of the Riemann problem for two-dimensional gas dynamics. *SIAM J. Sci. Comput.* **1993**, *14*, 1394–1414. [CrossRef]
44. Schulz-Rinne, C.W. Classification of the Riemann problem for two-dimensional gas dynamics. *SIAM J. Math. Anal.* **1993**, *24*, 76–88. [CrossRef]
45. Lax, P.D.; Liu, X.D. Solution of two-dimensional Riemann problems of gas dynamics by positive schemes. *SIAM J. Sci. Comput.* **1998**, *19*, 319–340. [CrossRef]
46. Zhang, R.; Zhang, M.; Shu, C.W. On the order of accuracy and numerical performance of two classes of finite volume WENO schemes. *Commun. Comput. Phys.* **2011**, *9*, 807–827. [CrossRef]
47. Pirozzoli, S. Numerical methods for high-speed flows. *Annu. Rev. Fluid Mech.* **2011**, *43*, 163–194. [CrossRef]

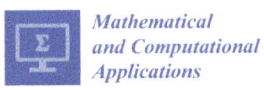

Article
Modified Representations for the Close Evaluation Problem

Camille Carvalho

Department of Applied Mathematics, University of California Merced, 5200 North Lake Road, Merced, CA 95343, USA; ccarvalho3@ucmerced.edu

Abstract: When using boundary integral equation methods, we represent solutions of a linear partial differential equation as layer potentials. It is well-known that the approximation of layer potentials using quadrature rules suffer from poor resolution when evaluated closed to (but not on) the boundary. To address this challenge, we provide modified representations of the problem's solution. Similar to Gauss's law used to modify Laplace's double-layer potential, we use modified representations of Laplace's single-layer potential and Helmholtz layer potentials that avoid the close evaluation problem. Some techniques have been developed in the context of the representation formula or using interpolation techniques. We provide alternative modified representations of the layer potentials directly (or when only one density is at stake). Several numerical examples illustrate the efficiency of the technique in two and three dimensions.

Keywords: boundary integral equations; layer potential identities; density subtractions; quadrature rules

Citation: Carvalho, C. Modified Representations for the Close Evaluation Problem. *Math. Comput. Appl.* **2021**, *26*, 69. https://doi.org/10.3390/mca26040069

Received: 28 June 2021
Accepted: 23 September 2021
Published: 28 September 2021

Publisher's Note: MDPI stays neutral with regard to jurisdictional claims in published maps and institutional affiliations.

Copyright: © 2021 by the author. Licensee MDPI, Basel, Switzerland. This article is an open access article distributed under the terms and conditions of the Creative Commons Attribution (CC BY) license (https://creativecommons.org/licenses/by/4.0/).

1. Introduction

One can represent the solution of partial differential boundary-value problems using boundary integral equation methods, which involves integral operators defined on the domain's boundary called layer potentials. Using layer potentials, the solution can be evaluated anywhere in the domain without restriction to a particular mesh. For that reason boundary integral equations have found broad applications, including in fluid mechanics, electromagnetics, and plasmonics [1–8].

The close evaluation problem refers to the nonuniform error produced by high-order quadrature rules used to discretize layer potentials. This phenomenon arises when computing the solution close to the boundary (i.e., at close evaluation points). It is well understood that this growth in error is due to the fact that the integrands of the layer potentials become increasingly peaked as the evaluation point approaches the boundary (nearly singular behavior), leading in limited cases to an $O(1)$ error [9].

There exists a plethora of manners to address the close evaluation problem: using extraction methods based on Taylor series expansions [10], regularizing the nearly singular behavior of the integrand and adding corrections [11,12], compensating quadrature rules via interpolation [13], using Quadrature By Expansion related techniques (QBX) [9,14–19], using adaptive methods [20], using singularity subtraction techniques and interpolation [21–23], or using asymptotic approximations [24–26], to name a few. Most techniques rely on either providing corrections to the *kernel* (related to the fundamental solution of the PDE at stake), or to the *density* (solution of the boundary integral equation).

In the latter category, it is well-known that Laplace's double-layer potential can be straightforwardly modified via a *density subtraction technique* based on Gauss' law (e.g., [27]). This modification alleviates the close evaluation problem, and provides a better approximation for any given numerical method. However this identity technique is specific to Laplace's double-potential. Other identities have been derived for other problems, such as for the elastostatic problem [28].

In this paper, we provide modified representations of layer potentials, and we give guidance to address the close evaluation problem in two and three dimensions. In particular, we modify Laplace's single-layer potential (representing the solution of the exterior Neumann Laplace problem) and Helmholtz layer potentials (in the context of a sound-soft scattering problem). With some given quadrature rule, the resulted modified representations allow us to obtain better approximations compared to standard representations. The proposed modifications are based on subtracting specific solutions (or *auxiliary functions*) of the PDE at stake. The use of auxiliary functions have been developed in the context of Boundary Regularized Integral Equation Formulation (BRIEF) [29–31] to regularize the representation formula on the boundary, or in the context of density interpolation techniques [21,23,32] to regularize layer potentials (generalization of density subtractions). Those techniques commonly consider multiple auxiliary functions, and may require to solve additional problems to find such functions. The proposed work concentrates on regularizing nearly singular integrals using explicitly one analytic auxiliary function, and when representing the solution with layer potentials involving only one density (no representation formula). We provide several examples of auxiliary functions (and compare them), and provide guidelines to find them. The proposed modified representations are simple and easy to implement, and allow one to straightforwardly gain accuracy in evaluating the solution, especially when computational resources are limited. This work provides valuable insights into Laplace and Helmholtz layer potentials. Additionally this can also be applied to modify boundary integral equations to avoid weakly singular integrals.

The paper is organized as follows: Section 2 presents some context and motivation for the proposed modified representations. Section 3 establishes the modified representations and general guidelines to find appropriate auxiliary functions. Sections 4 and 5 illustrate the efficiency of the modified representations for Laplace and Helmholtz in two and three dimensions, off and on boundary. Finally, Section 6 presents our concluding remarks, Appendices A and B provide a brief summary of the Nyström methods used in two and three dimensions, and Appendix C details some proofs for Section 3.

2. Motivation for Modified Representations

Consider a *domain* $D \subset \mathbb{R}^d$, $d = 2, 3$, that is a bounded simply connected open set with smooth boundary (of class $\mathcal{C}^2$), and a linear elliptic partial differential equation of the form $\mathcal{L}u = 0$. It is common to represent the solution v of that PDE using the so-called representation formula (e.g., Theorem 6.5 in [33], Theorem 3.1 in [34]). In particular for v satisfying $\mathcal{L}v = 0$ in D, we have the following identities:

$$\int_{\partial D} \partial_{n_y} G(x,y) v(y) d\sigma_y - \int_{\partial D} G(x,y) \partial_{n_y} v(y) d\sigma_y = \begin{cases} -v(x) & x \in D, \\ -\frac{1}{2} v(x) & x \in \partial D, \\ 0 & x \in E := \mathbb{R}^d \setminus \bar{D}, \end{cases} \quad (1)$$

where G denotes the fundamental solution of considered PDE, n_y is the unit outward normal of D at y, and $d\sigma_y$ is the integration surface element. For instance, (1) holds true for $\mathcal{L} := \Delta$ and $\mathcal{L} := \Delta + k^2$, the Laplace and the Helmholtz equation, respectively. The goal of this paper is to use (1) with well-chosen v to modify the representation of the solution of boundary value problems associated with $\mathcal{L}$. Let us illustrate the strategy with, for example, the Exterior Neumann Laplace problem:

$$\begin{vmatrix} \text{Find } u \in \mathcal{C}^2(E) \cap \mathcal{C}^1(\bar{E} := \mathbb{R}^d \setminus D) \text{ such that:} \\ \Delta u = 0 \quad \text{in } E, \quad \partial_n u = g \quad \text{on } \partial D, \quad \lim_{|x| \to \infty} u(x) = o(1), \end{vmatrix} \quad (2)$$

with some smooth data g (with null average). The solution of Problem (2) can be represented using Green's formula [34,35]:

$$\begin{aligned} u(x) &= \int_{\partial D} \partial_{n_y} G(x,y) u(y) \, d\sigma_y - \int_{\partial D} G(x,y) \partial_{n_y} u(y) \, d\sigma_y, \quad x \in E, \\ &= \int_{\partial D} \partial_{n_y} G(x,y) u(y) \, d\sigma_y - \int_{\partial D} G(x,y) g(y) \, d\sigma_y, \quad x \in E, \end{aligned} \quad (3)$$

where

$$G(x,y) = \begin{cases} -\dfrac{1}{2\pi} \log|x-y| & \text{for } d=2, \\ \dfrac{1}{4\pi} \dfrac{1}{|x-y|} & \text{for } d=3, \end{cases} \qquad (4)$$

and the trace on the boundary satisfies the boundary integral equation of the second kind:

$$\frac{1}{2} u(x^*) - \int_{\partial D} \partial_{n_y} G(x^*, y) u(y) \, d\sigma_y = \int_{\partial D} G(x^*, y) g(y) \, d\sigma_y, \quad x^* \in \partial D. \qquad (5)$$

The fundamental solution G is singular when $y = x^*$. For $x \in \mathbb{R}^d \setminus \partial D$, assume we can write $x = x^* \pm \ell n_{x^*}$ with n_{x^*} the unit outward normal at x^*, and $\ell > 0$ the distance from the boundary. Then G is *nearly singular* at $y = x^*$ when $|x - y| = \ell \ll 1$ (i.e., when x is close to the boundary). A layer potential is said to be a weakly singular integral (resp. a nearly singular integral) when its kernel (G or $\partial_n G$ in the cases above) is singular at $y = x^*$ (resp. nearly singular at $y = x^*$). There exist high-order quadrature rules to approximate weakly singular integrals with very high accuracy (e.g., [36–39]). However, high accuracy is lost for nearly singular integrals: this is the so-called close evaluation problem. Assuming we have solved (5), we can modify (3) using (1) to address the close evaluation problem. Taking the difference we obtain

$$u(x) = \int_{\partial D} \partial_{n_y} G(x,y)[u(y) - v(y)] \, d\sigma_y - \int_{\partial D} G(x,y)[g(y) - \partial_{n_y} v(y)] \, d\sigma_y, \quad x \in E. \qquad (6)$$

If one finds v such that $v(x^*) = u(x^*)$ and $\partial_{n_{x^*}} v(x^*) = g(x^*)$, where $x^* \in \partial D$ denotes the closest boundary point of the evaluation point x ($x = x^* + \ell n_{x^*}$), then (6) does not suffer from the close evaluation problem.

Similarly, one can represent the solution of Problem (2) using *a single-density representation* given by the single-layer potential:

$$u(x) = \int_{\partial D} G(x,y) \rho(y) \, d\sigma_y, \quad x \in D, \qquad (7)$$

with ρ a continuous density solution of the boundary integral equation of the second-kind:

$$-\frac{1}{2} \rho(x^*) + \int_{\partial D} \partial_{n_x^*} G(x^*, y) \rho(y) \, d\sigma_y = g(x^*), \quad x^* \in \partial D. \qquad (8)$$

Assuming we have solved (8) for ρ, subtracting (1) from (7) we obtain

$$u(x) = \int_{\partial D} G(x,y)[\rho(y) - \partial_{n_y} v(y)] \, d\sigma_y + \int_{\partial D} \partial_{n_y} G(x,y) v(y) \, d\sigma_y, \quad x \in E. \qquad (9)$$

If one finds v such that $v(x^*) = 0$ and $\partial_{n_{x^*}} v(x^*) = \rho(x^*)$, then (9) does not suffer from the close evaluation problem.

Representations (6) and (9) are attractive representations, and several works have provided guidelines on how to build appropriate solutions v. For (6) one can use Taylor-like functions $v(x) = u(x^*) \tilde{g}(x) + \partial_{n_{x^*}} u(x^*) \tilde{f}(x)$, with $\tilde{g}$ and $\tilde{f}$ solutions of some Laplace boundary value problems [29–31]. This technique has been first developed in the context of Boundary Regularized Integral Equation Formulation (BRIEF) (namely to solve (5) using the same subtraction technique on boundary) and applied to evaluate the solution near the boundary. For (9) one can use density interpolation methods [21,23,32]: $v = v(x^*, y) = \sum_{j=0}^{J} c_j(y) H_j(x^* - y)$ where $(H_j)_j$ satisfy the PDE (in the above case $(H_j)_j$ are harmonic functions). In both methods the chosen auxiliary functions v necessarily depend on the trace u (and/or normal trace $\partial_n u$), or the density ρ at the closest evaluation point. Furthermore they require to satisfy at least two conditions (two boundary value problems or two boundary conditions).

In this paper, we provide another construction of modified representations for single-density representations of Laplace and Helmholtz boundary value problems. The construction relies on auxiliary functions v that are independent of the density (solution of the boundary integral equation), and requires fewer constraints in the context of (7). As a consequence, our approach provides more freedom in choosing v. The proposed modified representations are also simple to implement and do not add significant computational costs. In what follows we provide modified representations for Laplace and Helmholtz in 2D and 3D, and provide several examples to illustrate the efficiency of the method.

3. Modified Representations

We present modified representations for single-density representations of Laplace and Helmholtz boundary value problems. In particular, we consider the interior Dirichlet Laplace problem (where one can represent the solution using the double-layer potential), the exterior Neuman Laplace problem (2) (using the single-layer potential (7)), and the sound-soft scattering problem.

3.1. Modified Representation for the Laplace Double-Layer Potential

The interior Dirichlet problem for Laplace consists of finding $u \in \mathcal{C}^2(D) \cap \mathcal{C}^1(\overline{D})$ such that

$$\Delta u = 0 \quad \text{in } D, \quad u = f \quad \text{on } \partial D, \tag{10}$$

with some smooth data f. The solution of Problem (10) can be represented as a double-layer potential [34,35]:

$$u(x) = \int_{\partial D} \partial_{n_y} G(x,y) \mu(y) \, d\sigma_y, \quad x \in D, \tag{11}$$

with G defined in (4), and μ a continuous density solution of the boundary integral equation:

$$-\frac{1}{2}\mu(x^*) + \int_{\partial D} \partial_{n_y} G(x^*,y) \mu(y) \, d\sigma_y = f(x^*), \quad x^* \in \partial D. \tag{12}$$

We now make use of (1) to modify (11). One can show the following (see Appendix C.1 for details):

Proposition 1. *Given* $x = x^* - \ell n_{x^*} \in D$ *with* $x^* \in \partial D$, *let* v *be a solution of Laplace's equation in* $D \subset \mathbb{R}^d$, $d = 2, 3$, *such that*

$$\text{v}(x^*) = 1, \quad \partial_{n_{x^*}} \text{v}(x^*) = 0. \tag{13}$$

The solution of the exterior Dirichlet Laplace problem (11) *admits the modified representation:*

$$u(x) = \int_{\partial D} \partial_{n_y} G(x,y) \mu(y) [1 - \text{v}(y)] \, d\sigma_y + \int_{\partial D} \partial_{n_y} G(x,y) [\mu(y) - \mu(x^*)] \text{v}(y) \, d\sigma_y$$
$$- \mu(x^*)\text{v}(x^*) + \mu(x^*) \int_{\partial D} G(x,y) \left[\partial_{n_y} \text{v}(y) - \partial_{n_{x^*}} \text{v}(x^*)\right] d\sigma_y - \mu(x^*)\partial_{n_{x^*}} \text{v}(x^*), \quad x \in D. \tag{14}$$

The modified representation (14) *has smoother integrands than* (11), *and it addresses the close evaluation problem, in the sense that nearly singular terms vanish as* $y \to x^*$.

From Proposition 1 we can now build auxiliary functions v independent of μ, and there exist plenty of candidates: constant, linear, based on Green's function ($\text{v}(y) = G(y, x_0)$ with $x_0 \in E$), quadratic ($\text{v}(y_1, y_2) = 1 + (y_1 - x_1^*)(y_2 - x_2^*)$, $\text{v}(y_1, y_2) = 1 + (y_1 - x_1^*)^2 - (y_2 - x_2^*)^2$), $\text{v}(y_1, y_2, y_3) = e^{y_3}(\sin y_1 + \sin y_2)$, etc. The solution $\text{v} \equiv 1$ naturally satisfies the conditions (13), and the modified representation (14) boils down to

$$u(x) = \int_{\partial D} \partial_{n_y} G(x,y) [\mu(y) - \mu(x^*)] \, d\sigma_y - \mu(x^*), \quad x \in D. \tag{15}$$

The modified representation (15) is well-known and widely used (e.g., [9,25,27]), it is the simplest representation that naturally addresses the close evaluation problem. Thus, we do not provide numerical results for this case. Rather, we concentrate on other layer potentials.

3.2. Modified Representation for the Laplace Single-Layer Potential

Going back to Problem (2), one can show the following (see Appendix C.2 for details):

Proposition 2. *Given $x = x^* + \ell n_{x^*} \in E$ with $x^* \in \partial D$, let v be a solution of Laplace's equation in $D \subset \mathbb{R}^d$, $d = 2, 3$, such that*

$$\partial_{n_{x^*}} v(x^*) = 1. \tag{16}$$

The solution of the exterior Neumann Laplace problem (2) admits the modified representation:

$$u(x) = \int_{\partial D} G(x,y)\rho(y)\left[1 - \partial_{n_y} v(y)\right] d\sigma_y + \int_{\partial D} G(x,y)[\rho(y) - \rho(x^*)]\partial_{n_y} v(y) \, d\sigma_y \\ + \rho(x^*) \int_{\partial D} \partial_{n_y} G(x,y)\rho(y)[v(y) - v(x^*)] \, d\sigma_y, \quad \forall x \in E. \tag{17}$$

The modified representation (17) has smoother integrands than (7).

Contrary to auxiliary functions provided in Taylor-like methods and density interpolation methods (discussed in Section 2), auxiliary functions v do not depend on ρ and rely on only one constraint (16). Therefore, there is a lot of freedom in choosing v: given u a solution of Laplace's equation, then one chooses $v := \frac{u}{\partial_{n_x^*} u(x^*)}$ (as long as $\partial_{n_x^*} u(x^*) \neq 0$). Candidates may then include:

- The linear function $v(y) = n_{x^*} \cdot y$;
- The function $v(y) = 2^{d-1}\pi G(y, x^* + n_{x^*})$ based on Green's function;
- The quadratic product function $v(y) = \frac{(y_1 - x_{0,1})(y_2 - x_{0,2})}{n_{x^*,1}(x_2^* - x_{0,2}) + n_{x^*,2}(x_1^* - x_{0,1})}$, $x_0 \in D$;
- The quadratic difference function $v(y) = \frac{1}{2}\frac{(y_1 - x_{0,1})^2 - (y_2 - x_{0,2})^2}{n_{x^*,1}(x_1^* - x_{0,1}) - n_{x^*,2}(x_2^* - x_{0,2})}$, $x_0 \in D$.

Note that the above candidates are valid in $\mathbb{R}^d$, one can also consider any of the quadratic functions above in $\mathbb{R}^3$ as a function of (y_i, y_j), $i, j = 1, 2, 3$, $j \neq i$. In Section 4, we will test (17) using several candidates v and make comparisons. The modified representation (17) adds two terms to compute compared to (7), it is the price to pay to gain accuracy at close evaluation points. We will make comparative tests to quantify this aspect.

3.3. Modified Representation for the Helmholtz Double- and Single-Layer Potentials

We consider in this case the sound-soft scattering problem:

$$\begin{vmatrix} \text{Find } u \in C^2(E) \cup C^1(\bar{E}) \text{ such that:} \\ \Delta u + k^2 u = 0 \quad \text{in } E, \quad u = f \quad \text{on } \partial D, \quad \lim_{R \to \infty} \int_{|y|=R} |\partial_n u - iku|^2 \, d\sigma_y = 0, \end{vmatrix} \tag{18}$$

with some smooth data f associated with the wavenumber k. Above, the last condition represents the Sommerfeld radiation condition. The solution of Problem (18) can be represented as a combination of double- and single-layer potentials [40]:

$$u(x) = \int_{\partial D} \left[\partial_{n_y} G^H(x,y) - ik G^H(x,y)\right] \mu(y) \, d\sigma_y, \quad x \in E, \tag{19}$$

with G^H defined by

$$G^H(x,y) = \begin{cases} \dfrac{i}{4} H_0^{(1)}(k|x-y|), & \text{for } d = 2, \\ \dfrac{1}{4\pi} \dfrac{e^{ik|x-y|}}{|x-y|}, & \text{for } d = 3, \end{cases} \tag{20}$$

with $H_0^{(1)}(\cdot)$ the Hankel function of the first kind, and μ a continuous density satisfying:

$$\frac{1}{2}\mu(x^*) + \int_{\partial D}\left[\partial_{n_y}G^H(x^*,y) - ikG^H(x^*,y)\right]\mu(y)\,d\sigma_y = f(x^*), \quad x^* \in \partial D. \qquad (21)$$

One obtain the following:

Proposition 3. *Given $x = x^* + \ell n_{x^*} \in E$ with $x^* \in \partial D$, let v be a solution of Helmholtz equation in $D \subset \mathbb{R}^d$, $d = 2, 3$, such that*

$$v(x^*) = 1, \quad \partial_{n_{x^*}}v(x^*) = ik. \qquad (22)$$

Then the solution of the sound-soft scattering problem (18) admits the modified representation:

$$\begin{aligned}
u(x) &= \int_{\partial D}\left[\partial_{n_y}G^H(x,y) - \partial_{n_y}v(y)G^H(x,y)\right][\mu(y) - \mu(x^*)]\,d\sigma_y \\
&+ \int_{\partial D}G^H(x,y)\left[\partial_{n_y}v(y) - ik\right]\mu(y)\,d\sigma_y \\
&+ \mu(x^*)\int_{\partial D}\partial_{n_y}G^H(x,y)[1 - v(y)]\,d\sigma_y, \quad \forall x \in E.
\end{aligned} \qquad (23)$$

The modified representation (23) has smoother integrands than (19).

The proof can be found in Appendix C.3. One can check in particular that plane waves $v(y) = e^{ikn_{x^*}\cdot(y-x^*)}$ do satisfy (22), whereas Green-based functions like $v(y) = G^H(y, x_* + n_{x^*})$ (up to some constant) cannot. We will use (23) with plane waves for the numerical examples.

4. Numerical Examples

The accuracy in approximating (11)–(15), (7)–(17), (19)–(23), respectively, relies on the resolution of the boundary integral Equations (12), (8) and (21), respectively. In what follows we assume that the boundary integral equations are sufficiently resolved. Given the density's resolution, we compare the representations and their modified ones through several examples. All the codes can be found in [41].

4.1. Exterior Neumann Laplace Problem

4.1.1. Example 1: Exterior Laplace in Two Dimensions

Since ∂D is a closed smooth boundary, we use the Periodic Trapezoid Rule (PTR) to approximate (7) and (17), where we will use several v according to Proposition 2. We consider an exact solution of Problem (2):

$$u_{\text{exact}}(x) = u_{\text{exact}}(x_1, x_2) = \frac{x_1 - x_{0,1}}{|x - x_0|^2}, \quad x_0 = (x_{0,1}, x_{0,2}) \in D,$$

which consists of choosing $g(x^*) = \partial_{n_{x^*}}u_{\text{exact}}(x^*)$, for any $x^* \in \partial D$.

All simulations are done outside of a kite-shaped domain using the Periodic Trapezoid Rule with $N = 128$ quadrature points for the following representations:

- **V0:** standard representation (7);
- **V1:** modified representation (17) with the linear function $v_1(y) = n_{x^*}\cdot y$;
- **V2:** modified representation (17) with the Green's function $v_2(y) = 2\pi G(y, x^* + n^*)$;
- **V3:** modified representation (17) with the quadratic function $v_3(y) = \frac{1}{2}\frac{y_1^2 - y_2^2}{n_{x^*,1}x_1^* - n_{x^*,2}x_2^*}$;
- **V4:** modified representation (17) with the quadratic function

$$v_4(y) = \frac{(y_1 - 5)(y_2 - 5)}{n_{x^*,1}(x_2^* - 5) + n_{x^*,2}(x_1^* - 5)}.$$

We solved (8) using the Nyström method based on the Periodic Trapezoid Rule (using Matlab classic *backslash*). The accuracy of all methods is limited by the accuracy of the resolution for ρ (in particular when considering moderate N). This can be assessed by looking the density's Fourier coefficients decay: in this case the coefficients decay is bounded by 10^{-5} for $N = 128$. The results in Figures 1 and 2 show that given ρ resolved, the approximation of the modified representations provide better results overall. Far from the boundary, all methods approximate well the solution. As the evaluation point gets closer to the boundary ($\ell \to 0$), V0 approximated by PTR suffers from the close evaluation problem and the error increases (see [9]). Note that the single-layer potential commonly suffers less from this phenomenon than the double-layer potential (e.g., [24]). Using the modified representations (V1–V4) allows to reduce the error by a couple of orders of magnitude for the close evaluation problem. All modified representations provide a satisfactory correction overall. We use a naïve (straightforward) implementation of (7) and (17) in Matlab, computed on a Mac mini SSD 512Go. We provide run times in Table 1 for various number of quadrature points. Run times do not count the time to compute the boundary integral equation for ρ (being the same for all methods).

Representation V0 is obviously cheaper (less terms to compute) than V1–V4, and V1 is cheaper than V2–V4 due to simpler terms: there are less operations to conduct to compute $v_1(y)$ than the other provided auxiliary functions.

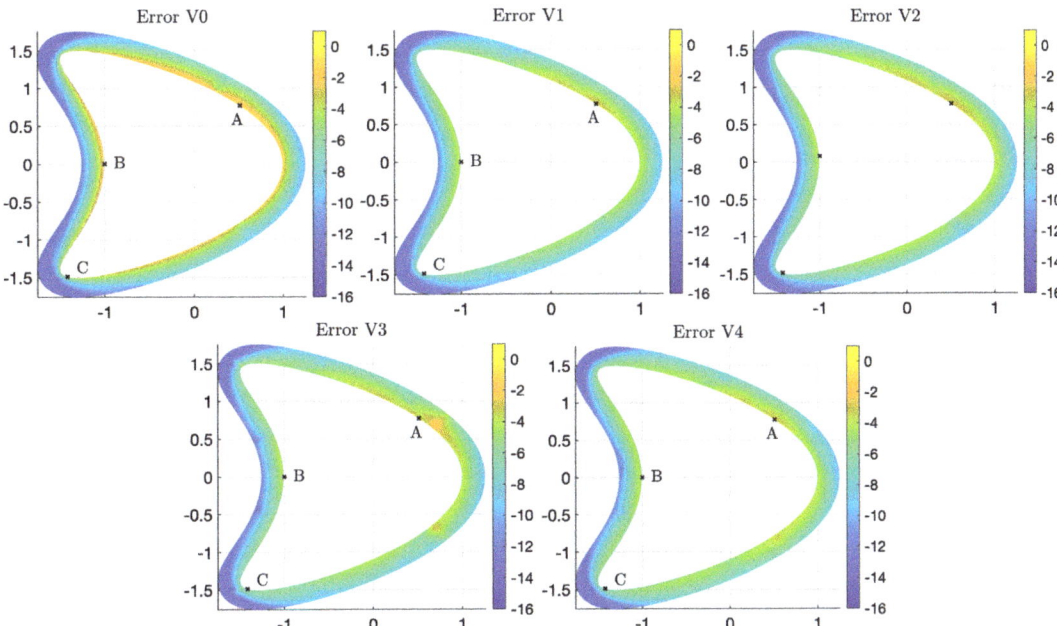

Figure 1. Laplace 2D single-layer. Plots of $\log_{10}$ of the error for the evaluation of the solution of (2) out of the kite domain defined by the boundary $y(t) = (\cos t + 0.65\cos(2t) - 0.65, 1.5\sin t)$, $t \in [0, 2\pi]$, for the Neumann data, $g = \partial_n u_{\text{exact}}$ with $x_0 = (0.1, 0.4)$, for representations V0, V1, V2, V3, V4 computed using PTR with $N = 128$. Computations are made on a boddy-fitted grid with $N \times 200$ grid points.

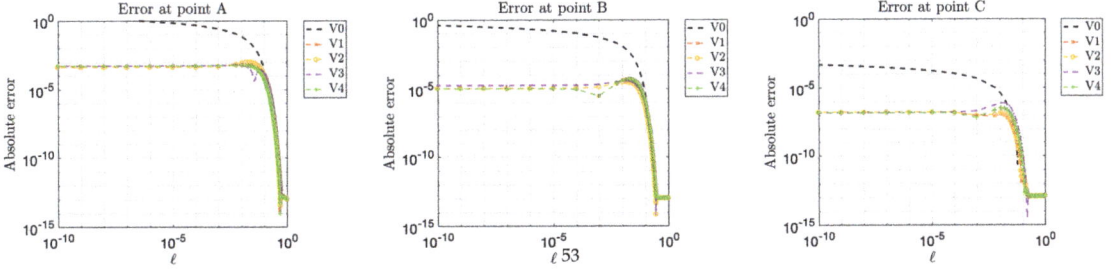

Table 1. Laplace 2D single-layer. CPU times (in seconds) for various number of quadrature points and representations. Times account for computing the solution at $N \times 12$ grid points ($\ell = 10^{-k}$, $k = [\![0, 11]\!]$) on a body-fitted grid.

Method	V0	V1	V2	V3	V4
$N = 128$	0.014	0.044	0.055	0.045	0.05
$N = 256$	0.056	0.07	0.112	0.08	0.081
$N = 512$	0.12	0.192	0.263	0.2	0.19

To better compare the methods, Figure 3 represents log plots of the maximum error with respect to the number of quadrature points N and for various distances ℓ from point A (indicated in Figure 1). The results show that modified representations allow to gain a couple of order of magnitude even for moderate N ($N < 100$). Additionally, the error using V0 decreases linearly with the number of quadrature points whereas it is cubic using modified representations. While there is no significant difference between the considered modified representations V1–V4, one may consider run times (and simplicity of auxiliary function v) to discuss competitiveness. Based on the above results, overall representation V1 seems to be the best choice for the best computational cost-accuracy trade-off. Let us emphasize that the focus of this paper is to highlight the efficacy and simplicity of the proposed modified representations, given a quadrature rule. Our results show that modified representations allow to naturally gain a couple of orders of magnitude in the error, addressing the close evaluation problem even for moderate computational resources. Additionally, the proposed auxiliary functions are independent of the density ρ. In the next section we investigate the efficacy of (17) in three dimensions.

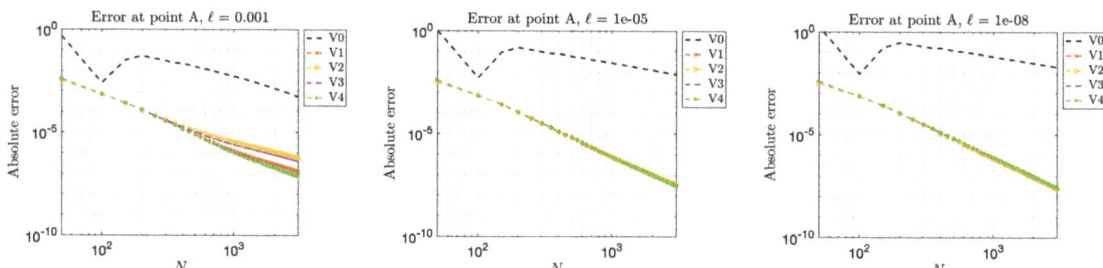

Figure 3. Laplace 2D single-layer. Log-log plots of the errors with respect to N made in computing the solution at some distance ℓ along the normal from point A plotted as black ×'s in Figure 1.

4.1.2. Example 2: Exterior Laplace in Three Dimensions

Given a *domain* $D \subset \mathbb{R}^3$ with smooth boundary, we assume ∂D to be an analytic, closed, and oriented surface that can be parameterized by $y = y(s,t)$ for $s \in [0, \pi]$ and $t \in [-\pi, \pi]$. Then one can write (7) as

$$u(x) = \int_{-\pi}^{\pi} \int_0^{\pi} G(x, y(s,t)) J(s,t) \rho(y(s,t)) \sin(s) ds dt, \quad (24)$$

with $J(s,t) = |y_s(s,t) \times y_t(s,t)|/\sin(s)$ the Jacobian. We now work with a surface integral defined on a sphere, and we use a *three-step method* (see [26] for details) to approximate (7) and (17). This method corresponds to a modification of the product Gaussian quadrature rule (PGQ) [42], and it has been shown to be very effective for computing layer potentials in three dimensions at close evaluation points compared to other quadrature methods for nearly singular integrals [26]. It relies on (i) rotating the local coordinate system so that x^* corresponds to the north pole, (ii) use Periodic Trapezoid Rule with $2N$ quadrature points to approximate the integral with respect to t, (iii) use Gauss–Legendre with N quadrature points mapped to $(0, \pi)$ (and not $(-1, 1)$) to approximate the integral with respect to s. This leads to the approximation:

$$u(x) \approx \frac{\pi^2}{2N} \sum_{i=1}^{N} \sum_{j=1}^{2N} w_i \sin(s_i) F(s_i, t_j),$$

with $F(s_i, t_j) = G(x, y(s_i, t_j)) J(s_i, t_j) \rho(y(s_i, t_j))$, $t_j = -\pi + \pi(j-1)/N$, $j = 1, \cdots, 2N$, $s_i = \pi(z_i + 1)/2$, $i = 1, \cdots, N$ with $z_i \in (-1, 1)$ the N-point Gauss–Legendre quadrature rule abscissas with corresponding weights w_i for $i = 1, \cdots, N$. One proceeds similarly for (17). We consider an exact solution of Problem (2):

$$u_{\text{exact}}(x) = \frac{1}{|x - x_0|}, \quad x_0 \in D,$$

which consists of choosing $g(x^*) = \partial_{n_{x^*}} u_{\text{exact}}(x^*)$, for any $x^* \in \partial D$. The efficacy of the three-step method for various geometries (including effects of curvature) is presented in [26]. Naively implementing this method has the same computational cost as the PGQ method. We do not focus in this paper on fast implementations but do believe that it is possible to speed up this method using ideas that have been previously developed including the fast multipole method [20]. Then for simplicity, results will be computed on a sphere where the resolution of ρ does not require a lot of quadrature points. One can apply the technique for arbitrary closed smooth surfaces, but might be limited by the resolution of (8). All simulations are done outside of a sphere of radius 2 using the three-step method with $N = 16$ for the following representations:

- **V0:** standard representation (7);
- **V1:** modified representation (17) with the linear function $v_1(y) = n_{x^*} \cdot y$;
- **V2:** modified representation (17) with the Green's function $v_2(y) = 4\pi G(y, x^* + n^*)$;
- **V3:** modified representation (17) with the quadratic function $v_3(y) = \frac{1}{2} \frac{y_1^2 - y_2^2}{n_{x^*,1} x_1^* - n_{x^*,2} x_2^*}$;
- **V4:** modified representation (17) with the quadratic product function

$$v_4(y) = \frac{(y_1 - 5)(y_2 - 5)}{n_{x^*,1}(x_2^* - 5) + n_{x^*,2}(x_1^* - 5)}.$$

Note that there are other quadratic polynomials v (as a function of 2 variables instead of 3, see [22] where those polynomials serve as basis for interpolation method). We make here the choice to test using similar functions as in Section 4.1.1. We solve (8) using a Galerkin method and the product Gaussian quadrature rule [36,42–45] (see Appendix B for details).The accuracy in approximating V0–V4 is limited by the accuracy of the resolution for ρ. This can be assessed by looking at the coefficients' decay of the density spherical harmonic expansion. In this case the coefficients' decay has reached 10^{-15}. The results in Figure 4 show that given ρ resolved, the approximation of the modified representations provide better results overall, except for V2 where the error plateaus around 10^{-7} for small ℓ (providing less accurate results compared to standard representation V0). Note that the single-layer potential commonly suffers less from the close evaluation than the double-layer potential, and the chosen method provides already a good approximation. This is the reason why the error when considering V0 decays as ℓ decreases [26]. The modified representations allow to make it even better. To better assess the efficacy of the modified representations in three dimensions, Figure 5 represents log plots of the maximum error with respect to $N \in \{8, 16, 24, 32\}$ (the method uses $2N \times N$ quadrature points) and for various distances ℓ from point B. The results show that as $\ell \to 0$, V1–V4 allow to gain a couple of orders of magnitude in the error, even for a small N. Note that the error produced by three-step method does not seem to depend on N, and in this case there are more variations with respect to the choice of auxiliary function v than in two dimensions. Here, V1 (the linear function) is the best representation, producing the smallest errors (and the fastest to compute as indicated in Table 2). Again, the three-step method has been designed to treat nearly-singular integrals. It is the reason why the method provides already satisfactory results (given the resolution of ρ). The modified representations allow to significantly gain even more accuracy in this case, even with limited computational resources.

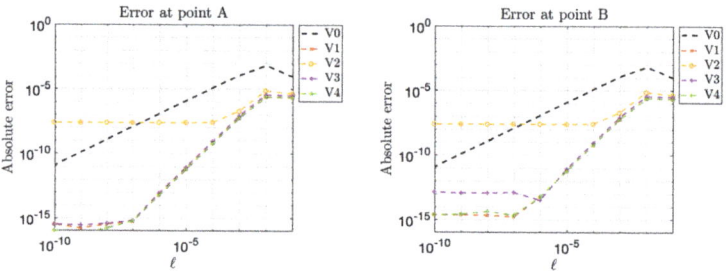

Figure 4. Laplace 3D single-layer. Log-log plots of the errors with respect to ℓ made in computing the solution of (2) for the Neumann data, $g(x^*) = -\frac{n_{x^*} \cdot (x^* - x_0)}{|x^* - x_0|^3}$ with $x_0 = (0,0,0)$, outside of a sphere a radius 2, along the normal of point A = $(-0.0065, -0.0327, 1.9997)$ (**left**), of point B = $(-0.3526, -1.7728, 0.8561)$ (**right**).

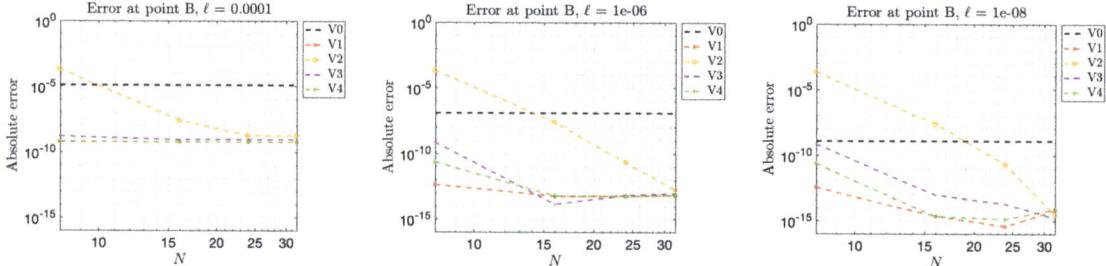

Figure 5. Laplace 3D single-layer. Log-log plots of the errors with respect to N made in computing the solution (as described in Figure 4) at some distance ℓ along the normal from point B = $(-0.3526, -1.7728, 0.8561)$.

Table 2. Laplace 3D single-layer. CPU times (in seconds) for various number of quadrature points and representations for computing the solution (as described in Figure 4) from points A and B, for $\ell = 10^{-k}, k = [\![0, 11]\!]$.

Method	V0	V1	V2	V3	V4
N = 8	0.028	0.029	0.032	0.031	0.046
N = 16	0.143	0.146	0.148	0.150	0.142
N = 24	0.352	0.344	0.346	0.35	0.356

4.2. Scattering Problem

Using Proposition 3, we compare (19) with the modified representation (23) obtained with $v(y) = e^{ikn_{x^*} \cdot (y - x^*)}$:

$$u(x) = \int_{\partial D} \left[\partial_{n_y} G^H(x,y) - ik(n_y \cdot n_{x^*}) e^{ik(n_{x^*} \cdot (y - x^*))} G^H(x,y) \right] [\mu(y) - \mu(x^*)] d\sigma_y$$
$$+ ik \int_{\partial D} [(n_y \cdot n_{x^*}) e^{ik(n_{x^*} \cdot (y - x^*))} - 1] G^H(x,y) \mu(y) d\sigma_y \qquad (25)$$
$$+ \mu(x^*) \int_{\partial D} \partial_{n_y} G^H(x,y) [1 - e^{ik(n_{x^*} \cdot (y - x^*))}] d\sigma_y, \quad x \in \mathbb{R}^d \setminus \bar{D}.$$

4.2.1. Example 3: Scattering in Two Dimensions

We consider an exact solution of Problem (18):

$$u_{\text{exact}}(x) = \frac{i}{4} H_0^{(1)}(k|x - x_0|), \quad x_0 \in D,$$

which consists of choosing $f(x^*) = u_{\text{exact}}(x^*)$, for any $x^* \in \partial D$. All simulations are done outside of a star-shaped domain using the Periodic Trapezoid Rule with $N = 256$ quadrature points and $k = 15$ for the following representations:

- **V0:** standard representation (19);
- **V1:** modified representation (25) (i.e., (23) with the plane wave function $v_1(y) = e^{ikn_{x^*} \cdot (y - x^*)}$).

We solved (21) using Kress product quadrature rule [40] (see Appendix A). The quadrature rule is well adapted to approximate kernels with a logarithmic singularity. The accuracy of both methods is limited by the resolution for μ (the Fourier coefficients' decay of the density is bounded by 10^{-6} for $N = 256$ and $k = 15$). The results in Figures 6 and 7 show that given μ resolved, the approximation of the modified representation provides better results overall. Similarly to Laplace's examples, both methods approximate well the solution far from the boundary. As the evaluation point gets closer to the boundary ($\ell \to 0$), V0 approximated with PTR suffers from the close evaluation problem leading to large errors (see [9]). Using the modified representation V1 allows to reduce the error by a couple of order of magnitude for the close evaluation problem.

Figure 8 represents log plots of the maximum error with respect to the number of quadrature points $N \in [\![50, 3000]\!]$ and for various distances ℓ from point A (indicated in Figure 6). The results show that for any number of quadrature points, the error when considering V0 explodes as we approach the boundary (error larger than 10^5) while the error with V1 remains bounded (of the order of 10^{-2} in the case presented above). In this case standard rerpresentation V0 strongly suffers from the close evaluation problem, however the modified representation V1 significantly reduces the error. Even when standard quadrature rules fail to compute the standard representation, the proposed modified one regularizes the solution and provides satisfactory results without significant additional computational time (as shown in Table 3).

Table 3. Helmholtz 2D. CPU times (in seconds) for various number of quadrature points and representations. Times account for computing the solution for $N \times 12$ grid points (for $\ell = 10^{-k}$, $k = [\![0, 11]\!]$) on a body-fitted grid.

Method	$N = 128$	$N = 256$	$N = 512$
V0	0.18	0.27	0.71
V1	0.21	0.33	0.89

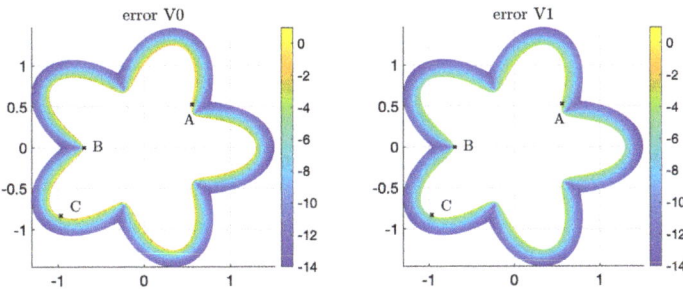

Figure 6. Helmholtz 2D. Plots of $\log_{10}$ of the error for the evaluation of the solution of (18) out of the star domain defined by the boundary $y(t) = (1 + 0.3 \cos 5t) * (\cos t, \sin t)$, $t \in [0, 2\pi]$, for the Dirichlet data, $f(x^*) = \frac{i}{4} H_0^{(1)}(15|x^* - x_0|)$ with $x_0 = (0.2, 0.8)$, for representations V0, V1, computed using PTR with $N = 256$.

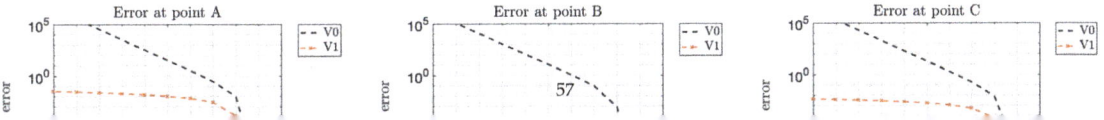

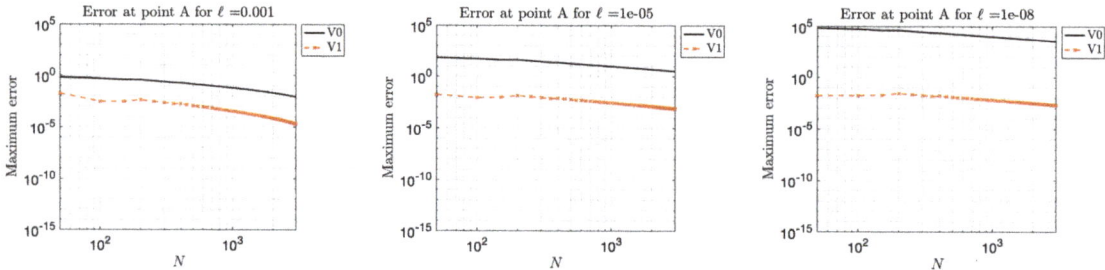

Figure 8. Helmholtz 2D. Log-log plots of the errors with respect to N made in computing the solution at some distance ℓ along the normal from point A plotted as black ×'s in Figure 6.

4.2.2. Example 4: Scattering in Three Dimensions

We consider an exact solution of (10):

$$u_{\text{exact}}(x) = \frac{1}{4\pi} \frac{e^{ik|x-x_0|}}{|x-x_0|}, \quad x_0 \in D,$$

which consists of choosing $f(x^*) = u_{\text{exact}}(x^*)$, for any $x^* \in \partial D$.

All simulations are done outside of an ellipsoid parameterized by $y(s,t) = (2\cos(t)\sin(s), \sin(t)\sin(s), 2\cos(s))$, $(s,t) \in [0,\pi] \times [-\pi,\pi]$, and using the three-step method with various N. This is in order to investigate the technique in the context of limited resolution, namely the coefficients' decay of the density spherical harmonic expansion does not reach machine precision. We consider $k = 5$ and the following representations:

- **V0:** standard representation (19);
- **V1:** modified representation (25).

We solved (21) using Galerkin method and the product Gaussian quadrature rule (see Appendix B for details). The accuracy of both methods is limited by the accuracy of the resolution for μ. This limitation can be checked for instance by looking at the density spherical harmonics coefficients' decay: for $k = 5$, the resolution will be capped around 10^{-2} for $N = 16$, 10^{-4} for $N = 24$, and 10^{-7} for $N = 32$. The results in Figure 9 show that given μ resolved, standard representation incurs bigger errors at close evaluation points while the modified representation provides better results overall. Here, the resolution of the boundary integral equation was fairly limited. Figure 10 represents log plots of the maximum error with respect to $N \in [\![8,32]\!]$ (the method uses $2N \times N$ quadrature points) and for various distances ℓ from the boundary from point A. While the three-step method has been designed to treat nearly-singular integrals and provided satisfactory results for Laplace's problems, the method here requires more quadrature points to achieve accuracy due to the wavenumber (see Section 4.2.3 for more details). The standard representation V0 suffers from both the close evaluation problem and the poor density resolution. The modified representation V1 allows to gain accuracy even with limited resolution (without significant additional computational time as indicated in Table 4).

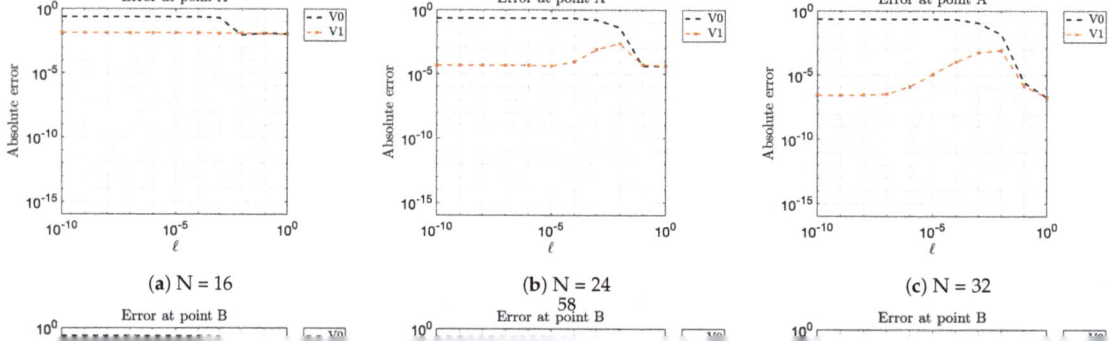

Table 4. Helmholtz 3D. CPU times (in seconds) for various number of quadrature points and representations. Times account for computing the solution from points A and B, for $\ell = 10^{-k}$, $k = [\![0, 11]\!]$.

Method	$N = 8$	$N = 16$	$N = 20$
V0	0.027	0.15	0.313
V1	0.03	0.15	0.314

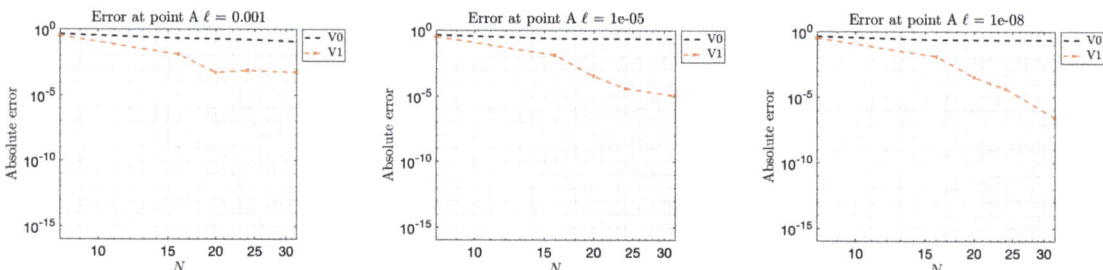

Figure 10. Helmholtz 3D. Log-plot of the maximum error for computing the solution as described in Figure 9 with ∂D being the ellipsoid parameterized by $y(s,t) = (2\cos(t)\sin(s), \sin(t)\sin(s), 2\cos(s))$, $(s,t) \in [0,\pi] \times [-\pi,\pi]$, at some distance ℓ along the normal from point A= $(-0.7664, 0.0607, 1.8433)$.

4.2.3. High Frequency Behavior

It is well-known that for a fixed number of quadrature points N, accuracy is lost for larger wavenumbers k. Figures 11 and 12 represent the high frequency behavior for the Examples 3 and 4, for various k and N. We consider the same quadrature rules, exact solution u_{exact}, boundary shapes, as in Sections 4.2.1 and 4.2.2, but we vary k and/or N. The modified representation annihilates some oscillatory behavior by subtracting plane waves along the normal of the evaluation points. It then allows a better approximation for a wider range of wavenumbers (until the number of quadrature points is not enough), and results in a greater wavenumber stability. The results in Figures 11 and 12 confirm this phenomenon.

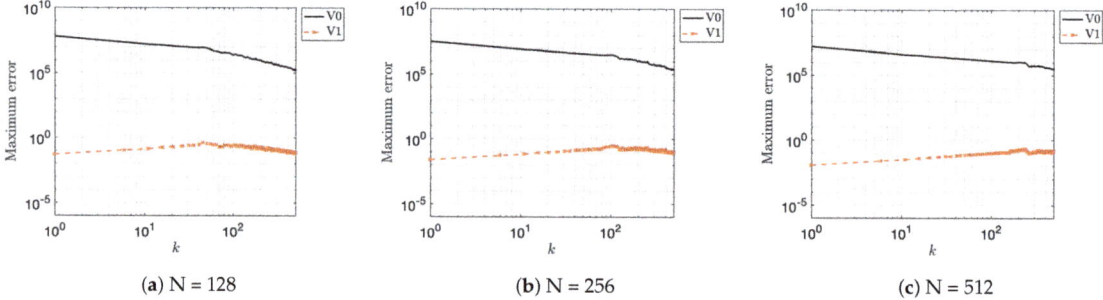

Figure 11. Helmholtz 2D. Log-Log of the maximum error in computing the solution of Problem (18) as described in Section 4.2.1, with respect to the wavenumber k, for various number of quadrature points N.

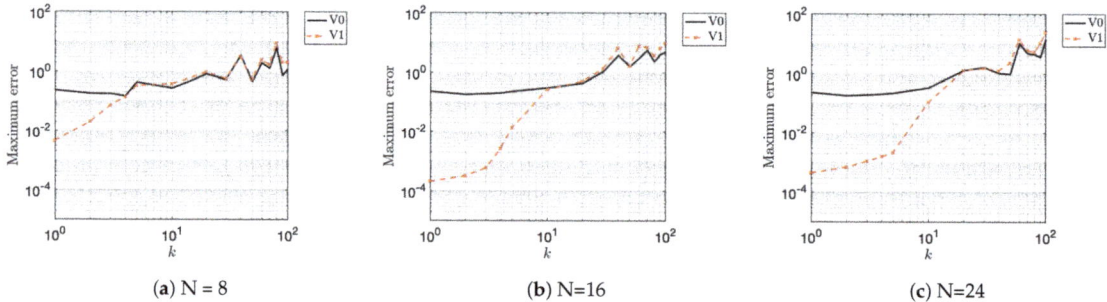

Figure 12. Helmholtz 3D. Log-Log of the maximum error in computing the solution of Problem (18) as described in Section 4.2.2, with respect to the wavenumber k, for various number of quadrature points N.

5. Modified Boundary Integral Equations

We have used (1) to modify the representation of solution of boundary value problems close to (but not on) the boundary. One could also use (1) to avoid weakly singular integrals in the boundary integral equation as done in BRIEF [31]. In the section we present a modified representation of (21).

Proposition 4. *Given $x^* \in \partial D$, let v be a solution of Helmholtz equation in $D \subset \mathbb{R}^d$, $d = 2, 3$, satisfying conditions (22). Then the boundary integral Equation (21) admits the modified representation:*

$$\int_{\partial D} \left[\partial_{n_y} G^H(x^*, y) - \partial_{n_y} v(y) G^H(x^*, y)\right] [\mu(y) - \mu(x^*)] \, d\sigma_y$$
$$+ \int_{\partial D} G^H(x^*, y) \left[\partial_{n_y} v(y) - ik\right] \mu(y) \, d\sigma_y + \mu(x^*) \int_{\partial D} \partial_{n_y} G^H(x^*, y) [1 - v(y)] \, d\sigma_y \quad (26)$$
$$= f(x^*), \quad \forall x^* \in \partial D.$$

The modified representation (26) has smoother integrands than (21).

The proof can be found in Appendix C.3. Using again $v(y) = e^{ikn_{x^*} \cdot (y-x^*)}$, Proposition 4 gives us the modified boundary integral equation:

$$\int_{\partial D} \left[\partial_{n_y} G^H(x^*, y) - ik(n_y \cdot n_{x^*}) e^{ikn_{x^*} \cdot (y-x^*)} G^H(x^*, y)\right] [\mu(y) - \mu(x^*)] \, d\sigma_y$$
$$+ ik \int_{\partial D} G^H(x^*, y) \left[(n_y \cdot n_{x^*}) e^{ikn_{x^*} \cdot (y-x^*)} - 1\right] \mu(y) \, d\sigma_y \quad (27)$$
$$+ \mu(x^*) \int_{\partial D} \partial_{n_y} G^H(x^*, y) \left[1 - e^{ikn_{x^*} \cdot (y-x^*)}\right] d\sigma_y = f(x^*), \quad x^* \in \partial D.$$

Equation (27) has no singular integrals (in the sense its integrands have vanishing singularities), in particular it could be approximated using standard quadrature rules such as PTR in two dimensions. Going back to Examples 3 and 4 presented in Sections 4.2.1 and 4.2.2, we now compare the approximation of the representations (19)–(25) where the density μ has been computed via (21)–(27). We then have four representations:

- **V0**: standard representation (19) with previous approximation of (21);
- **V1**: modified representation (25) with previous approximation of (21);
- **V2**: standard representation (19), approximation of (27) using PTR as Nyström method (2D), using product Gaussian quadrature rule (3D);
- **V3**: modified representation (25), approximation of (27) using PTR as Nyström method (2D), using product Gaussian quadrature rule (3D).

Figure 13 represents the results in two dimensions and illustrates how the resolution of μ limits the approximation of the solution of (18). Far from the boundary the error made using V2–V3 cannot be better than order 10^{-6}. This limitation is due to the poor resolution of μ using Nyström method based on PTR to approximate (25). This can be assessed by looking at the density Fourier coefficients' decay, which caps at 10^{-6} for $N = 256$. However, as the evaluation point gets closer to the boundary ($\ell \to 0$), V3 yields competitive (sometimes better) results. Additionally, the use of Nyström PTR allows to reduce CPU times as indicated in Table 5. The modified boundary integral Equation (27) can be approximated using standard quadrature rules such as Periodic Trapezoid Rule (note that Nyström PTR was not possible to use to solve for (21) due to singular integrals). Its resolution may be limited but it offers interesting corrections for the close evaluation problem using simple quadrature rules as well as faster solvers.

Table 5. Helmholtz 2D. CPU times (in seconds) for various number of quadrature points to compute the solution of the boundary integral equation.

Method	$N = 128$	$N = 256$	$N = 512$
(21) with Kress product rule	0.12	0.45	1.70
(27) with PTR	0.09	0.302	1.16

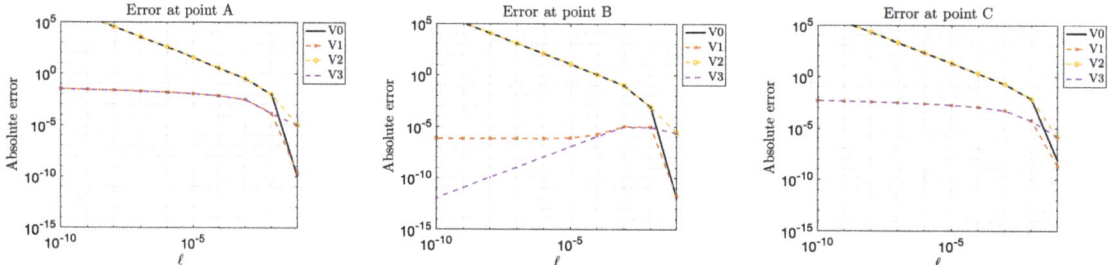

Figure 13. Helmholtz 2D. Log-Log plot of the error along the normal for the solution of (18) out of the star domain defined by the boundary $y(t) = (1.55 + 0.4\cos 5t) * (\cos t, \sin t)$, $t \in [0, 2\pi]$, for the Dirichlet data, $f(x^*) = \frac{i}{4}H_0^{(1)}(15|x^* - x_0|)$ with $x_0 = (0.2, 0.8)$, at the three points A,B,C plotted as black ×'s in Figure 6.

The results in Figure 14 show that the resolution of the solution using both methods yields the same accuracy in three dimensions. The product Gaussian quadrature rule is an open quadrature at the singular point $y = x^*$ (see Appendix B). Thus, the modification introduced in (25) does not affect the approximation. The product Gaussian quadrature rule is a well-used, efficient, easy to implement method, but one could consider a closed quadrature rule to study the effect of (27) more closely.

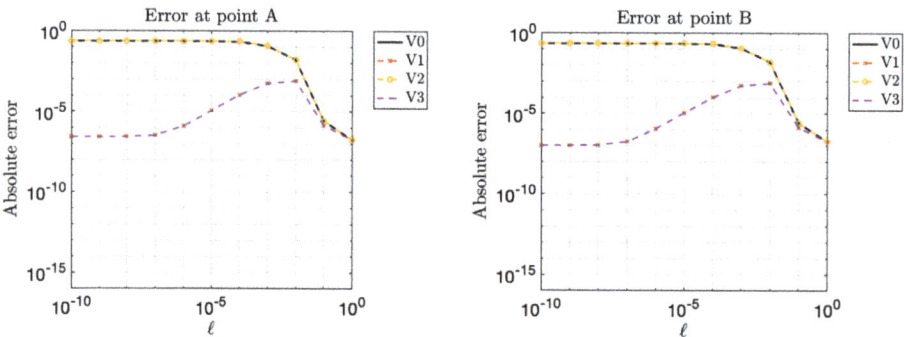

Figure 14. Helmholtz 3D. Log-Log plot of the error for the problem described in Figure 9 using $N = 32$, and for the four representations (standard or modified, off and on boundary).

6. Conclusions

In this paper, we have provided modified representations for Laplace and Helmholtz layer potentials to address the close evaluation problem in several boundary value problems. Similar to Gauss' law, we take advantage of one auxiliary function, satisfying the partial differential equation at stake. A similar technique has been used in the context of BRIEF and density interpolation. Our approach provides guidelines on how to develop them independently of the density, and valuable insights into the layer potentials inherent nearly singular behavior. Several examples in two and three dimensions have been presented and demonstrated the efficiency of the modified representations. Given a quadrature rule, the modified representation of the solution provides a better approximation by several orders of magnitude even with limited computational resources. This assumes that the density, solution of the boundary integral equation, is sufficiently well-resolved. The modified boundary integral equation has no singular behaviors anymore, and allows us to use standard quadrature rules that do not treat singularities.

We have provided general modified representations, one can use them with any solution of their choice as long as they follow the provided guidelines to address the close evaluation. One can use this technique to modify any other wave problems, including sound-hard, penetrable obstacles. Future work includes applying those techniques to plasmonic scattering problems [46,47], deriving an asymptotic analysis to quantify the limit behavior of the error as the evaluation point approaches the boundary, as well as extensions to other partial differential equations such as Stokes problems and others.

Funding: This research was funded by the NSF grant number DMS 1819052.

Acknowledgments: The author would like to thank Shilpa Khatri, Arnold D. Kim, Marc Bonnet, Zoïs Moitier, and the reviewers for fruitful discussions and feedback.

Conflicts of Interest: The author declares no conflict of interest. The funders had no role in the design of the study; in the collection, analyses, or interpretation of data; in the writing of the manuscript, or in the decision to publish the results.

Appendix A. Kress Product Quadrature

In this section, we provide a brief summary about the Kress product quadrature rule [40] used to compute the density μ, solution of (21), in two dimensions. Denoting the parameterization of ∂D as $y(t)$, $t \in (0, 2\pi)$, and denoting $x^* = y(t^*)$, we compactly rewrite (21)

$$\frac{1}{2}\mu(t^*) + \int_0^{2\pi} K(t, t^*)\mu(t)\,dt = f(t^*), \tag{A1}$$

with the abuse of notation $K(t, t^*) = \left(\partial_{n_y} G^H(x^*, y(t)) - ikG^H(x^*, y(t))\right)|y'(t)|$, $\mu(t) = \mu(y(t))$, and $f(t) = f(y(t))$. The Kress product quadrature rule is well adapted for weakly singular integrals involving kernel with a logarithmic singularity. To that aim one rewrites:

$$K(t, t^*) = K_1(t, t^*)\log\left(4\sin^2\left(\frac{t^* - t}{2}\right)\right) + K_2(t, t^*),$$

with smooth functions K_1, K_2 (the expression of K_1, K_2 can be found in [40]). Then one discretizes the integral using $N = 2n$ quadrature points as follows:

$$\int_0^{2\pi} K(t, t^*)\mu(t)\,dt \approx \sum_{k=0}^{2n-1}\left(R_k^{(n)}(t^*)K_1(t^*, t_k) + \frac{\pi}{n}K_2(t^*, t_k)\right)\mu(t_k),$$

with $t_k = \frac{\pi k}{n}$, $k = 0, \ldots, 2n-1$, and $R_k^{(n)}(t^*)$ the weights

$$R_k^{(n)}(t^*) = -\frac{2\pi}{n}\sum_{j=1}^{n-1}\frac{1}{j}\cos(j(t^* - t_k)) - \frac{\pi}{n^2}\cos(n(t^* - t_k)), \quad k = 0, \ldots, 2n-1.$$

Appendix B. Galerkin Approximation

In this section, we provide a brief summary about the Galerkin approximation used to compute the solutions of (8), (12), (21) and (27) in three dimensions. First, we compactly write (8), (12), (21) and (27) as

$$\mathcal{K}[\psi] = F, \tag{A2}$$

with ψ denoting the density (i.e., μ, ρ), and F denoting the Dirichlet or Neumann data. We introduce the approximation for ψ

$$\psi(y(\theta, \varphi)) \approx \sum_{n=0}^{N-1}\sum_{m=-n}^{n} Y_{nm}(\theta, \varphi)\hat{\psi}_{nm}, \tag{A3}$$

with $y(\theta, \varphi)$, $\theta \in (0, \pi)$, $\varphi \in (-\pi, \pi)$ a parameterization of the boundary ∂D, $\{Y_{nm}(\theta, \varphi)\}_{n,m}$ the orthonormal set of spherical harmonics. For $x^* \in \partial D$, we write $x^* = y(\theta^*, \varphi^*)$. Note that N in (A3) corresponds also to the same order of the quadrature rule used to approximate (8), (12), (21) and (27). Substituting (A3) into (A2) and taking the inner product with $Y_{n'm'}(\theta^*, \varphi^*)$, we obtain the Galerkin equations

$$\sum_{n=0}^{N-1}\sum_{m=-n}^{n} \langle Y_{n'm'}, \mathscr{K}[Y_{nm}]\rangle \hat{\psi}_{nm} = \langle Y_{n'm'}, F\rangle. \tag{A4}$$

We construct the $N^2 \times N^2$ linear system for the unknown coefficients, $\hat{\psi}_{n'm'}$ resulting from (A4) evaluated for $n' = 0, \cdots, N-1$ with corresponding values of m'. To compute the inner products, $\langle Y_{n'm'}, \mathscr{K}[Y_{nm}]\rangle$ and $\langle Y_{n'm'}, F\rangle$, we use the product Gaussian quadrature rule for spherical integrals [42]. This corresponds to approximate the integral with respect to φ using N Gauss–Legendre quadrature points, and the integral with respect to θ using a $2N$ Periodic Trapezoid Rule points. One can proceed as in the three-step method (see Section 4.1.2, and [26] for more details), by adding a rotation of the local coordinate system so that x^* corresponds to the north pole, and by using the N Gauss–Legendre quadrature points mapped to $(0, \pi)$ and not $(-1, 1)$.

For (12) we have

$$\mathscr{K}[Y_{nm}](\theta^*, \varphi^*) = -\frac{1}{2}Y_{nm}(\theta^*, \varphi^*) + \int_{-\pi}^{\pi}\int_{0}^{\pi} \partial_{n_y} G^L(\theta^*, \varphi^*, \theta, \varphi) J(\theta, \varphi) \sin(\theta) Y_{nm}(\theta, \varphi) d\theta d\varphi.$$

For (8) we make use of the adjoint $\mathscr{K}^*$ of $\mathscr{K}$. Using Gauss' law we write $\sum_{n=0}^{N-1}\sum_{m=-n}^{n} \langle \mathscr{K}^*[Y_{n'm'}], Y_{nm}\rangle \hat{\psi}_{nm} = \langle Y_{n'm'}, F\rangle$ with

$$\mathscr{K}^*[Y_{n'm'}](\theta, \varphi) = \int_{-\pi}^{\pi}\int_{0}^{\pi} \partial_{n_x^*} G^L(\theta^*, \varphi^*, \theta, \varphi) J(\theta^*, \varphi^*) \sin(\theta^*)[Y_{n'm'}(\theta^*, \varphi^*) - Y_{n'm'}(\theta, \varphi)] d\theta^* d\varphi^*.$$

For (21) we have

$$\mathscr{K}[Y_{nm}](\theta^*, \varphi^*) = \frac{1}{2}Y_{nm}(\theta^*, \varphi^*) + \int_{-\pi}^{\pi}\int_{0}^{\pi}\left[\partial_{n_y} G^H(\theta^*, \varphi^*, \theta, \varphi) - ikG^H(\theta^*, \varphi^*, \theta, \varphi)\right]$$
$$J(\theta, \varphi)\sin(\theta)Y_{nm}(\theta, \varphi)d\theta d\varphi,$$

and for (27) we have

$$\mathscr{K}_m[Y_{nm}](\theta^*, \varphi^*) = \int_{-\pi}^{\pi}\int_{0}^{\pi}\left[\partial_{n_y} G^H(\theta^*, \varphi^*, \theta, \varphi) - ik(n_y \cdot n_{x^*})e^{ik(n_{x^*}\cdot(y(\theta,\varphi)-y(\theta^*,\varphi^*)))}G^H(\theta^*, \varphi^*, \theta, \varphi)\right]$$
$$J(\theta, \varphi)\sin(\theta)[Y_{nm}(\theta, \varphi) - Y_{nm}(\theta^*, \varphi^*)]d\theta d\varphi$$
$$+ ik\int_{-\pi}^{\pi}\int_{0}^{\pi}[(n_y \cdot n_{x^*})e^{ik(n_{x^*}\cdot(y(\theta,\varphi)-y(\theta^*,\varphi^*)))} - 1]G^H(\theta^*\varphi^*, \theta, \varphi)J(\theta, \varphi)\sin(\theta)Y_{nm}(\theta, \varphi)d\theta d\varphi$$
$$+ Y_{nm}(\theta^*, \varphi^*)\int_{-\pi}^{\pi}\int_{0}^{\pi}[1 - e^{ik(n_{x^*}\cdot(y(\theta,\varphi)-y(\theta^*,\varphi^*)))}]\partial_{n_y}G^H(\theta^*\varphi^*, \theta, \varphi)J(\theta, \varphi)\sin(\theta)d\theta d\varphi.$$

Appendix C. Proof of Modified Representations

Appendix C.1. Modified Double-Layer Potential (14)

Given v solution of Laplace's equation in $D \subset \mathbb{R}^d$, $d = 2, 3$, and for $x \in D$ we write $x = x^* - \ell n_{x^*}$, with $x^* \in \partial D$. Then we write (11) as:

$$u(x) = \int_{\partial D} \partial_{n_y} G(x, y)\mu(y)[1 - v(y)]\, d\sigma_y + \int_{\partial D} \partial_{n_y} G(x, y)\mu(y)v(y)\, d\sigma_y$$
$$= \int_{\partial D} \partial_{n_y} G(x, y)\mu(y)[1 - v(y)]\, d\sigma_y + \int_{\partial D} \partial_{n_y} G(x, y)[\mu(y) - \mu(x^*)]v(y)\, d\sigma_y$$
$$+ \mu(x^*)\int_{\partial D} \partial_{n_y} G(x, y)v(y) - G(x, y)\partial_{n_y}v(y)\, d\sigma_y + \mu(x^*)\int_{\partial D} G(x, y)\partial_{n_y}v(y)\, d\sigma_y$$

Using (1) the third term becomes $-\mu(x^*)v(x^*)$. Then

$$u(x) = \int_{\partial D} \partial_{n_y} G(x,y)\mu(y)[1-v(y)]\,d\sigma_y + \int_{\partial D} \partial_{n_y} G(x,y)[\mu(y)-\mu(x^*)]v(y)\,d\sigma_y - \mu(x^*)v(x^*)$$
$$+ \mu(x^*)\int_{\partial D} G(x,y)[\partial_{n_y}v(y) - \partial_{n_{x^*}}v(x^*)]\,d\sigma_y + \mu(x^*)\partial_{n_{x^*}}v(x^*)\int_{\partial D} G(x,y)\,d\sigma_y$$

which is (14), after using (1) for the last term.

Appendix C.2. Proof of Proposition 2

In this section, we derive (17). Given v solution of Laplace's equation in $D \subset \mathbb{R}^d$, $d = 2,3$, and for $x \in E$ we write $x = x^* + \ell n_{x^*}$, with $x^* \in \partial D$. Then we write (7) as:

$$u(x) = \int_{\partial D} G(x,y)\rho(y)\left[1 - \partial_{n_y}v(y)\right]d\sigma_y + \int_{\partial D} G(x,y)\rho(y)\partial_{n_y}v(y)\,d\sigma_y$$
$$= \int_{\partial D} G(x,y)\rho(y)\left[1 - \partial_{n_y}v(y)\right]d\sigma_y + \int_{\partial D} G(x,y)[\rho(y) - \rho(x^*)]\partial_{n_y}v(y)\,d\sigma_y$$
$$+ \rho(x^*)\int_{\partial D} G(x,y)\partial_{n_y}v(y) - \partial_{n_y}G(x,y)v(y)\,d\sigma_y + \rho(x^*)\int_{\partial D}\partial_{n_y}G(x,y)v(y)\,d\sigma_y$$

Using (1), the third term vanishes. Then

$$u(x) = \int_{\partial D} G(x,y)\rho(y)\left[1 - \partial_{n_y}v(y)\right]d\sigma_y + \int_{\partial D} G(x,y)[\rho(y) - \rho(x^*)]\partial_{n_y}v(y)\,d\sigma_y$$
$$+ \rho(x^*)\int_{\partial D}\partial_{n_y}G(x,y)[v(y) - v(x^*)]\,d\sigma_y + \rho(x^*)v(x^*)\int_{\partial D}\partial_{n_y}G(x,y)\,d\sigma_y$$

The last term vanishes using (1) then one obtains (17).

Appendix C.3. Proof of Propositions 3, 4

In this section, we derive (23), (26). Given v solution of the Helmholtz equation in $D \subset \mathbb{R}^d$, $d = 2,3$, and for $x \in E$ we write $x = x^* + \ell n_{x^*}$, with $x^* \in \partial D$. Then we write (19) as:

$$u(x) = \int_{\partial D}\left[\partial_{n_y}G^H(x,y) - \partial_{n_y}v(y)G^H(x,y)\right]\mu(y)\,d\sigma_y + \int_{\partial D}\left[\partial_{n_y}v(y) - ik\right]G^H(x,y)\mu(y)\,d\sigma_y$$
$$= \int_{\partial D}\left[\partial_{n_y}G^H(x,y) - \partial_{n_y}v(y)G^H(x,y)\right][\mu(y) - \mu(x^*)]\,d\sigma_y + \mu(x^*)\int_{\partial D}\left[\partial_{n_y}G^H(x,y)v(y) - G^H(x,y)\partial_{n_y}v(y)\right]d\sigma_y \quad \text{(A5)}$$
$$+ \int_{\partial D}\left[\partial_{n_y}v(y) - ik\right]G^H(x,y)\mu(y)\,d\sigma_y + \mu(x^*)\int_{\partial D}\partial_{n_y}G^H(x,y)[(1-v(y)]\,d\sigma_y$$

Using (1), the third term vanishes, then one obtains (23). One proceeds similarly starting with (21): one can show that the layer potentials in (21) correspond to (A5) for $x = x^* \in \partial D$. Finally, (1) gives that the third term boils down to $-\frac{1}{2}\mu(x^*)v(x^*)$, which finishes the proof.

References

1. Akselrod, G.M.; Argyropoulos, C.; Hoang, T.B.; Ciracì, C.; Fang, C.; Huang, J.; Smith, D.R.; Mikkelsen, M.H. Probing the mechanisms of large Purcell enhancement in plasmonic nanoantennas. *Nat. Photonics* **2014**, *8*, 835–840. [CrossRef]
2. Barnett, A.H.; Wu, B.; Veerapaneni, S. Spectrally accurate quadratures for evaluation of layer potentials close to the boundary for the 2D Stokes and Laplace equations. *SIAM J. Sci. Comput.* **2015**, *37*, B519–B542. [CrossRef]
3. Keaveny, E.E.; Shelley, M.J. Applying a second-kind boundary integral equation for surface tractions in Stokes flow. *J. Comput. Phys.* **2011**, *230*, 2141–2159. [CrossRef]
4. Marple, G.R.; Barnett, A.; Gillman, A.; Veerapaneni, S. A fast algorithm for simulating multiphase flows through periodic geometries of arbitrary shape. *SIAM J. Sci. Comput.* **2016**, *38*, B740–B772. [CrossRef]
5. Mayer, K.M.; Lee, S.; Liao, H.; Rostro, B.C.; Fuentes, A.; Scully, P.T.; Nehl, C.L.; Hafner, J.H. A label-free immunoassay based upon localized surface plasmon resonance of gold nanorods. *ACS Nano* **2008**, *2*, 687–692. [CrossRef]
6. Novotny, L.; Van Hulst, N. Antennas for light. *Nat. Photonics* **2011**, *5*, 83–90. [CrossRef]

7. Sannomiya, T.; Hafner, C.; Voros, J. In situ sensing of single binding events by localized surface plasmon resonance. *Nano Lett.* **2008**, *8*, 3450–3455. [CrossRef]
8. Smith, D.J. A boundary element regularized Stokeslet method applied to cilia-and flagella-driven flow. *Proc. R. Soc. Lond. A* **2009**, *465*, 3605–3626. [CrossRef]
9. Barnett, A.H. Evaluation of layer potentials close to the boundary for Laplace and Helmholtz problems on analytic planar domains. *SIAM J. Sci. Comput.* **2014**, *36*, A427–A451. [CrossRef]
10. Schwab, C.; Wendland, W. On the extraction technique in boundary integral equations. *Math. Comput.* **1999**, *68*, 91–122. [CrossRef]
11. Beale, J.T.; Lai, M.C. A method for computing nearly singular integrals. *SIAM J. Numer. Anal.* **2001**, *38*, 1902–1925. [CrossRef]
12. Beale, J.T.; Ying, W.; Wilson, J.R. A simple method for computing singular or nearly singular integrals on closed surfaces. *Commun. Comput. Phys.* **2016**, *20*, 733–753. [CrossRef]
13. Helsing, J.; Ojala, R. On the evaluation of layer potentials close to their sources. *J. Comput. Phys.* **2008**, *227*, 2899–2921. [CrossRef]
14. Af Klinteberg, L.; Tornberg, A.-K. A fast integral equation method for solid particles in viscous flow using quadrature by expansion. *J. Comput. Phys.* **2016**, *326*, 420–445. [CrossRef]
15. Af Klinteberg, L.; Tornberg, A.-K. Error estimation for quadrature by expansion in layer potential evaluation. *Adv. Comput. Math.* **2017**, *43*, 195–234. [CrossRef]
16. Epstein, C.L.; Greengard, L.; Klöckner, A.K. On the convergence of local expansions of layer potentials. *SIAM J. Numer. Anal.* **2013**, *51*, 2660–2679. [CrossRef]
17. Klöckner, A.; Barnett, A.; Greengard, L.; O'Neil, M. Quadrature by expansion: A new method for the evaluation of layer potentials. *J. Comput. Phys.* **2013**, *252*, 332–349. [CrossRef]
18. Rachh, M.; Klöckner, A.; O'Neil, M. Fast Algorithms for Quadrature by Expansion I: Globally Valid Expansions. *J. Comput. Phys.* **2017**, *345*, 706–731. [CrossRef]
19. Wala, M.; Klöckner, A. A Fast Algorithm for Quadrature by Expansion in Three Dimensions. *J. Comput. Phys.* **2019**, *388*, 655–689. [CrossRef]
20. Greengard, L.; O'Neil, M.; Rachh, M.; Vico, F. Fast multipole methods for the evaluation of layer potentials with locally-corrected quadratures. *J. Comput. Phys. X* **2021**, *10*, 100092.
21. Pérez-Arancibia, C. A plane-wave singularity subtraction technique for the classical Dirichlet and Neumann combined field integral equations. *Appl. Numer. Math.* **2018**, *123*, 221–240. [CrossRef]
22. Pérez-Arancibia, C.; Faria, L.; Turc, C. Harmonic density interpolation methods for high-order evaluation of Laplace layer potentials in 2D and 3D. *J. Comput. Phys.* **2019**, *376*, 411–434. [CrossRef]
23. Pérez-Arancibia, C.; Turc, C.; Faria, L. Planewave density interpolation methods for 3D Helmholtz boundary integral equations. *SIAM J. Sci. Comput.* **2019**, *41*, A2088–A2116. [CrossRef]
24. Carvalho, C.; Khatri, S.; Kim, A.D. Asymptotic analysis for close evaluation of layer potentials. *J. Comput. Phys.* **2018**, *355*, 327–341. [CrossRef]
25. Carvalho, C.; Khatri, S.; Kim, A.D. Asymptotic approximation for the close evaluation of double-layer potentials. *SIAM J. Sci. Comput.* **2020**, *42*, A504–A533. [CrossRef]
26. Khatri, S.; Kim, A.D.; Cortes, R.; Carvalho, C. Close evaluation of layer potentials in three dimensions. *J. Comput. Phys.* **2020**, *423*, 109798. [CrossRef]
27. Hwang, W.S. A regularized boundary integral method in potential theory. *Comput. Methods Appl. Mech. Eng.* **2013**, *259*, 9. [CrossRef]
28. Liu, Y.J.; Rudolphi, T.J. New identities for fundamental solutions and their applications to non-singular boundary element formulations. *Comput. Mech.* **1999**, *24*, 286–292. [CrossRef]
29. Klaseboer, E.; Sun, Q.; Chan, D.Y. Non-singular boundary integral methods for fluid mechanics applications. *J. Fluid Mech.* **2012**, *696*, 78. [CrossRef]
30. Sun, Q.; Klaseboer, E.; Khoo, B.C.; Chan, D.Y. A robust and non-singular formulation of the boundary integral method for the potential problem. *Eng. Anal. Bound. Elem.* **2014**, *1*, 117–123. [CrossRef]
31. Sun, Q.; Klaseboer, E.; Khoo, B.-C.; Chan, D.Y. Boundary regularized integral equation formulation of the Helmholtz equation in acoustics. *R. Soc. Open Sci.* **2015**, *2*, 140520. [CrossRef] [PubMed]
32. Faria, L.M.; Pérez-Arancibia, C.; Bonnet, M. General-purpose kernel regularization of boundary integral equations via density interpolation. *Comput. Methods Appl. Mech. Eng.* **2021**, *378*, 113703. [CrossRef]
33. Kress, R. *Linear Integral Equations*; Springer: New York, NY, USA, 1989.
34. Colton, D.; Kress, R. *Integral Equation Methods in Scattering Theory*; SIAM: Philadelphia, PA, USA, 2013.
35. Guenther, R.B.; Lee, J.W. *Partial Differential Equations of Mathematical Physics and Integral Equations*; Dover Publications: New York, NY, USA, 1996.
36. Atkinson, K.E. *The Numerical Solution of Integral Equations of the Second Kind*; Cambridge University Press: Cambridge, UK, 1997.
37. Bremer, J.; Gimbutas, Z.; Rokhlin, V. A nonlinear optimization procedure for generalized gaussian quadratures. *SIAM J. Sci. Comput.* **2010**, *32*, 1761–1788. [CrossRef]
38. Bruno, O.P.; Kunyansky, L.A. A fast, high-order algorithm for the solution of surface scattering problems: basic implementation, tests, and applications. *J. Comput. Phys.* **2001**, *169*, 80–110. [CrossRef]

39. Ganesh, M.; Graham, I. A high-order algorithm for obstacle scattering in three dimensions. *J. Comput. Phys.* **2004**, *198*, 211–242. [CrossRef]
40. Kress, R. Boundary integral equations in time-harmonic acoustic scattering. *Math. Comput. Model.* **1991**, *15*, 229–243. [CrossRef]
41. Carvalho, C. Subtraction-Techniques Codes. Available online: https://doi.org/10.5281/zenodo.5523373 (accessed on 25 September 2021).
42. Atkinson, K.E. Numerical integration on the sphere. *ANZIAM J.* **1982**, *23*, 332–347. [CrossRef]
43. Atkinson, K.E. The numerical solution Laplace's equation in three dimensions. *SIAM J. Numer. Anal.* **1982**, *19*, 263–274. [CrossRef]
44. Atkinson, K.E. Algorithm 629: An integral equation program for Laplace's equation in three dimensions. *ACM Trans. Math. Softw.* **1985**, *11*, 85–96. [CrossRef]
45. Atkinson, K.E. A survey of boundary integral equation methods for the numerical solution of Laplace's equation in three dimensions. In *Numerical Solution of Integral Equations*; Springer: Boston, MA, USA, 1990; pp. 1–34.
46. Ammari, H.; Millien, P.; Ruiz, M.; Zhang, H. Mathematical analysis of plasmonic nanoparticles: The scalar case. *Arch. Ration. Mech. Anal.* **2017**, *2*, 597–658. [CrossRef]
47. Helsing, J.; Karlsson, A. An extended charge-current formulation of the electromagnetic transmission problem. *SIAM J. Appl. Math.* **2020**, *80*, 951–976. [CrossRef]

Article

Multi-Physics Inverse Homogenization for the Design of Innovative Cellular Materials: Application to Thermo-Elastic Problems

Matteo Gavazzoni [1,*], Nicola Ferro [2,*], Simona Perotto [2] and Stefano Foletti [1]

[1] Dipartimento di Meccanica, Politecnico di Milano, Via La Masa 1, I-20156 Milano, Italy; stefano.foletti@polimi.it
[2] MOX, Dipartimento di Matematica, Politecnico di Milano, Piazza L. da Vinci 32, I-20133 Milano, Italy; simona.perotto@polimi.it
* Correspondence: matteo.gavazzoni@polimi.it (M.G.); nicola.ferro@polimi.it (N.F.)

Abstract: We present a new algorithm to design lightweight cellular materials with required properties in a multi-physics context. In particular, we focus on a thermo-elastic setting by promoting the design of unit cells characterized both by an isotropic and an anisotropic behavior with respect to mechanical and thermal requirements. The proposed procedure generalizes the microSIMPATY algorithm to a thermo-elastic framework by preserving all the good properties of the reference design methodology. The resulting layouts exhibit non-standard topologies and are characterized by very sharp contours, thus limiting the post-processing before manufacturing. The new cellular materials are compared with the state-of-art in engineering practice in terms of thermo-elastic properties, thus highlighting the good performance of the new layouts which, in some cases, outperform the consolidated choices.

Keywords: topology optimization; cellular materials; multi-physics; homogenization; anisotropic mesh adaptation

Citation: Gavazzoni, M.; Ferro, N.; Perotto, S.; Foletti, S. Multi-Physics Inverse Homogenization for the Design of Innovative Cellular Materials: Application to Thermo-Elastic Problems. *Math. Comput. Appl.* **2022**, 27, 15. https://doi.org/10.3390/mca27010015

Received: 7 January 2022
Accepted: 9 February 2022
Published: 15 February 2022

Publisher's Note: MDPI stays neutral with regard to jurisdictional claims in published maps and institutional affiliations.

Copyright: © 2022 by the authors. Licensee MDPI, Basel, Switzerland. This article is an open access article distributed under the terms and conditions of the Creative Commons Attribution (CC BY) license (https://creativecommons.org/licenses/by/4.0/).

1. Introduction

Cellular materials represent an effective solution for structural applications where conventional monolithic materials fail to satisfy the design constraints [1]. The fast advancements in additive manufacturing technologies experienced in the last few years have further amplified the interest toward metamaterials. In addition, the possibility to employ a large variety of bulk materials in manufacturing processes (e.g., metals, polymers, ceramics [2–4]) has enabled the design of new metamaterials, featuring innovative combinations of physical effective properties. The possibility to blend different materials in order to reach diverse objectives proved to have a great impact in all the contexts where multi-functionality is required. For example, in [5–8], biocompatible 3D-printed metal bone implants promoting bone ingrowth are proposed by properly tailoring the material microstructure in order to reproduce the elastic modulus and the permeability of the human bone. Other applications range from thermal-cloaking systems fitly combining microstructure geometry and orientation [9,10] to lattice-based heat exchangers, where good thermal conductivity and convection properties are exploited to enhance the devices' performance [11,12].

From a modeling viewpoint, the proposal of innovative multifunctional cellular materials can benefit from the most recent advancements in topology optimization [13], properly combined with direct and inverse homogenization processes [14–16]. Several optimization approaches can be exploited in the context of metamaterial design. The layout of the employed microstructures can be selected a priori, starting from consolidated dictionaries of unit cells [16–21], or designed from scratch to match the expected effective properties [22–28]. In this context, a single- or a multi-objective topology optimization at the microscale can drive the design of new unit cells matching target properties at the macroscale, potentially in a multi-physics framework. For instance, the optimization of homogenized elastic properties is tackled in [29–31] with the aim of maximizing the bulk (or shear) modulus. To this aim, the authors control specific components of the homogenized elastic tensor or resort to the minimization of the compliance of a given structural part. Other works focus on a multi-physics optimization (for instance, by considering elastic, thermal, and electrical properties) by providing microstructures optimized with respect to diverse objectives and physics [32–35].

Nevertheless, it is well-known that standard topology optimization techniques suffer from typical issues that may compromise the effective performance and manufacturability of the new layouts. Among the most recurrent, we mention the possible presence of intermediate densities, the non-smooth contours of the final design and the generation of unit cells which turn out to be unprintable since presenting too thin struts. All these drawbacks are strictly related to the selected computational grid: a coarse mesh promotes jagged boundaries and a diffused void/material interface; vice versa, an extremely fine mesh leads to a non-affordable computational effort and fosters the generation of too complex structures. Filtering offers a possible remedy to address all these concerns by alternating smoothing with sharpening phases to be properly tuned. Such a tuning is not a trivial task and may often lead to non-optimal design solutions [13,33,36].

The selection of a computational mesh customized to the design problem has been proved to be instrumental in order to limit the main issues of topology optimization. For instance, in [37], the combination of a standard density-based method for topology optimization with an anisotropic mesh adaptation procedure has been used to get rid of intermediate densities, irregular boundaries, and thin struts in the design of structures at the macroscale. The proposed algorithm, named SIMPATY (SIMP with mesh AdaptiviTY), is based on a robust mathematical tool, namely an a posteriori estimator for the discretization error, and it leads to final designs characterized by reliable mechanical properties as well as by free-form features. The same procedure has been successfully exploited at the microscale, with the proposal of the microSIMPATY algorithm [26]. So far, this procedure has been used for the design of unit cells with optimized mechanical properties in a linear elasticity setting [27,38].

In this work, we propose a new pipeline for the design of new cellular materials by extending the microSIMPATY algorithm to a multi-physics context. The objective is to obtain lightweight metamaterials with prescribed requirements on the elastic and thermal conductivity properties, which are characterized by a ready-to-print topology. The design strategy here developed is confined to a 2D setting and has to be meant as a proof-of-concept, preliminary to a 3D implementation. However, to corroborate the effectiveness of the proposed methodology, we perform a cross-comparison between the new cells and the standard ones in thermo-elastic applications.

The paper is organized as follows. Section 2 represents the core of the paper. It provides the physical problem constraining the optimization process, outlines the main theoretical tools to perform the optimization, and formalizes the thermo-elastic design procedure in the MultiP-microSIMPATY algorithm. Three design cases are considered in Section 3 to apply the MultiP-microSIMPATY algorithm to diverse scenarios. Section 4 further analyzes the results in the previous section by comparing the new designs with the state-of-the-art. Finally, Section 5 outlines the most remarkable contributions of the work together with some future perspectives.

2. Methods

In this paper, we refer to a multi-physics framework, in order to provide new layouts for the design of cellular materials. A standard issue consists in optimizing the microscale to ensure desired properties at the macroscale. To deal with this two-scale setting, it is crucial to properly transfer the physical characterization of the micro- to the macroscale. Direct and inverse homogenization represent widespread solutions in such a direction [14,39–41].

In particular, the direct approach incorporates the microscopic effects into a homogenized macroscopic model. As a consequence, the microscopic behavior is known, whereas we have to identify the (homogenized) macroscopic characterization. Vice versa, inverse homogenization starts from desired macroscopic physical properties and designs the microscale in order to match such features, thus swapping the role played by known and unknown scales with respect to direct homogenization.

In this paper, we focus on a two-dimensional setting and on linear thermo-elastic properties, so that, at the macroscale, the reference models are the linear elasticity equation [42] and the linear thermal conduction problem, as identified by the standard stress–strain (σ-ε) and heat flux–temperature ($\mathbf{q}$-θ) relations, given by

$$\sigma(\mathbf{u}) = \begin{bmatrix} \sigma_{11}(\mathbf{u}) \\ \sigma_{22}(\mathbf{u}) \\ \sigma_{12}(\mathbf{u}) \end{bmatrix} = \begin{bmatrix} E_{1111} & E_{1122} & E_{1112} \\ E_{2211} & E_{2222} & E_{2212} \\ E_{1211} & E_{1222} & E_{1212} \end{bmatrix} \begin{bmatrix} \varepsilon_{11}(\mathbf{u}) \\ \varepsilon_{22}(\mathbf{u}) \\ 2\varepsilon_{12}(\mathbf{u}) \end{bmatrix} = \mathbf{E}\,\varepsilon(\mathbf{u}), \quad (1)$$

and

$$\mathbf{q}(\theta) = \begin{bmatrix} q_1(\theta) \\ q_2(\theta) \end{bmatrix} = \begin{bmatrix} k_{11} & k_{12} \\ k_{21} & k_{22} \end{bmatrix} \begin{bmatrix} \dfrac{\partial \theta}{\partial x_1} \\ \dfrac{\partial \theta}{\partial x_2} \end{bmatrix} = \mathbf{k}\,\nabla\theta, \quad (2)$$

respectively, where $\mathbf{E}$ and $\mathbf{k}$ are the stiffness and the conductivity tensors characterizing the considered solid material.

We observe that in view of the homogenization procedures, the constitutive laws (1) and (2) have to be properly modified to include the effects of the microscale, into

$$\sigma^H(\mathbf{u}) = \mathbf{E}^H\,\varepsilon(\mathbf{u}), \quad \mathbf{q}^H(\theta) = \mathbf{k}^H\,\nabla\theta,$$

where $\mathbf{E}^H$ and $\mathbf{k}^H$ denote the homogenized stiffness and thermal conductivity tensors, as detailed in the next section.

2.1. Inverse Homogenization

Inverse homogenization is the procedure that allows us to design microstructures with prescribed properties at the macroscale. The required features are mathematically commuted into a goal functional $\mathcal{J}$ and into suitable constraints driving a topology optimization process to be solved in the unit cell $Y \subset \mathbb{R}^2$ whose periodic repetition yields the cellular material [26,27,29,38,43]. According to a density-based approach, a standard way to perform such an optimization leads us to define an auxiliary scalar field, ρ, that models the relative material density at the microscale. A priori, it is assumed that $\rho = 1$ labels the material, while $\rho = 0$ identifies the void. However, since density $\rho \in L^\infty(Y,[0,1])$ can take all the values in $[0,1]$, it is standard to penalize the intermediate values (i.e., intermediate material densities) that are not physically consistent. To this aim, we resort to the SIMP method, which modifies the reference state equations by weighting the constitutive laws with a suitable power of the density [13].

In general, the optimization problem we are interested to solve is

$$\min_{\rho \in L^\infty(Y,[0,1])} \mathcal{J}(z(\rho),\rho) : \begin{cases} a_\rho(z(\rho),w) = F_\rho(w) \quad \forall w \in W \\ \mathbf{L}_B \leq \mathbf{C}(z(\rho),\rho) \leq \mathbf{U}_B, \end{cases} \quad (3)$$

where $z = z(\rho)$ denotes the state variable depending on the density field, $a_\rho(\cdot,\cdot)$ together with $F_\rho(\cdot)$ defines the state equation constraining the topology optimization, W is a suitable function space [44], and the box inequality includes specific design and physical requirements, with $\mathbf{C}(\cdot,\cdot)$ being the vector gathering the quantities to be controlled through the corresponding lower and upper bounds, $\mathbf{L}_B$ a $\mathbf{U}_B$.

In the analysis below, we pick the objective functional $\mathcal{J}$ as

$$\mathcal{M}(\rho) = \int_Y \rho \, dY \tag{4}$$

since we are interested in minimizing the total mass, $\mathcal{M}$, of the cellular structure, i.e., to design lightweight materials.

According to a standard homogenization procedure, as the state equation, we select the linear elasticity model at the microscale weighed by the density function, which describes the Y-periodic displacement field fluctuations, $\mathbf{u}^{*,ij}$, induced by the test displacement fields $\mathbf{u}^{0,ij}$, with $\mathbf{u}^{0,11} = [x,0]^T$, $\mathbf{u}^{0,22} = [0,y]^T$ and $\mathbf{u}^{0,12} = [y,0]^T$. This leads us to identify the forms $a_\rho(\cdot,\cdot)$ and $F_\rho(\cdot)$ in (3) with

$$\begin{aligned} a_\rho^{E,ij}\left(\mathbf{u}^{*,ij}(\rho),\mathbf{v}\right) &= \frac{1}{|Y|}\int_Y \rho^p\, \sigma(\mathbf{u}^{*,ij}):\varepsilon(\mathbf{v})dY, \\ F_\rho^{E,ij}(\mathbf{v}) &= \frac{1}{|Y|}\int_Y \rho^p\, \sigma(\mathbf{u}^{0,ij}):\varepsilon(\mathbf{v})dY, \end{aligned} \tag{5}$$

respectively, with $p \in \mathbb{R}^+$, $ij \in I = \{11, 22, 12\}$, and where the superscript E refers to the elasticity setting. The state equation associated with (5) is completed with fully periodic conditions on the cell boundary ∂Y, according to the asymptotic homogenization theory. Thus, W in (3) coincides with the space $\mathcal{U}_\#^2 = [H_\circ^1(Y)]^2$ of the $H^1(Y)$-vector functions satisfying periodic boundary conditions along ∂Y.

To include also the thermal component in the topology optimization, we further constrain the process with the ρ-weighed thermal conductivity model at the microscale, which are characterized by the forms

$$\begin{aligned} a_\rho^{k,m}(\theta^{*,m}(\rho),v) &= \frac{1}{|Y|}\int_Y \rho^s\, \mathbf{q}(\theta^{*,m}):\nabla v\, dY, \\ F_\rho^{k,m}(v) &= \frac{1}{|Y|}\int_Y \rho^s\, \mathbf{q}(\theta^{0,m}):\nabla v\, dY, \end{aligned} \tag{6}$$

with $s \in \mathbb{R}^+$, $m \in J = \{1,2\}$, and where the superscript k refers to the thermal framework. Analogously to (5), we complete the thermal state equation identified by (6) with periodic boundary conditions along ∂Y, so that $\theta^{*,m}$ and $v \in \mathcal{U}_\#^1 = H_\circ^1(Y)$, where $\theta^{*,m}$ denotes the temperature fluctuations associated with the test temperature fields $\theta^{0,m}$ (namely, $\theta^{0,1} = x$ and $\theta^{0,2} = y$).

The two problems at the microscale, (5) and (6), are instrumental to define the homogenized elastic tensor, $\mathbf{E}^H$, and the homogenized thermal conductivity tensor, $\mathbf{k}^H$, which will be involved in the box constraints in (3). The component-wise definition of $\mathbf{E}^H$ and $\mathbf{k}^H$ is

$$E_{ijkl}^H = \frac{1}{|Y|}\int_Y \rho^p \left[\sigma(\mathbf{u}^{0,ij}) - \sigma(\mathbf{u}^{*,ij}(\rho))\right] : \left[\varepsilon(\mathbf{u}^{0,kl}) - \varepsilon(\mathbf{u}^{*,kl}(\rho))\right] dY, \tag{7}$$

$$k_{mn}^H = \frac{1}{|Y|}\int_Y \rho^s \left[\mathbf{q}(\theta^{0,m}) - \mathbf{q}(\theta^{*,m}(\rho))\right] : \left[\nabla\theta^{0,n} - \nabla\theta^{*,n}(\rho)\right] dY, \tag{8}$$

respectively, with $ij, kl \in I$ and $m, n \in J$. In particular, the two-sided inequality in (3) will be exploited to promote diverse mechanical and thermal behaviors along the different spatial directions. To this aim, we constrain the two ratios E_{2222}^H/E_{1111}^H and k_{22}^H/k_{11}^H so that they vary in suitable ranges. This choice allows us to penalize the mechanical and the thermal contributions in a different way along the two directions, as shown in the numerical assessment. An additional two-sided control is enforced on the first and the last diagonal terms, E_{1111}^H and E_{1212}^H, of the homogenized elastic tensor, as well as on the first diagonal term, k_{11}^H, of the homogenized thermal conductivity tensor.

To sum up, the optimization setting we are led to deal with coincides with the following problem:

$$\min_{\rho \in L^\infty(Y,[0,1])} \mathcal{M}(\rho) : \begin{cases} a_\rho^{E,ij}\left(\mathbf{u}^{*,ij}(\rho), \mathbf{v}\right) = F_\rho^{E,ij}(\mathbf{v}) & \forall \mathbf{v} \in \mathcal{U}_\#^2,\ ij \in I \\ a_\rho^{k,m}\left(\theta^{*,m}(\rho), v\right) = F_\rho^{k,m}(v) & \forall v \in \mathcal{U}_\#^1,\ m \in J \\ E_{1111}^{\text{low}} \leq E_{1111}^H \leq E_{1111}^{\text{up}} \\ E_{1212}^{\text{low}} \leq E_{1212}^H \leq E_{1212}^{\text{up}} \\ \left(\dfrac{E_{2222}}{E_{1111}}\right)^{\text{low}} \leq \dfrac{E_{2222}^H}{E_{1111}^H} \leq \left(\dfrac{E_{2222}}{E_{1111}}\right)^{\text{up}} \\ k_{11}^{\text{low}} \leq k_{11}^H \leq k_{11}^{\text{up}} \\ \left(\dfrac{k_{22}}{k_{11}}\right)^{\text{low}} \leq \dfrac{k_{22}^H}{k_{11}^H} \leq \left(\dfrac{k_{22}}{k_{11}}\right)^{\text{up}} \\ \rho_{\min} \leq \rho \leq 1 \end{cases} \quad (9)$$

where all the bound values, $(\cdot)^{\text{low}}$ and $(\cdot)^{\text{up}}$, will be set according to the application at hand and in order to avoid an unfeasible solution (inappropriate constraints might lead to an empty solution space). Finally, the last inequality in (9) is meant to ensure the well-posedness of both the elasticity and the thermal problems (5) and (6), $\rho_{\min}$ being a suitable value in $(0, 1)$ (see Section 3 for more details).

2.2. Discretization on Anisotropic Adapted Meshes

With a view to the solution of problem (9), all the quantities involved in the state equations, as well as in the constraints, have to be discretized on a suitable tessellation of the unit cell Y. For this purpose, we resort to a computational mesh $\mathcal{T}_h = \{K\}$ customized to the problem at hand and characterized by stretched elements (i.e., a so-called anisotropic adapted mesh). Mesh $\mathcal{T}_h$ is employed to discretize both the test and the trial functions in the state equations, as well as the density function ρ, by means of a finite element scheme [44]. The anisotropic reference setting is the one proposed in [45]. In particular, the anisotropic features of each element K coincide with the lengths, $\lambda_{1,K}$, $\lambda_{2,K}$, and the directions, $\mathbf{r}_{1,K}$, $\mathbf{r}_{2,K}$, of the semi-axes of the ellipse circumscribed to K through the standard affine map, $T_K : \hat{K} \to K$, between the reference element $\hat{K}$ and the triangle K.

Concerning the adaptation procedure, we resort to a metric-based approach driven by an a posteriori estimator for the discretization error associated with the density function ρ. Among the error estimators available in the literature [46,47], we refer to an a posteriori recovery-based error analysis. Following the seminal work by O.C. Zienkewicz and J.Z. Zhu [48], we control the H^1-seminorm of the discretization error on the density, $e_\rho = \rho - \rho_h$. The selection of such an estimator is motivated by the fact that the density ρ exhibits strong gradients (i.e., large values for the H^1-seminorm) across the material–void interface. This feature will yield meshes whose elements are crowded along the boundaries of the structure, thus promoting the design of very smooth layouts. To this aim, we exactly integrate the so-called recovered error, $\mathbf{E}_\nabla = P(\nabla \rho_h) - \nabla \rho_h$, namely,

$$\begin{aligned} |e_\rho|_{H^1(Y)}^2 &= \|\nabla e_\rho\|_{L^2(Y)}^2 = \int_Y |\nabla \rho - \nabla \rho_h|^2 dY \\ &\simeq \|\mathbf{E}_\nabla\|_{L^2(Y)}^2 = \int_Y |P(\nabla \rho_h) - \nabla \rho_h|^2 dY, \end{aligned} \quad (10)$$

where ρ_h denotes the finite element discretization of ρ in the space V_h^r of the piecewise polynomials of degree $r \in \mathbb{N}$ associated with $\mathcal{T}_h$. The operator $P : [V_h^{r-1}]^2 \to [V_h^s]^2$ in (10), with $s \in \mathbb{N}$, denotes the recovered gradient, which, in general, provides a more accurate estimate of the exact gradient $\nabla \rho$ with respect to the discrete gradient $\nabla \rho_h$. Several recipes are available in the literature to define P [49–52]. In particular, we select operator $P : [V_h^0]^2 \to [V_h^0]^2$ as the area-weighted average of $\nabla \rho_h$ over the patch of the elements, $\Delta_K = \{T \in \mathcal{T}_h : T \cap K \neq \emptyset\}$, associated with K; i.e., we opt for

$$P(\nabla \rho_h)(\mathbf{x}) = \frac{1}{|\Delta_K|} \sum_{T \in \Delta_K} |T| \left. \nabla \rho_h \right|_T \quad \forall \mathbf{x} \in K, \tag{11}$$

with $|\omega|$ the area of the generic domain $\omega \subset \mathbb{R}^2$, where we have set the degree of the finite element space for ρ_h to $r = 1$. Space V_h^1 is also adopted to discretize the components of the displacement vectors $\mathbf{u}^{*,ij}$ as well as the temperature fields $\theta^{*,m}$ in (9), with $ij \in I$ and $m \in J$.

According to [53,54], we here adopt the anisotropic generalization of (10). This estimator essentially exploits the anisotropic counterpart of the definition of the H^1-seminorm [45], based on the symmetric semidefinite positive matrix G_{Δ_K}, with entries

$$\left[G_{\Delta_K}(\nabla g)\right]_{i,j} = \sum_{T \in \Delta_K} \int_T \frac{\partial g}{\partial x_i} \frac{\partial g}{\partial x_j} \, dT \quad i, j = 1, 2, \tag{12}$$

with $g \in H^1(Y)$, and where it is understood $x_1 = x$ and $x_2 = y$. Thus, the squared H^1-seminorm $|e_\rho|^2_{H^1(Y)}$ is evaluated by the (global) error estimator $\eta^2 = \sum_{K \in \mathcal{T}_h} \eta_K^2$, where

$$\eta_K^2 = \frac{1}{\lambda_{1,K} \lambda_{2,K}} \sum_{i=1}^{2} \lambda_{i,K}^2 \left(\mathbf{r}_{i,K}^T G_{\Delta_K}(\mathbf{E}_\nabla) \mathbf{r}_{i,K}\right), \tag{13}$$

defines the local error estimator. The contribution between brackets coincides with the projection of the squared L^2-norm of the recovered error along the anisotropic directions, while the scaling factor $(\lambda_{1,K} \lambda_{2,K})^{-1}$ guarantees the consistency with the isotropic case (for more details, see [53]).

The new adapted mesh is generated after commuting the error estimator η_K into a new mesh spacing (the metric), $\mathcal{M}$, consisting of the triplet $\{\lambda_{1,K}^{adapt}, \lambda_{2,K}^{adapt}, \mathbf{r}_{1,K}^{adapt}\}$, where the direction $\mathbf{r}_{2,K}^{adapt}$ is automatically defined being $\mathbf{r}_{1,K}^{adapt} \cdot \mathbf{r}_{2,K}^{adapt} = 0$, for each element $K \in \mathcal{T}_h$. This operation is performed by taking into account three different criteria, namely, (i) the minimization of the mesh cardinality $\#\mathcal{T}_h$; (ii) an accuracy requirement on the discretization error $|e_\rho|_{H^1(Y)}$ (i.e, on the error estimator η), controlled up to a user-defined tolerance TOL; (iii) the equidistribution of the error throughout the mesh elements (i.e., $\eta_K^2 = \text{TOL}^2/\#\mathcal{T}_h$). These three criteria lead us to solve a constrained minimization problem on each triangle $K \in \mathcal{T}_h$. The solution to this local optimization problem can be analytically derived, as proved in [55], being

$$\lambda_{1,K}^{adapt} = g_2^{-1/2} \left(\frac{\text{TOL}^2}{2 \, \#\mathcal{T}_h \, |\hat{\Delta}_K|}\right)^{1/2}, \quad \mathbf{r}_{1,K}^{adapt} = \mathbf{g}_2,$$

$$\lambda_{2,K}^{adapt} = g_1^{-1/2} \left(\frac{\text{TOL}^2}{2 \, \#\mathcal{T}_h \, |\hat{\Delta}_K|}\right)^{1/2}, \quad \mathbf{r}_{2,K}^{adapt} = \mathbf{g}_1 \tag{14}$$

where g_1, g_2 and $\mathbf{g}_1$, $\mathbf{g}_2$ are the eigenvalues and the eigenvectors of the scaled matrix $\hat{G}_{\Delta_K}(\mathbf{E}_\nabla) = G_{\Delta_K}(\mathbf{E}_\nabla)/|\Delta_K|$, with $g_1 \geq g_2 > 0$.

Finally, the metric $\mathcal{M} = \{\lambda_{1,K}^{adapt}, \lambda_{2,K}^{adapt}, \mathbf{r}_{1,K}^{adapt}\}_{K \in \mathcal{T}_h}$ has to be changed into a quantity associated with the vertices of $\mathcal{T}_h$ received as an input by the selected mesh generator. A standard choice consists in an arithmetic mean formula applied to the patch of elements associated with each vertex in $\mathcal{T}_h$ [54].

The anisotropic mesh adaptation based on the metric (14) is customized to a topology optimization problem in the algorithm SIMPATY, proposed in [37]. This procedure has been successfully employed for the design of structures at the macroscale [37,56,57] as well as for the design of new metamaterials with the proposal of the algorithm microSIMPATY [26,27]. Moreover, a combination of topology optimization at the macroscale and at the microscale is carried out in [38]. In particular, a multiscale topology optimization process is used for the design of orthotic devices for 3D printing manufacturing with the proposal of patient-specific innovative solutions.

It has been verified that the adoption of an adapted anisotropic mesh leads to free-form layouts characterized by very smooth boundaries both at the macroscale and at the microscale, mitigating some of the well-known drawbacks of standard topology optimization, such as the massive employment of filtering, the staircase effect, and the generation of too complex structures [13,33,36]. However, in [57], it has been observed that the presence of deformed elements inside the structures makes the finite element analysis less reliable. To overcome this issue, the authors suggest a hybrid approach. Thus, the mesh is kept isotropic, with a uniform diameter h^{iso} in the full-material regions, $\{\mathbf{x} \in Y : \rho_h(\mathbf{x}) > \rho^{th}\}$ with ρ^{th} as a user-defined threshold, whereas the stretched triangles are preserved along the material–void interface. Actually, these hybrid meshes ensure an effective balance between the smoothness of the structure and robust engineering performances (the interested reader can find a quantitative investigation of the benefits of the hybrid approach in terms of accuracy in (Section 5, [57])). For this reason, we resort to hybrid meshes in the sequel.

2.3. Multi-Physics Optimization Algorithm

In this section we propose the multi-physics adaptive inverse homogenization procedure, which generalizes the algorithm proposed in [26]. The discretization of the state equations associated with (5) and (6) is performed with the open-source finite element solver FreeFEM [58], which provides the ideal environment to implement an anisotropic mesh adaptation procedure in Section 2.2 through the built-in mesh generator BAMG (Bidimensional Anisotropic Mesh Generator).

The developed multi-physics optimization procedure is listed in the pseudocode in Algorithm 1. The main loop (lines 3–12) includes an optimization step, a filtering phase, and the mesh adaptation. At each global iteration k, the optimization problem is solved (line 4, function optimize) by taking into account all the constraints on the components of the elastic and of the thermal conductivity tensors in (9). To this aim, we use the interior point algorithm IPOPT [59], although any other optimization tool can be selected [60]. IPOPT requires as input the functional $\mathcal{J}$ to be minimized; the vector $\mathcal{C}$ gathering the constrained quantities in the optimization procedure; the two vectors $\mathbf{c}^l$ and $\mathbf{c}^u$ of the lower and upper bounds for the components in $\mathcal{C}$; the array $\mathcal{G}$ collecting the derivative of the functional $\mathcal{J}$ and of the constraints $\mathcal{C}$ with respect to ρ, computed by the adjoint Lagrangian approach (for more details, we refer to [27]); the initial guess ρ_h^k to start the optimization process; the accuracy TOPT for the minimization problem; the maximum number of iterations IT to stop the optimization. In particular, in the numerical assessment of Section 3, we set TOPT $= 10^{-5}$, and IT $= 100$ for k $= 0$ and IT $= 10$ for all the successive iterations. The higher value for IT for k $= 0$ takes into account that the initial guess ρ_h^0 can be completely arbitrary with respect to the minimum to be reached. On the contrary, a smaller value for IT is sufficient for k > 0, since the initial guess, ρ_h^k, coincides with the output of a previous optimization step.

Algorithm 1 MultiP-microSIMPATY

1: **Input**: CTOL, kmax, c^l, c^u, ρ_h^0, TOPT, IT, kfmax, τ, β, $\mathcal{T}_h^0$, TOL, HYB
2: Set: k = 0, errC = 1+CTOL;
3: **while** errC > CTOL & k < kmax **do**
4: $\quad\rho_h^{k+1}$ = optimize($\mathcal{J}$, $\mathcal{C}$, c^l, c^u, $\mathcal{G}$, ρ_h^k, TOPT, IT);
5: $\quad$**if** k < kfmax **then**
6: $\quad\quad\rho_h^{k+1}$ = helmholtz(ρ_h^{k+1}, τ);
7: $\quad\quad\rho_h^{k+1}$ = heaviside(ρ_h^{k+1}, β);
8: $\quad$**end if**
9: $\quad\mathcal{T}_h^{k+1}$ = adapt($\mathcal{T}_h^k$, ρ_h^{k+1}, TOL, HYB);
10: $\quad$errC = $\left|\#\mathcal{T}_h^{k+1} - \#\mathcal{T}_h^k\right|/\#\mathcal{T}_h^k$;
11: $\quad$k = k+1;
12: **end while**
13: $\mathcal{T}_h = \mathcal{T}_h^k$;
14: $\rho_h = \rho_h^k$;
15: $\left[\mathbf{E}^H, \mathbf{k}^H\right]$ = homogenize(ρ_h);
16: **return** $\mathcal{T}_h$, ρ_h, $\mathbf{E}^H$, $\mathbf{k}^H$

Function optimize returns the density ρ_h^{k+1}, which is successively processed by means of Helmholtz and Heaviside filters (lines 6–7, functions helmholtz and heaviside) [61,62]. The two filtering operations work in a complementary way. The Helmholtz partial differential equation is instrumental to remove too thin features, although promoting intermediate densities along the layout contour. In more detail, it consists of a low-pass filter based on a diffusion kernel with radius $\tau \in \mathbb{R}^+$. On the contrary, the Heaviside filter, coinciding with a β-dependent regularization of the Heaviside function with $\beta \in \mathbb{R}^+$, penalizes the intermediate material densities, also due to the Helmholtz filter, thus increasing the sharpness of the material/void interface. The combined filtering takes place for the first kfmax global iterations only. This choice leads to start the mesh adaptation procedure with a density field, which is free from too complex features, while exhibiting a clear alternation between void and material. The filtering phase becomes redundant when the optimization loop approaches the minimum, so that mesh adaptation alone suffices to ensure well-defined structures. In the next section, filtering parameters τ and β are set equal to 0.02 and 5 respectively, while kfmax = 25.

The next step coincides with the mesh adaptation procedure detailed in Section 2.2 and here represented by function adapt (line 9). The input parameter TOL establishes the accuracy of the error estimator η through the predicted metric in (14). Parameter HYB is a boolean flag that, in correspondence with the full material, switches the employment of an isotropic mesh on or off.

The main loop is controlled by a check on the stagnation of the relative difference between the cardinality of two consecutive meshes (line 10), up to a maximum number of global iterations kmax (line 3). The choices TOL = 10^{-5} and kmax= 100 are preserved throughout all the numerical assessment below.

Algorithm MultiP-microSIMPATY returns the final adapted mesh $\mathcal{T}_h$, the optimized density ρ_h, and the homogenized elastic and conductivity tensors, $\mathbf{E}^H$ and $\mathbf{k}^H$, which are computed by function homogenize (line 15), based on (7) and (8).

We remark that the procedure itemized in Algorithm 1 is fully general and it can be applied in a straightforward way to different multi-physics contexts after properly modifying the formulation in (9).

3. Results

We analyze three different cases of microstructure design according to (9). In order to highlight the interplay between the different (thermal and mechanical) physics involved, we consider configurations where the thermal conductivity and the elastic stiffness requirements act along different directions. For instance, a high shear stiffness combined with a high thermal conductivity along the x-direction orient the material along two opposite directions, with the prescription of a conflict configuration.

The whole verification below shares common choices for some physical quantities and discretization parameters. In particular, the unit cell $Y \subset \mathbb{R}^2$ is identified with the unitary square, $Y = (0,1)^2$. Moreover, we set the Young modulus, E_Y, and the Poisson ratio, ν, to 1 and 0.3, respectively, and we consider an isotropic solid material with unitary thermal conductivity by setting $k_{11} = k_{22} = 1$ and $k_{12} = k_{21} = 0$. These choices allow us to obtain normalized homogenized mechanical and thermal properties for the cellular structures. Following [26], both the SIMP-powers, p and s, in (7) and (8) are chosen equal to 4 to penalize intermediate densities.

Concerning the discretization frame, we choose a random density field, ρ_h^0, as the initial guess for the optimization process, in order to avoid any bias. In particular, ρ_h^0 is defined on an initial structured mesh characterized by 30 subdivisions per side, and with values ranging from $\rho_{min} = 10^{-4}$ to 1 (see Figure 1 for an example).

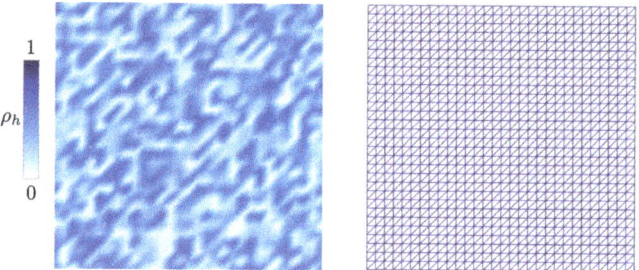

Figure 1. Initial guess ρ_h^0 (**left**) and corresponding mesh $\mathcal{T}_h^0$ (**right**).

Finally, to ensure a reliable finite element analysis, we resort to the hybrid mesh adaptation procedure (HYB = 1 in function adapt). In particular, we choose the threshold value $\rho^{th} = 0.9$ to manage the alternation between isotropic and anisotropic elements, and the isotropic tessellation is characterized by the uniform diameter $h^{iso} = 0.03$ (approximately 1/30 of the design domain dimension).

After the optimization, we perform a verification step to check the actual mechanical and thermal properties of the material yielded by a periodic repetition of the optimized unit cell. To this aim, we use the Abaqus software (Abaqus, Dassault Systèmes Simulia Corp, Johnston, RI, USA). The layouts provided by Algorithm 1 are imported in Abaqus after a thresholding, which neglects the density smaller than 0.75. The obtained geometry is remeshed on a uniform isotropic triangular mesh with an average size equal to 0.01, while the displacement and temperature fields are discretized with quadratic finite elements, completed with periodic boundary conditions. The verification here performed can be considered as a preliminary step toward the integration of the MultiP-microSIMPATY algorithm into a common workflow for structural analysis.

3.1. Design Case 1

The main goal of this first optimization process is to design a lightweight unit cell characterized by isotropic mechanical homogenized properties and, vice versa, anisotropic thermal homogenized features. This problem can be cast in setting (9), after making the following choices for the constraints:

$$\begin{cases} 0.05 \leq E^H_{1111} \leq 0.08 \\ 0.055 \leq E^H_{1212} \leq 0.080 \\ 1 \leq \dfrac{E^H_{2222}}{E^H_{1111}} \leq 2 \\ 0.01 \leq k^H_{11} \leq 1.00 \\ 0.00 \leq \dfrac{k^H_{22}}{k^H_{11}} \leq 0.58. \end{cases} \quad (15)$$

The isotropic mechanical behavior and the anisotropic thermal properties are enforced by the constraints in (15)$_3$ and (15)$_5$. In particular, we expect ratios E^H_{2222}/E^H_{1111} and k^H_{22}/k^H_{11} to coincide with the corresponding lower and upper bounds, respectively. Moreover, since a control on the ratios does not ensure E^H_{1111}, E^H_{2222}, k^H_{11}, and k^H_{22} to be in a physically admissible range of values, we further constrain the optimization through the box inequalities (15)$_1$ and (15)$_4$. Finally, a control on the component E^H_{1212} of the homogenized stiffness tensor closes the minimization problem, thus further restricting the solution space.

For the values set for the input parameters, the MultiP-microSIMPATY algorithm converges in 51 global iterations. Figure 2 shows the layout and the associated anisotropic adapted mesh at three different iterations.

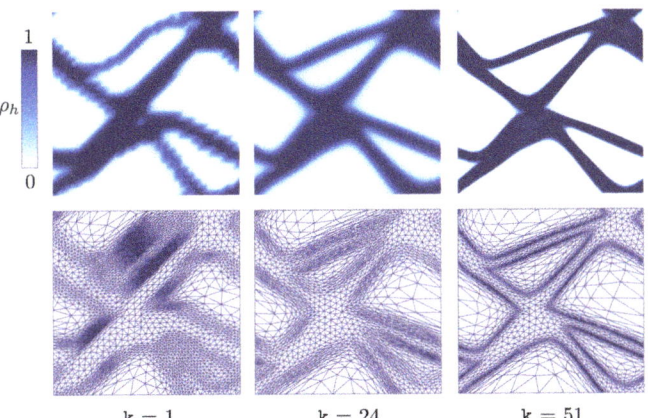

Figure 2. Design Case 1: density field (**top**) and associated anisotropic adapted mesh (**bottom**) for three different global iterations.

We remark that the final topology of the layout is already detected at the first iteration, although the quality of the solution is improved throughout the optimization process. In particular, at the first iteration (k = 1), we observe a significant staircase effect together with the presence of intermediate densities along the microstructure interface. At the end of the filtering phase (k = 24), the jagged boundaries are fully smoothed, despite the intermediate densities still blurring the design. The spreading effect along the material/void interface is gradually reduced when switching off the filtering, i.e., for k > 24, as shown by the last column in Figure 2. Thus, the final optimized solution (k = 51) shows an extremely sharp transition from material to void and smooth boundaries, which make the structure ready for printing or manufacturing, with a limited need for post-processing.

Concerning the adapted mesh, we recognize the effect of the hybrid approach, which combines stretched elements to discretize the strong gradients of the density field, coarse anisotropic triangles outside the structure, and isotropic elements in correspondence with the material.

Additional quantitative information on the MultiP-microSIMPATY algorithm is provided by Table 1 and by the diagrams in Figure 3, which show the evolution of the objective function and of the constrained quantities (top), together with the trend of the mesh cardinality (bottom), over the global iterations. Notice that the values of the constraints have been normalized between 0 and 1 (see the highlighted area in the top panel of Figure 3). It is evident that the mass exhibits a completely different trend when compared with the constrained quantities. The value of the objective function oscillates with values between 0.325 and 0.475 over the first 35 iterations, and it eventually converges toward a stable phase. On the contrary, all the constrained quantities are characterized by mild oscillations. In particular, k_{11}^H remains essentially constant over the whole optimization process. The plot of the ratios E_{2222}^H/E_{1111}^H and k_{22}^H/k_{11}^H confirms that the two inequalities are in conflict so that the active constraints are the lower and upper bound, respectively. Moreover, from the values in Table 1, it can be observed that the stiffness component along the x-direction, E_{1111}^H, reaches a value that is about 25% lower than the corresponding c^l. This can be ascribed to the presence of very thin struts generated by the severe thresholding ($\rho_h < 0.75$) applied before performing the analyses in Abaqus.

Table 1. Design cases 1, 2, and 3: values of the constraints and of the objective functional computed with Abaqus software, together with the lower and the upper bounds, c^l and c^u, involved in the optimization.

	E_{1111}^H	E_{1212}^H	$\dfrac{E_{2222}^H}{E_{1111}^H}$	k_{11}^H	$\dfrac{k_{22}^H}{k_{11}^H}$	$\mathcal{M}$
			Design Case 1			
c^u	0.080	0.080	2.000	1.000	0.580	
c	0.038	0.056	1.299	0.199	0.566	0.292
c^l	0.050	0.055	1.000	0.010	0.000	
			Design Case 2			
c^u	0.350	0.150	2.000	1.000	2.000	
c	0.250	0.086	0.299	0.317	0.597	0.412
c^l	0.230	0.080	0.300	0.300	0.000	
			Design Case 3			
c^u	0.150	0.100	1.100	0.400	1.100	
c	0.151	0.083	1.074	0.260	1.002	0.415
c^l	0.100	0.080	1.000	0.250	1.000	

The evolution of the topology in Figure 2 is consistent with the trend in Figure 3 (top panel). The topology does not essentially vary during the optimization process, according to the almost constant trend of the constraints. On the other hand, the highly oscillatory trend of $\mathcal{M}$ in the first optimization stage is related to the effect of the smoothing and of the sharpening operations, which are confined to the first 24 iterations. From k = 25, only the minimization process and the mesh adaptation contribute to a mass variation, with less striking changes.

Finally, in Figure 4 (left), we show the 3×3-cell material generated by a periodic repetition of the optimized unit cell.

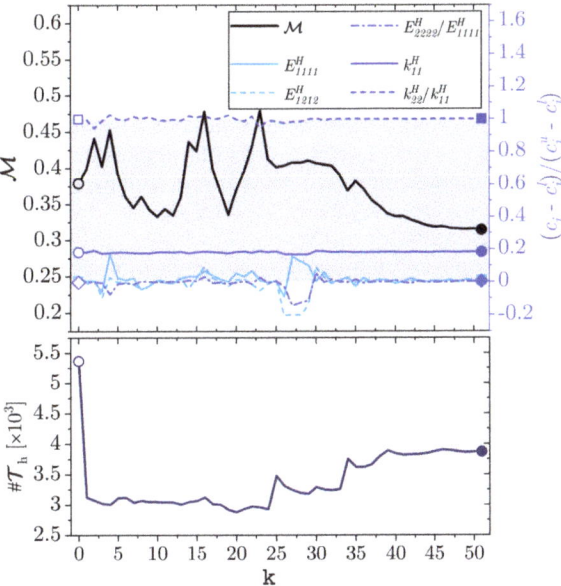

Figure 3. Design Case 1: evolution of the objective functional $\mathcal{M}$ and of the constraints c_i (**top**); trend of the mesh cardinality $\#\mathcal{T}_h$ (**bottom**) throughout the global iterations k.

Figure 4. Design Cases 1, 2, and 3 (**left–right**): 3×3-cell meta-material.

3.2. Design Case 2

The second MultiP-microSIMPATY run aims at designing a microstructure that provides high stiffness and thermal conductivity along the x-direction and a high shear stiffness. As for the Design Case 1, these requirements might originate a set of conflicting constraints. In fact, the two former demands are expected to orient the material along the x-direction, while the latter requirement prescribes also the presence of material along the diagonal of the cell Y, which could react by tension to shear loading. This design setting is formalized by problem (9) when completed by the following set of constraints:

$$\begin{cases} 0.23 \leq E^H_{1111} \leq 0.35 \\ 0.08 \leq E^H_{1212} \leq 0.15 \\ 0.3 \leq \dfrac{E^H_{2222}}{E^H_{1111}} \leq 2.0 \\ 0.3 \leq k^H_{11} \leq 1.0 \\ 0 \leq \dfrac{k^H_{22}}{k^H_{11}} \leq 2. \end{cases} \quad (16)$$

We highlight that the bounds for the stiffness tensor components to be promoted, E^H_{1111} and E^H_{1212}, are set by taking into account the mass minimization goal, i.e., by keeping them considerably lower than 1.

Algorithm 1 stops in 56 iterations due to mesh stagnation. Figure 5 gathers the density field distribution together with the associated anisotropically adapted computational mesh at iterations $k = 5, 20$, and 56. At the fifth iteration, the cell presents very thin struts that are progressively erased by the combined action of the Helmholtz and the Heaviside filters. For $k = 20$, the topology essentially coincides with the final optimized one, although the layout still exhibits intermediate density values along the boundaries. The structure contours become sharper and sharper throughout the next iterations when filtering is switched off and thanks to the mesh adaptation procedure.

Concerning the final topology, we observe that most of the material is aligned along the two main diagonals of Y. This guarantees high shear stiffness, while ensuring a low stiffness along the y-direction, so that the lower bound for E^H_{2222}/E^H_{1111} is reached. On the other side, the requirements on E^H_{1111} and k^H_{11} are taken into account by the two thinner struts along the x-direction, which improves the corresponding stiffness and the thermal conductivity. Figure 4 (center) provides a sketch of the metamaterial associated with the optimized cell in a 3×3 cellular pattern.

For a more quantitative characterization of the optimized structure in terms of mass and reached constraints, we refer to Table 1. We notice that to address the conflict among the several requirements, the optimization process pushes all the constrained quantities toward the lower bound of the corresponding range, while increasing the mass of the structure if compared, for instance, with the previous design case.

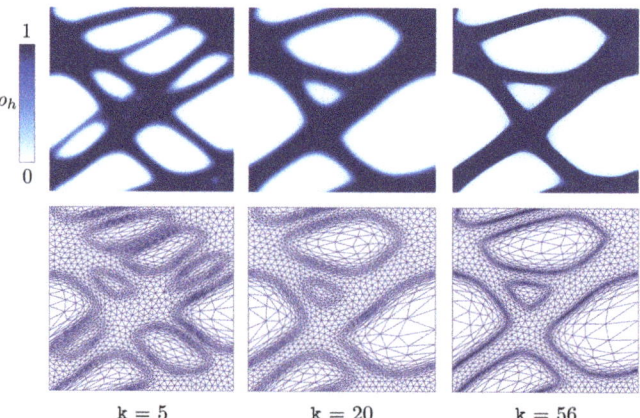

Figure 5. Design Case 2: density field (**top**) and associated anisotropic adapted mesh (**bottom**) for three different global iterations.

3.3. Design Case 3

As a third design, we carry out the optimization of a microcell characterized by similar stiffness and thermal conductivity along the x- and y-directions and by a high shear stiffness. This leads to solve problem (9) when the following constraints are enforced:

$$\begin{cases} 0.10 \leq E^H_{1111} \leq 0.15 \\ 0.08 \leq E^H_{1212} \leq 0.10 \\ 1.0 \leq \dfrac{E^H_{2222}}{E^H_{1111}} \leq 1.1 \\ 0.25 \leq k^H_{11} \leq 0.40 \\ 1.0 \leq \dfrac{k^H_{22}}{k^H_{11}} \leq 1.1. \end{cases} \quad (17)$$

The limited range for the two ratios E^H_{2222}/E^H_{1111} and k^H_{22}/k^H_{11} is consistent with the request for comparable stiffness and thermal conductivities along the two directions, whereas the mass minimization goal justifies the tight variation for the other tensors components.

The MultiP-microSIMPATY algorithm resorts to 35 loops before satisfying the stopping criterion. Figure 6 shows the density field and the mesh for three different global iterations of the algorithm. As for the previous design cases, thin features are removed by filtering during the first 24 iterations, while intermediate densities are erased in the second part of the process by the mesh adaptation procedure. As a consequence, the final microstructure exhibits very sharp density gradients, so that little post-processing has to be applied. In the final layout, most of the material is allocated along the two main diagonals of the domain, which ensures the required high shear stiffness as well as the balance between stiffness and thermal conductivity with respect to the horizontal and vertical directions.

Table 1 offers some additional quantitative information regarding the optimized structure. All the box constraints are satisfied (with a slight violation for the component E^H_{1111}) in the presence of a structure mass comparable with the one obtained for Design Case 2 (about 40% with respect to the full material configuration). We refer to Figure 4 (right) for an example of the microcellular material associated with the optimized cell.

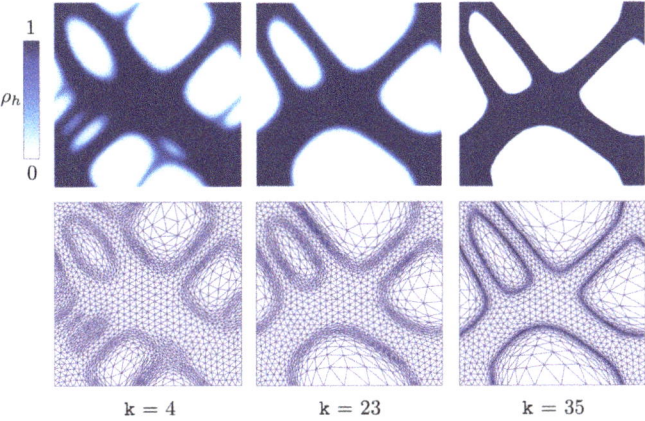

Figure 6. Design Case 3: density field (**top**) and associated anisotropic adapted mesh (**bottom**) for three different global iterations.

4. Discussion of Results

This section is meant to highlight the benefits of the MultiP-microSIMPATY algorithm. To this aim, we compare the layouts provided by the proposed methodology with unit cells available in engineering practice and with cellular materials designed by a standard inverse homogenization procedure, which does not exploit mesh adaptation.

4.1. Comparison with Off-The-Shelf Designs

This first investigation is carried out by comparing each of the three designs in the previous section with state-of-the-art unit cells in terms of mechanical and thermal performance, after setting a reference value for the overall mass. The quantities involved in such a comparison are the homogenized elastic modulus, E_x^H and E_y^H, associated with the direction x and y, which coincide with the inverse of the diagonal entries, C_{11}^H and C_{22}^H, of the compliance matrix $\mathbf{C}^H = (\mathbf{E}^H)^{-1}$; the homogenized shear modulus, G^H, equal to the inverse of the third diagonal entry of matrix $\mathbf{C}^H$; the homogenized thermal conductivities, k_{11}^H and k_{22}^H, along the x- and y-direction. The results of this analysis are summarized in Table 2.

Table 2. Comparison between the MultiP-microSIMPATY optimized structures and off-the-shelf designs in terms of homogenized elastic and thermal properties, for comparable volume fraction values.

		E_x^H	E_y^H	G^H	k_{11}^H	k_{22}^H
		\multicolumn{5}{c|}{Design Case 1}				
D1		0.012	0.015	0.056	0.200	0.113
A		0.009	0.009	0.075	0.163	0.163
B		0.095	0.042	0.059	0.198	0.131
		\multicolumn{5}{c|}{Design Case 2}				
D2		0.126	0.039	0.082	0.317	0.126
C		0.341	0.116	0.002	0.432	0.125
		\multicolumn{5}{c|}{Design Case 3}				
D3		0.070	0.070	0.082	0.260	0.261
L		0.188	0.188	0.072	0.255	0.255

Concerning Design Case 1, we perform two comparisons. Since the geometry provided by MultiP-microSIMPATY is similar to a square cell rotated by 45°, we choose simple squares (A and B) characterized by the same rotation as state-of-the-art unit cells. The basic squares in layout A fully couple mechanical and thermal features, thus excluding this cell for the purpose addressed in the first design case. This justifies the selection of cell B where the reinforcing horizontal strut mimics the very thin diagonal member connecting the adjacent sides in the proposed layout (D1). From a structural perspective, the horizontal strut in B increases the nodal connectivity and reacts with tension/compression to a load applied along the x-axis. This fact is confirmed by the non-isotropic elastic behavior of the material (compare the values E_x^H and E_y^H). Regarding thermal conduction, the strut promotes heat transfer along the horizontal direction, as highlighted by the discrepancy between k_{11}^H and k_{22}^H. In the optimized layout D1, the thin member is instead slightly inclined and does not connect two opposite nodes. Thus, the elastic modulus along the two directions is similar, since the strut reacts by bending to a load applied along the

x-axis. Moreover, the thin member promotes the heat transfer along the x-direction, thus decoupling the ratios E_y^H/E_x^H and k_{22}^H/k_{11}^H.

The unit cell D2 has been designed to ensure high stiffness and conductivity along the x-direction as well as a high shear modulus. As reference layout, we consider a square cell characterized by a rectangular cavity. This choice offers us a trivial solution to optimize stiffness and conductivity along direction x. The optimization performed by MultiP-microSIMPATY is corroborated by the values of G^H. In fact, cell D2 is characterized by a shear modulus, which is approximately 40 times higher when compared with the reference layout, although the values of E_x^H and k_{11}^H for cell D2 are, on average, 30% lower with respect to cell C.

Finally, the Design Case 3 aims at ensuring equal elastic modulus and conductivity along the x- and y-directions, as well as a high shear modulus. The paradigm for an isotropic stretch-based lattice, namely the standard triangular cell (L), is assumed as the off-the-shelf layout. A comparison between the corresponding values in Table 2 shows a 15% increment in the shear modulus of cell D3. In addition, both cells D3 and L exhibit the requested isotropic behavior in terms of the selected mechanical and thermal properties.

4.2. Comparison with Standard Inverse Homogenization

This section is meant to verify the benefits led by mesh adaptation in the context of thermo-elastic inverse homogenization, which is in accordance with the preliminary remarks in Section 3.

To this aim, we carry out a comparison between the MultiP-microSIMPATY algorithm and a standard inverse homogenization procedure. This comparison is performed in terms of mass. We expect that the employment of mesh adaptation leads to efficiently allocate the available material, thus promoting the mass minimization. As a reference standard approach, we implement a non-adaptive version of Algorithm 1, where the adaptation loop (lines 3–12) is replaced by the single call

$$\rho_h = \texttt{optimize}(\tilde{\mathcal{J}}, \tilde{\mathcal{C}}, \mathbf{c}^l, \mathbf{c}^u, \tilde{\mathcal{G}}, \rho_h^0, \texttt{TOPT}, \texttt{IT}).$$

We refer to this variant of Algorithm 1 as MultiP-microSIMP. In this case, the optimization is performed on the filtered density, so that the goal functional, the constraints, and the associated derivatives are modified accordingly (this justifies the new notation $\mathcal{Q} \to \tilde{\mathcal{Q}}$, with $\mathcal{Q} = \mathcal{J}, \mathcal{C}, \mathcal{G}$, where $\tilde{\mathcal{Q}}$ refers to quantities dependent on the filtered density). This choice is recurrent in topology optimization [61,62]. As far as all the parameters required by the optimization are concerned, we preserve the same values as in Section 2.3, while the computational mesh coincides with a 50×50 structured mesh.

Figure 7 compares the optimized layouts delivered by MultiP-microSIMP (top) and MultiP-microSIMPATY (bottom) for the three design cases in Section 3. The topologies characterizing the three cells vary when resorting to mesh adaptation. In general, MultiP-microSIMPATY provides more complex layouts, which however are still manufacturable. The presence of intermediate densities in the cells yielded by MultiP-microSIMP is highligthed by the blurred structure contours, promoted by the massive employment of filtering. Table 3 quantitatively assesses the optimization performance of the two algorithms by collecting the mass of the corresponding unit cells, together with the percentage mass reduction ensured by MultiP-microSIMPATY. On average, a mass saving of approximately 10% is guaranteed by the sharp detection of the material/void interface, i.e., by the removal of intermediate densities.

The use of filtering deserves further discussion. In particular, we prove the redundancy of the filtering phase after a sufficiently large number of global optimization iterations. To this aim, we run Algorithm 1 for $\texttt{kfmax} = 25$ and $\texttt{kfmax} = \texttt{kmax}$ (i.e., smoothing and sharpening filters in lines 6–7 are applied at each global iteration). Figure 8 compares the output associated with these two choices. The final topology provided by both the procedures is the same. This confirms that filtering is instrumental only in the identification of the final layout, and this takes place during the first iterations. From the top-left panel, the slightly diffusive action of the selected filtering is also evident, giving rise to intermediate densities along the layout boundaries. On the other hand, the removal of filtering allows mesh adaptation to sharply detect gradients from material to void, thus increasing the quality of the final output (compare the two panels on the left panel). The improvement in terms of boundary detection is confirmed also by the final adapted mesh, which captures the steep gradients of the density with thinner refined areas (compare the two panels on the right).

Figure 7. Comparison between the optimized cells delivered by MultiP.microSIMP (**top**) and by MultiP-microSIMPATY (**bottom**) for the Design Cases 1, 2, and 3 (from left to right).

Table 3. Comparison between the optimized cells delivered by MultiP-microSIMPATY and a standard inverse homogenization algorithm in terms of mass.

	D1	D2	D3
MultiP-microSIMP	0.330	0.443	0.486
MultiP-microSIMPATY	0.292	0.412	0.415
Mass reduction [%]	11.5%	7.0%	14.6%

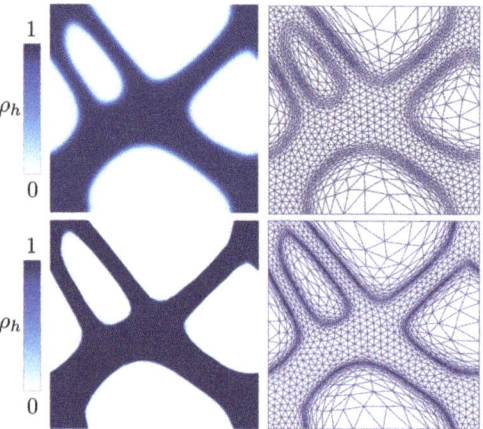

Figure 8. Effect of filtering for the MultiP-microSIMPATY algorithm: density field (**left**) and associated anisotropic adapted mesh (**right**) when filtering is applied during the whole optimization process (**top**) and in the first 25 iterations only (**bottom**).

Finally, we highlight that the presence of blurred interfaces may raise issues in the extraction of the final geometry after the optimization procedure. In fact, the extracted geometry strongly depends on the cut-off threshold, with a possible significant alteration of the overall mass and the expected thermo-elastic properties.

5. Conclusions and Perspectives

In this paper, we provide a new methodology for the design of cellular materials optimized by means of multi-physics inverse homogenization, which was discretized on customized computational meshes. The inverse homogenization problem is modeled by a standard density-based topology optimization at the microscale; the grid is generated by exploiting an anisotropic a posteriori error estimator that drives a mesh adaptation procedure. These two phases are iteratively coupled in the MultiP-microSIMPATY algorithm in order to deliver layouts characterized by clear-cut contours. In particular, the goal of the analyzed test cases is the design of lightweight structures with prescribed elastic and thermal properties, according to a multi-physics framework.

The main results of this work can be outlined as follows:

(i) The MultiP-microSIMPATY algorithm provides original design solutions, complying also with conflicting requirements;

(ii) The good performance of microSIMPATY has been confirmed also in a thermo-elastic context. Standard issues typical of topology optimization, such as the presence of intermediate densities, of jagged boundaries, and of too complex structures, is mitigated by the employment of a mesh customized to the design process (see Figure 7 and Table 3);

(iii) The new cellular materials have been successfully compared with consolidated solutions in terms of mechanical and thermal properties (see Table 2);

(iv) Filtering can be considerably limited thanks to the use of mesh adaptation. This turns into an improvement in terms of accuracy of the optimization process (see Figure 8);

(v) The employment of an anisotropic mesh adaptation provides advantages with a view to a manufacturing phase. Indeed, the unit cells designed by MultiP-microSIMPATY exhibit very smooth geometries which demand for a very limited post-processing;

(vi) The procedure here settled turns out to be fully general with respect to the selected multi-physics context.

Possible future developments include the extension of the MultiP-microSIMPATY design procedure to a 3D setting. The proposed methodology could also be exploited in a multiscale topology optimization framework [38], inspired by the many possible applications in engineering practice (including medicine, aerospace, automotive, and architecture). In such a context, with a view to the manufacturing step, another issue that deserves further investigation is represented by the handling of the transition area between different cellular materials. Finally, innovative techniques, such as model reduction or machine learning, still represent topics of high relevance in topology optimization for a future examination [63–65].

Author Contributions: Conceptualization, M.G., N.F., S.P. and S.F.; methodology, M.G., N.F. and S.P.; software, M.G. and N.F.; formal analysis, N.F. and S.P.; investigation, M.G., N.F., S.P. and S.F.; resources, S.P. and S.F.; data curation, M.G. and N.F.; writing—original draft preparation, M.G.; writing—review and editing, N.F. and S.P.; visualization, M.G. and N.F.; supervision, S.P. and S.F.; project administration, S.P. and S.F.; funding acquisition, S.P. and S.F. All authors have read and agreed to the published version of the manuscript.

Funding: This research received no external funding.

Acknowledgments: This research is part of the activity of the METAMatLab at Politecnico di Milano. The first and the last authors acknowledge the Italian Ministry of Education, University and Research for the support provided through the "Department of Excellence LIS4.0-Lightweight and Smart Structures for Industry 4.0" Project. The second author thanks the Istituto Nazionale di Alta Matematica (INdAM) for the awarded grant. Finally, the third author acknowledges the research project GNCS-INdAM 2020 "Tecniche Numeriche Avanzate per Applicazioni Industriali".

Conflicts of Interest: The authors declare no conflict of interest.

References

1. Gibson, L.J.; Ashby, M.F. *Cellular Solids: Structure and Properties*, 2nd ed.; Cambridge Solid State Science Series; Cambridge University Press: Cambridge, UK, 1997.
2. Rashed, M.G.; Ashraf, M.; Mines, R.A.; Hazell, P.J. Metallic microlattice materials: A current state of the art on manufacturing, mechanical properties and applications. *Mater. Des.* **2016**, *95*, 518–533. [CrossRef]
3. Bauer, J.; Hengsbach, S.; Tesari, I.; Schwaiger, R.; Kraft, O. High-strength cellular ceramic composites with 3D microarchitecture. *Proc. Natl. Acad. Sci. USA* **2014**, *111*, 2453–2458. [CrossRef] [PubMed]
4. Schaedler, T.A.; Carter, W.B. Architected Cellular Materials. *Annu. Rev. Mater. Res.* **2016**, *46*, 187–210. [CrossRef]
5. Ahmadi, S.M.; Campoli, G.; Amin Yavari, S.; Sajadi, B.; Wauthle, R.; Schrooten, J.; Weinans, H.; Zadpoor, A.A. Mechanical behavior of regular open-cell porous biomaterials made of diamond lattice unit cells. *J. Mech. Behav. Biomed. Mater.* **2014**, *34*, 106–115. [CrossRef]
6. Yan, C.; Hao, L.; Hussein, A.; Young, P. Ti-6Al-4V triply periodic minimal surface structures for bone implants fabricated via selective laser melting. *J. Mech. Behav. Biomed. Mater.* **2015**, *51*, 61–73. [CrossRef]
7. Taniguchi, N.; Fujibayashi, S.; Takemoto, M.; Sasaki, K.; Otsuki, B.; Nakamura, T.; Matsushita, T.; Kokubo, T.; Matsuda, S. Effect of pore size on bone ingrowth into porous titanium implants fabricated by additive manufacturing: An in vivo experiment. *Mater. Sci. Eng. C* **2016**, *59*, 690–701. [CrossRef]
8. Arabnejad, S.; Burnett Johnston, R.; Pura, J.A.; Singh, B.; Tanzer, M.; Pasini, D. High-strength porous biomaterials for bone replacement: A strategy to assess the interplay between cell morphology, mechanical properties, bone ingrowth and manufacturing constraints. *Acta Biomater.* **2016**, *30*, 345–356. [CrossRef]
9. Bandaru, P.R.; Vemuri, K.P.; Canbazoglu, F.M.; Kapadia, R.S. Layered thermal metamaterials for the directing and harvesting of conductive heat. *AIP Adv.* **2015**, *5*, 053403. [CrossRef]
10. Liu, D.P.; Chen, P.J.; Huang, H.H. Realization of a thermal cloak-concentrator using a metamaterial transformer. *Sci. Rep.* **2018**, *8*, 1–11. [CrossRef]
11. Attarzadeh, R.; Rovira, M.; Duwig, C. Design analysis of the "Schwartz D" based heat exchanger: A numerical study. *Int. J. Heat Mass Transf.* **2021**, *177*, 121415. [CrossRef]
12. Kaur, I.; Singh, P. State-of-the-art in heat exchanger additive manufacturing. *Int. J. Heat Mass Transf.* **2021**, *178*, 121600. [CrossRef]
13. Bendsøe, M.P.; Sigmund, O. *Topology Optimization*; Springer: Berlin/Heidelberg, Germany, 2004.
14. Sigmund, O. Materials with prescribed constitutive parameters: An inverse homogenization problem. *Int. J. Solids Struct.* **1994**, *31*, 2313–2329. [CrossRef]
15. Andreassen, E.; Andreasen, C.S. How to determine composite material properties using numerical homogenization. *Comp. Mater. Sci.* **2014**, *83*, 488–495. [CrossRef]
16. Allaire, G.; Geoffroy-Donders, P.; Pantz, O. Topology optimization of modulated and oriented periodic microstructures by the homogenization method. *Comput. Math. Appl.* **2019**, *78*, 2197–2229. [CrossRef]
17. Vigliotti, A.; Pasini, D. Mechanical properties of hierarchical lattices. *Mech. Mat.* **2013**, *62*, 32–43. [CrossRef]
18. Wang, Y.; Xu, H.; Pasini, D. Multiscale isogeometric topology optimization for lattice materials. *Comput. Methods. Appl. Mech. Eng.* **2017**, *316*, 568–585. [CrossRef]
19. Cheng, L.; Liu, J.; Liang, X.; To, A.C. Coupling lattice structure topology optimization with design-dependent feature evolution for additive manufactured heat conduction design. *Comput. Methods Appl. Mech. Eng.* **2018**, *332*, 408–439. [CrossRef]
20. Panesar, A.; Abdi, M.; Hickman, D.; Ashcroft, I. Strategies for functionally graded lattice structures derived using topology optimisation for Additive Manufacturing. *Addit. Manuf.* **2018**, *19*, 81–94. [CrossRef]
21. Moussa, A.; Rahman, S.; Xu, M.; Tanzer, M.; Pasini, D. Topology optimization of 3D-printed structurally porous cage for acetabular reinforcement in total hip arthroplasty. *J. Mech. Behav. Biomed. Mater.* **2020**, *105*, 103705. [CrossRef]
22. Coelho, P.G.; Fernandes, P.R.; Guedes, J.M.; Rodrigues, H.C. A hierarchical model for concurrent material and topology optimisation of three-dimensional structures. *Struct. Multidiscip. Optim.* **2008**, *35*, 107–115. [CrossRef]
23. Nakshatrala, P.B.; Tortorelli, D.A.; Nakshatrala, K.B. Nonlinear structural design using multiscale topology optimization. Part I: Static formulation. *Comput. Methods Appl. Mech. Eng.* **2013**, *261/262*, 167–176. [CrossRef]
24. Djourachkovitch, T.; Blal, N.; Hamila, N.; Gravouil, A. Multiscale topology optimization of 3D structures: A micro-architectured materials database assisted strategy. *Comput. Struct.* **2021**, *255*, 106574. [CrossRef]
25. Ferrer, A.; Oliver, J.; Cante, J.C.; Lloberas-Valls, O. Vademecum-based approach to multi-scale topological material design. *Adv. Model. Simul. Eng. Sci.* **2016**, *3*, 23. [CrossRef] [PubMed]
26. Ferro, N.; Micheletti, S.; Perotto, S. Density-Based Inverse Homogenization with Anisotropically Adapted Elements. *Lect. Notes Comput. Sci. Eng.* **2020**, *132*, 211–221.
27. di Cristofaro, D.; Galimberti, C.; Bianchi, D.; Ferrante, R.; Ferro, N.; Mannisi, M.; Perotto, S. Adaptive topology optimization for innovative 3D printed metamaterials. In Proceedings of the WCCM—ECCOMAS 2020 Conference, Volume 1200-Modeling and Analysis of Real World and Industry Applications, Online, 11–15 January 2021.
28. Auricchio, F.; Bonetti, E.; Carraturo, M.; Hömberg, D.; Reali, A.; Rocca, E. A phase-field-based graded-material topology optimization with stress constraint. *Math. Model. Methods Appl. Sci.* **2020**, *30*, 1461–1483. [CrossRef]
29. Huang, X.; Radman, A.; Xie, Y.M. Topological design of microstructures of cellular materials for maximum bulk or shear modulus. *Comput. Mater. Sci.* **2011**, *50*, 1861–1870. [CrossRef]
30. Xia, L.; Breitkopf, P. Concurrent topology optimization design of material and structure within FE2 nonlinear multiscale analysis framework. *Comput. Methods Appl. Mech. Eng.* **2014**, *278*, 524–542. [CrossRef]
31. Wang, Y.; Luo, Z.; Zhang, N.; Kang, Z. Topological shape optimization of microstructural metamaterials using a level set method. *Comput. Mater. Sci.* **2014**, *87*, 178–186. [CrossRef]

32. Torquato, S.; Hyun, S.; Donev, A. Optimal design of manufacturable three-dimensional composites with multifunctional characteristics. *J. Appl. Phys.* **2003**, *94*, 5748–5755. [CrossRef]
33. de Kruijf, N.; Zhou, S.; Li, Q.; Mai, Y.W. Topological design of structures and composite materials with multiobjectives. *Int. J. Solids Struct.* **2007**, *44*, 7092–7109. [CrossRef]
34. Challis, V.J.; Roberts, A.P.; Wilkins, A.H. Design of three dimensional isotropic microstructures for maximized stiffness and conductivity. *Int. J. Solids Struct.* **2008**, *45*, 4130–4146. [CrossRef]
35. Vineyard, E.; Gao, X.L. Topology and shape optimization of 2-d and 3-d micro-architectured thermoelastic metamaterials using a parametric level setmethod. *CMES-Comput. Model. Eng. Sci.* **2021**, *127*, 819–854.
36. Sigmund, O.; Petersson, J. Numerical instabilities in topology optimization: A survey on procedures dealing with checkerboards, mesh-dependencies and local minima. *Struct. Optim.* **1998**, *16*, 68–75. [CrossRef]
37. Micheletti, S.; Perotto, S.; Soli, L. Topology optimization driven by anisotropic mesh adaptation: Towards a free-form design. *Comput. Struct.* **2019**, *214*, 60–72. [CrossRef]
38. Ferro, N.; Perotto, S.; Bianchi, D.; Ferrante, R.; Mannisi, M. Design of cellular materials for multiscale topology optimization: Application to patient-specific orthopedic devices. *Struct. Multidiscip. Optim.* **2021**, *28*, 2021. [CrossRef]
39. Hassani, B.; Hinton, E. A review of homogenization and topology optimization I—homogenization theory for media with periodic structure. *Comput. Struct.* **1998**, *69*, 707–717. [CrossRef]
40. Hassani, B.; Hinton, E. A review of homogenization and topology optimization II—analytical and numerical solution of homogenization equations. *Comput. Struct.* **1998**, *69*, 719–738. [CrossRef]
41. Terada, K.; Hori, M.; Kyoya, T.; Kikuchi, N. Simulation of the multi-scale convergence in computational homogenization approaches. *Int. J. Solids Struct.* **2000**, *37*, 2285–2311. [CrossRef]
42. Gould, P.L. *Introduction to Linear Elasticity*; Springer: New York, NY, USA, 1994.
43. Noël, L.; Duysinx, P. Shape optimization of microstructural designs subject to local stress constraints within an XFEM-level set framework. *Struct. Multidiscip. Optim.* **2017**, *55*, 2323–2338. [CrossRef]
44. Ern, A.; Guermond, J.L. *Theory and Practice of Finite Elements*; Springer: New York, NY, USA, 2004.
45. Formaggia, L.; Perotto, S. New anisotropic a priori error estimates. *Numer. Math.* **2001**, *89*, 641–667. [CrossRef]
46. Ainsworth, M.; Oden, J.T. *A Posteriori Error Estimation in Finite Element Analysis*; John Wiley & Son: New York, NY, USA, 2000.
47. Bangerth, W.; Rannacher, R. *Adaptive Finite Element Methods for Differential Equations*; Birkhäuser Verlag: Basel, Germany, 2003.
48. Zienkiewicz, O.C.; Zhu, J.Z. A simple error estimator and adaptive procedure for practical engineerng analysis. *Int. J. Numer. Methods Eng.* **1987**, *24*, 337–357. [CrossRef]
49. Zienkiewicz, O.C.; Zhu, J.Z. The superconvergent patch recovery and a posteriori error estimates. I: The recovery technique. *Int. J. Numer. Meth. Eng.* **1992**, *33*, 1331–1364. [CrossRef]
50. Rodríguez, R. Some remarks on Zienkiewicz-Zhu estimator. *Numer. Methods Partial. Differ. Equations* **1994**, *10*, 625–635. [CrossRef]
51. Maisano, G.; Micheletti, S.; Perotto, S.; Bottasso, C.L. On some new recovery-based a posteriori error estimators. *Comput. Methods Appl. Mech. Eng.* **2006**, *195*, 4794–4815. [CrossRef]
52. Li, X.D.; Wiberg, N.E. A posteriori error estimate by element patch post-processing, adaptive analysis in energy and L_2 norms. *Comput. Struct.* **1994**, *53*, 907–919. [CrossRef]
53. Micheletti, S.; Perotto, S. Anisotropic adaptation via a Zienkiewicz-Zhu error estimator for 2D elliptic problems. In *Numerical Mathematics and Advanced Applications*; Kreiss, G., Lötstedt, P., Målqvist, A., Neytcheva, M., Eds.; Springer: Berlin/Heidelberg, Germany, 2010; pp. 645–653.
54. Farrell, P.E.; Micheletti, S.; Perotto, S. An anisotropic Zienkiewicz-Zhu-type error estimator for 3D applications. *Int. J. Numer. Meth. Eng.* **2011**, *85*, 671–692. [CrossRef]
55. Micheletti, S.; Perotto, S. Reliability and efficiency of an anisotropic Zienkiewicz-Zhu error estimator. *Comput. Methods Appl. Mech. Eng.* **2006**, *195*, 799–835. [CrossRef]
56. Ferro, N.; Micheletti, S.; Perotto, S. Compliance-stress constrained mass minimization for topology optimization on anisotropic meshes. *SN Appl. Sci.* **2020**, *2*, 1–11. [CrossRef]
57. Ferro, N.; Micheletti, S.; Perotto, S. An optimization algorithm for automatic structural design. *Comput. Methods Appl. Mech. Eng.* **2020**, *372*, 113335. [CrossRef]
58. Hecht, F. New development in FreeFem++. *J. Numer. Math.* **2012**, *20*, 251–265. [CrossRef]
59. Wächter, A.; Lorenz, T. Biegler. On the implementation of an interior-point filter line-search algorithm for large-scale nonlinear programming. *Math. Program.* **2006**, *106*, 25–57. [CrossRef]
60. Svanberg, K. The method of moving asymptotes-a new method for structural optimization. *Int. J. Numer. Meth. Eng.* **1987**, *24*, 359–373. [CrossRef]
61. Lazarov, B.S.; Sigmund, O. Filters in topology optimization based on Helmholtz-type differential equations. *Int. J. Numer. Meth. Eng.* **2011**, *86*, 765–781. [CrossRef]
62. Sigmund, O. Morphology-based black and white filters for topology optimization. *Struct. Multidiscip. Optim.* **2007**, *33*, 401–424. [CrossRef]
63. Caicedo, M.; Mroginski, J.L.; Toro, S.; Raschi, M.; Huespe, A.; Oliver, J. High performance reduced order modeling techniques based on optimal energy quadrature: Application to geometrically non-linear multiscale inelastic material modeling. *Arch. Comput. Methods Eng.* **2019**, *26*, 771–792. [CrossRef]
64. Ferro, N.; Micheletti, S.; Perotto, S. POD-assisted strategies for structural topology optimization. *Comput. Math. Appl.* **2019**, *77*, 2804–2820. [CrossRef]

65. Chi, H.; Zhang, Y.; Tang, T.L.E.; Mirabella, L.; Dalloro, L.; Song, L.; Paulino, G.H. Universal machine learning for topology optimization. *Comput. Methods. Appl. Mech. Eng.* **2021**, *375*, 112739. [CrossRef]

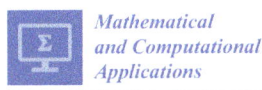

Article

Benchmarking Regridding Libraries Used in Earth System Modelling

Sophie Valcke *, Andrea Piacentini and Gabriel Jonville

Unité Mixte de Recherche 5318 «Climat Environnement Couplages et Incertitudes», Centre Européen de Recherche et Formation Avancée en Calcul Scientifique, Centre National de la Recherche Scientifique, CEDEX 1, 31057 Toulouse, France; piacentini@cerfacs.fr (A.P.); jonville@cerfacs.fr (G.J.)
* Correspondence: valcke@cerfacs.fr; Tel.: +33-(0)5-61-19-30-76

Citation: Valcke, S.; Piacentini, A.; Jonville, G. Benchmarking Regridding Libraries Used in Earth System Modelling. *Math. Comput. Appl.* **2022**, *27*, 31. https://doi.org/10.3390/mca27020031

Academic Editors: Simona Perotto, Gianluigi Rozza and Antonia Larese

Received: 19 January 2022
Accepted: 26 March 2022
Published: 1 April 2022

Publisher's Note: MDPI stays neutral with regard to jurisdictional claims in published maps and institutional affiliations.

Copyright: © 2022 by the authors. Licensee MDPI, Basel, Switzerland. This article is an open access article distributed under the terms and conditions of the Creative Commons Attribution (CC BY) license (https://creativecommons.org/licenses/by/4.0/).

Abstract: Components of Earth system models (ESMs) usually use different numerical grids because of the different environments they represent. Therefore, a coupling field sent by a source model has to be regridded to be used by a target model. The regridding has to be accurate and, in some cases, conservative, in order to ensure the consistency of the coupled model. Here, we present work done to benchmark the quality of four regridding libraries currently used in ESMs, i.e., SCRIP, YAC, ESMF and XIOS. We evaluated five regridding algorithms with four different analytical functions for different combinations of six grids used in real ocean or atmosphere models. Four analytical functions were used to define the coupling fields to be regridded. This benchmark calculated some of the metrics proposed by the CANGA project, including the mean, maximum, RMS misfit, and global conservation. The results show that, besides a few very specific cases that present anomalous values, the regridding functionality in YAC, ESMF and XIOS can be considered of high quality and do not present the specific problems observed for the conservative SCRIP remapping. The evaluation of the computing performance of those libraries is not included in the current work but is planned to be performed in the coming months. This exercise shows that benchmarking can be a great opportunity to favour interactions between users and developers of regridding libraries.

Keywords: regridding; remapping; interpolation; Earth system modelling; code coupling; coupler; coupling library; coupled models; ocean-atmosphere general circulation models

1. Introduction

Component models assembled in Earth system models (ESMs) usually have different grids because of the different environments that they represent, e.g., in an ocean model, the latitude–longitude grid convergence singularity can be conveniently displaced over a continent. Therefore, the coupling fields sent by a source component model have to be transformed for use by a target component on its grid. The first step is to define the addresses and weights of the source grid points that will contribute to the calculation of the coupling field on the target grid. The second step is regridding, i.e., the multiplication of the source grid values by the regridding weights to express the coupling field on the target grid. This spatial transformation is called regridding, remapping, or interpolation.

Different libraries exist for regridding in ESMs, offering different algorithms. We briefly describe here the two-dimensional (2D) algorithms used. With a nearest neighbour algorithm, the values of the nearest neighbours on the source grid, possibly weighted by their distance to the target point, are associated to each target grid point. A first-order non-conservative approximation uses, for each target point, the values of the coupling field at the four enclosing source grid points, as in a bilinear algorithm. Different algorithms are implemented for higher-order (non-conservative) regridding: one widely used schema is the bicubic interpolation, which uses the values of the four enclosing source neighbours but also the values of the field gradients in each direction and the cross gradient in the diagonal direction. In a first-order conservative remapping, the value for each target cell is computed as a weighted sum of the source cell values, with the contribution of a source cell being proportional to the fraction of the target cell intersected by the source cell. This method should be applied when it is important to conserve the area-integrated value of the coupling field, for example to conserve the energy associated with heat fluxes or water associated precipitation fields. The basis of a second-order conservative remapping is the same as the first-order conservative remapping but additional terms proportional to the gradients of the source field are applied.

The OASIS3-MCT (Ocean Atmosphere Sea Ice Soil 3—Model Coupling Toolkit) coupler [1] includes the SCRIP (Spherical Coordinate Remapping and Interpolation Package) library [2] for its regridding operations. A detailed analysis of the quality of the SCRIP library conservative remapping was realised in [3,4]. The impact of the different normalisation options and of a Lambert azimuthal projection above a certain latitude have been analysed for different types of grids. The general conclusion is that the SCRIP first-order conservative remapping may give satisfactory results for some types of grids for the different normalisation options; however, in some cases, only if the Lambert projection is activated and, in other cases, only if it is not. Furthermore, conservative regridding involving a Gaussian reduced grid always shows some problems, whether or not the Lambert projection is activated. This analysis motivated the exploration of other regridding libraries currently available for Earth system modelling, for a possible future interfacing in OASIS3-MCT. The regridding libraries analysed are the ones mostly used in Earth system modelling today, i.e., ATLAS, MOAB-Tempest Remap, YAC, ESMF and XIOS. The results of this exploration are presented in this paper and additional details can be found in [5]. Here we also show results for the SCRIP library, as a basis for comparison, but do not investigate specific problems when they arise, as the current objective is to evaluate alternative regridding libraries.

ATLAS [6] is an open-source library written in C++, currently being developed at the European Centre for Medium-Range Weather Forecast (ECMWF). It provides grids, mesh generation, and parallel data structures targeting numerical weather prediction or climate model developments. It is designed as an object-oriented modular library, with the capability to take advantage of the most recent computer architectures. It is meant to provide, among many other features, a set of parallel interpolation methods and is oriented toward the use of an internally consistent set of predefined grids and meshes. At the time of our evaluation, ATLAS provided nearest neighbour, linear, cubic and finite-element regridding methods but did not include any conservative remapping.

MOAB-Tempest Remap [7], which is also written in C++, is used in the energy exascale Earth system model (E3SM) [8], a state-of-the-art Earth system modelling project funded by the Department of Energy (DOE) in the United States. Through Fortran-compatible interfaces, it offers online conservative regridding based on a scalable advancing-front intersection algorithm, which allows to compute the supermesh defined by the intersection of the source and target grid cells. The supermesh is then used to assemble the higher-order, conservative, and monotonicity-preserving regridding weights.

YAC, Yet Another Coupler [9,10], is developed as a joint initiative between the German Climate Computing Center (DKRZ) and the Max Planck Institute for Meteorology (MPI-M). YAC is coded in C and a Fortran interface is also provided. Although targeting the German ICON (ICOsahedral Nonhydrostatic) model, the software provides multiple regridding methods, e.g., linear, nearest neighbour, first and second order conservative, and hybrid cubic Bernstein–Bézier patch [11] (see also Section 2.1.3) for the coupling of physical fields defined on regular and irregular grids on the sphere without a priori assumption about the particular grid structure or grid element types.

ESMF, the Earth System Modelling Framework [12,13], is an open-source software for coupling model components to form weather, climate, coastal, and other Earth science related applications. Today, ESMF is developed and governed by a set of partners in the USA that include the National Aeronautics and Space Administration (NASA), the National Oceanic and Atmospheric Administration (NOAA), the U.S. Navy, the National Center for Atmospheric Research (NCAR) and the national Earth System Prediction Capability (ESPC). Using ESMF, the scientist only codes the scientific part of their model into modular components and adapts it to the standard calling interface and standard data structures of the framework. Different modules, coded by either the scientists themselves or by others, can then be assembled into large scientific applications. ESMF offers a full interface to Fortran 90 and partial interface to C/C++ and Python. The ESMF software provides the underlying layers necessary for an efficient parallel execution of the scientific applications on different computer architectures, allowing for the coupling of the module to other components. ESMF supports regridding on combinations of 2D or 3D, spherical or cartesian coordinates with different regridding methods: nearest neighbour, bilinear, higher order, based on patch algorithm (see Section 2.1.3), and first and second order conservative.

XIOS [14], standing for XML-IO-Server, is an open-source library written in C++ with a Fortran interface developed at the Institut Pierre-Simon Laplace (IPSL) and dedicated to the management of I/O in climate codes. XIOS offers an impressive ensemble of online operations on model data (file rebuilding, time series, seasonal means, regridding, vertical interpolation, compression, etc.) based on external XML metadata definition, in order to minimize the post-processing of the data. Its regridding utility offers first and second order conservative remapping (but no non-conservative algorithms) on any type of grids used in Earth system modelling. Recently, XIOS has also been used as a coupler, i.e., managing communication of data, not only between a component and a file, but also between two components.

In order to compare these libraries, several aspects have to be considered. In a preliminary analysis, we enquired about the available regridding methods and evaluated the general software development environment, e.g., the coding language, project history, development plans, provision of support to external projects, and committed manpower. This first analysis led us to conclude that ATLAS and MOAB-Tempest Remap are certainly appealing libraries with good long-term perspectives regarding their development and support. However, their usage for regridding in OASIS3-MCT cannot be recommended at this point, as some basic capabilities were still missing in the version evaluated (0.21), in particular the handling of missing/masked values for MOAB-Tempest Remap or conservative regridding for ATLAS [15].

Therefore, we pushed further the analysis for YAC, ESMF, and XIOS and decided to benchmark the quality of their regridding. We also analyzed SCRIP as a basis for comparison, using criteria proposed by Coupling Approaches for Next-Generation Architectures (CANGA) project [16]. CANGA is a joint effort funded by the United States Department of Energy's Office of Science under the Scientific Discovery Through Advanced Computing (SciDAC) program that targets new high-performance coupling approaches for Earth system models on next-generation computers. Following CANGA, aspects to consider when evaluating a regridding library are the sensitivity (i.e., the algorithmic invariance of the scheme to the underlying mesh topology), the global conservation of integral quantities, the consistency (i.e., the preservation of discretization order and accuracy), the monotonicity (i.e., the preservation of global solution bounds), the dissipation (i.e., the smoothing of local solution maxima and minima that has to be minimal), the scalability, and the performance of the library. CANGA proposes metrics to quantify these aspects and we implemented the calculation of some of these metrics in our benchmark. The benchmark characteristics are detailed in Section 2.1, while its specific use for evaluating SCRIP, YAC, ESMF and XIOS is described in Section 2.2. In Section 3, we detail the benchmark results obtained for the four libraries. Finally, conclusions and perspectives of this work are presented in Section 4.

2. The Regridding Benchmark

Here, in Section 2.1, we describe the characteristics of the benchmark used to evaluate the regridding libraries that includes five algorithms, four different functions, and different combinations of six grids used in real ocean or atmosphere models. In Section 2.2, we provide some details on its application for the four libraries SCRIP, YAC, ESMF and XIOS.

2.1. The Benchmark Characteristics

2.1.1. Grids

The six grids considered in the benchmark are the following, given with their acronym used in the rest of the document and number of grid points:

- *torc*: the ocean model NEMO (Nucleus for European Modelling of the Ocean) [17], rotated-stretched logically-rectangular grid with 182×149 points horizontally;
- *nogt*: the ocean model NEMO, rotated-stretched logically-rectangular grid with 362×294 points horizontally;
- *bggd*: the atmosphere model LMDz (Laboratoire de Météorologie Dynamique zoom), [18] regular latitude–longitude grid with 144×143 points horizontally;
- *sse7*: the atmosphere model ARPEGE (Action de Recherche Petite Echelle Grande Echelle) [19], Gaussian reduced T127 with 24,572 points horizontally (unstructured, described with up to 7 vertices per cell);
- *icos*: the atmosphere model Dynamico [20], low-resolution unstructured icosahedral grid with 15,222 points horizontally;
- *icoh*: the atmosphere model Dynamico, high-resolution unstructured icosahedral grid with 2,016,012 points horizontally.

These grids are illustrated on Figure 1.

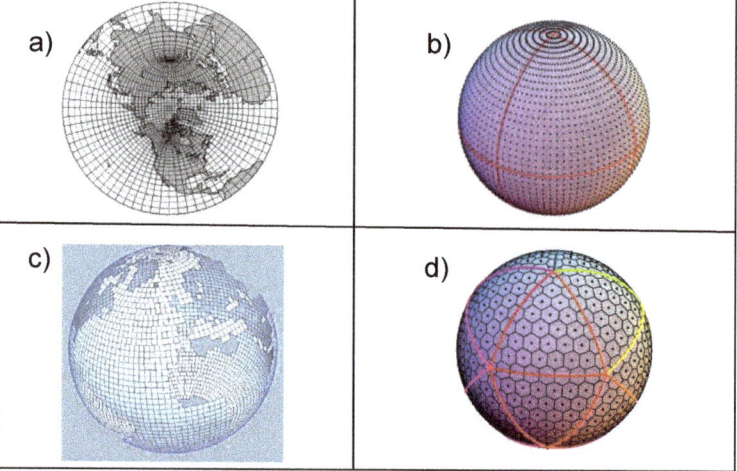

Figure 1. Illustration of the types of grids included in the benchmark: (**a**) rotated-stretched logically-rectangular (*torc, nogt*), (**b**) regular latitude–longitude (*bggd*), (**c**) Gaussian-reduced (*sse7*), and (**d**) icosahedral (*icos, icoh*).

The first five grids are relatively low-resolution grids. We decided to run the benchmark for the six pairs of these grids matching an ocean and an atmospheric grid and introduced the higher-resolution *icoh* grid only to test the impact of large resolution differences on the conservative regridding.

We note here that all grids used in this benchmark define a sea-land mask, with valid (non-masked) points over the ocean and not valid (masked) point over the land. In order to avoid non-matching sea-land masks between the ocean and the atmosphere grids, we adopted the following best practice that sets up a consistent atmosphere-ocean system and defines a well-posed coupled problem: The original sea-land mask of the ocean model is taken as is. For the atmosphere model, the fraction of water in each cell is defined by the conservative remapping of the ocean mask on the atmospheric grid. Then, the atmospheric coupling mask is adapted by associating a valid/active index to cells containing at least a surface fraction 1/1000 of water. Under 1/1000 of water, the atmospheric cell is considered to be completely masked. This method ensures that the total sea and land surfaces are the same in the ocean and atmosphere models, allowing global conservation of sea or land integrated quantities. It also minimizes the number of target grid points that does not receive a value with each specific regridding algorithm.

2.1.2. Analytical Functions

The four analytical functions used to define the coupling fields to be regridded, illustrated on Figure 2, are (see also Appendix A for their exact definition expressed in Fortran 90):

(a) *sinusoid*: a slowly varying standard sinusoid over the globe;
(b) *harmonic*: a more rapidly varying function with 16 maximums and 16 minimums in northern and southern bands;
(c) *vortex*: a slowly varying function with two added vortices, one in the Atlantic and one over Indonesia;
(d) *gulfstream*: the slowly varying standard sinusoid with a mimicked Gulf Stream.

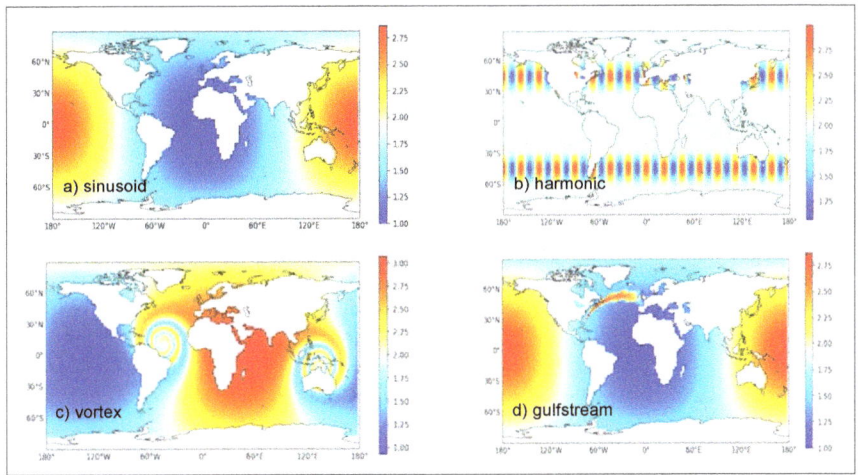

Figure 2. The four analytical functions defining the coupling field: (**a**) *sinusoid*, (**b**) *harmonic*, (**c**) *vortex*, (**d**) *gulfstream*.

2.1.3. Regridding Algorithms

The following algorithms were evaluated for the different regridding libraries, when available. The particularities of the algorithm for each library are described. We also specifically mention the option activated in the different regridding libraries to ensure that all valid target grid points receive a regridded value, even near the coasts.

1. Nearest neighbour

For all libraries, except for XIOS, which does not implement this algorithm, the value of the non-masked nearest neighbour on the source grid was assigned to each target grid point, i.e., only one neighbour was used.

For ESMF, the options allowing regridding on the cell centre locations of an unstructured grid (i.e., –*src_loc center* –*dst_loc center*) and the option ignoring degenerate cells in either the source or the destination grid (–*ignore_degenerate*) were activated. This option can be useful for the NEMO grids *torc* and *nogt*, which may have masked cells (i.e., not used in the regridding) collapsing into a point or line.

2. First order non-conservative

SCRIP uses a general scheme based on a local bilinear approximation. For non-masked target points that do not receive a value with the original bilinear algorithm, as can happen near the coast, the nearest non-masked source neighbour value was used, by default.

ESMF uses a standard bilinear algorithm. The same options as for the nearest neighbour regridding were activated (i.e., *–src_loc center –dst_loc center, –ignore_degenerate*). In addition, the option *–extrap_method nereststod* is turned on. Each target point that did not receive a value with the original algorithm used the closest unmasked source point to define its value (in order to reproduce the default behaviour of the SCRIP library).

We also note that with ESMF, grids can be described with the so-called *SCRIP* format or with an *unstructured* format. The *SCRIP* format (not to be confused with the SCRIP library itself) describes the grid with the latitude and the longitude of the centre and corners of each cell. The *unstructured* format describes the grid as an ensemble of elements and provides the element connectivity associating for each element a certain number of nodes in the list of nodes for which the latitude and longitude are provided.

For YAC, we activated an inverse-distance weighting of the vertex values of the source polygon enclosing the target point, and an average of the two nearest neighbours for target points falling outside any source polygon, so to ensure that all non-masked target grid points receive a regridded value.

XIOS does not implement any first-order non conservative regridding.

3. Second order non-conservative

For SCRIP, the bicubic regridding follows the general local bilinear remapping using the values of each vertex of the enclosing source cell and the values of the gradients in each local direction and in the cross direction. Again, the nearest non-masked source neighbour value is used for non-masked target points that do not receive any value with the original bicubic algorithm.

For YAC, the recently introduced hybrid cubic spherical Bernstein–Bézier (HCSBB) method [11] was used [10]. Compared to the patch algorithm used in ESMF (see below), the HCSBB method always results in an interpolated field that has a continuous first derivative. The source grid was first triangulated and the derivatives of the source field across the edges of the triangles were computed. Triangular patches were constructed from a blend of spherical Bernstein–Bézier polynomials using these derivatives, and then used to regrid each target point. Compared to the patch algorithm, this method uses a bigger stencil to compute each target point. The completion with 4-nearest non-masked neighbours is also activated for non-masked target points that do not receive any value with the original HCSBB algorithm.

For ESMF, the patch algorithm that is used is a technique commonly used in finite element modelling. Patch interpolation works by constructing multiple polynomial patches for the cells around the vertices of a source cell (e.g., for a square source cell four patches would be computed). For 2D grids, these polynomials are currently second degree 2D polynomials. The interpolated value at the destination point is the weighted average of all the patches for the source cell (e.g., the four patches for a square cell). This patch averaging prevents too strong overshoots and undershoots. The same options as for the first order non-conservative regridding (i.e., *–src_loc center –dst_loc center –ignore_degenerate –extrap_method nereststod*) were activated.

XIOS does not implement any second order non-conservative regridding.

4. First order conservative with FRACAREA and DESTAREA normalisations

In a first-order conservative remapping, the value for each target cell is computed as a weighted sum of source cell values, with the contribution of a source cell being proportional to the fraction of the target cell intersected by the source cell. In case of non-matching sea-land masks between the atmosphere and the ocean grids, different normalisation options exist. DESTAREA (DESTination AREA) uses the whole target cell area for the normalisation, whereas FRACAREA (FRACtional AREA) uses the intersected area of the target cell. DESTAREA ensures local conservation but may produce non-physical values while FRACAREA does not ensure local conservation but produces values that are physically consistent. We note also that the FRACAREA normalisation may give some good results for the wrong reasons, in the sense that the normalisation operation involving the intersected target cell area, as calculated by the library itself, may lead to the cancellation of error present in the weights before the normalisation. DESTAREA does not involve this error cancellation and therefore often reveals specific algorithmic problems. All libraries implement both normalisation options.

For conservative remappings, the SCRIP library assumes by default that the edges of the meshes follow a straight path in the longitude–latitude space. It is however possible, for the edge intersection calculation, to switch to a Lambert equivalent azimuthal projection above a certain latitude threshold if specified. We performed the benchmark tests either without any projection, or with a projection above 1.45 radians in latitude north. In the latter case, the results are denoted as *SCRIP-L* and in the former case, they are denoted as *SCRIP*. We mention here that, by default, target cells that do not intersect any non-masked source cells do not receive any value, even if this never happens in our tests thanks to the approach use to define the sea-land masks (see Section 2.1.1).

For conservative remapping, ESMF assumes by default that grid cells edges follow great circle paths along the sphere surface. The default normalisation is DESTAREA. To activate the FRACAREA normalisation, the option *–norm_type fracarea* was activated. The option *–ignore_degenerate* (see above) was also activated. In addition, the option *–ignore_unmapped*, i.e., do not do anything special for target point that does not receive a value with the original algorithm, was activated in order to reproduce the default behaviour of SCRIP.

With XIOS, the mesh edges can be described with great circle or latitude circles, and is automatically defined by the grid type. For unstructured and curvilinear grids (i.e., *torc, nogt, icos,* and *icoh* in our case), great circles are used. For longitude–latitude (i.e., *bggd* in our case), and Gaussian-reduced (i.e., *sse7*), latitude circles are used for the edges located on a latitude circle and great circles are used otherwise.

With YAC, the edges of the grid cells can be either defined with longitude and latitude circles or with great circles depending on the interface used. We used the interface defining the edges of the grid cells with great circles. We have to note here that this is not totally appropriate for the cell edges following a latitude circle as in the regular latitude-longitude grid *bggd* and in the Gaussian-reduced grid *sse7*.

5. Second order conservative with FRACAREA normalisation

As stated above, the basis of a second-order conservative remapping is the same as for the first-order conservative remapping but additional terms proportional to the gradients of the source field are applied. While remaining conservative, this remapping ensures that field details are reconstructed and that different target cells entirely located under the same source cell receive different values. This difference between the first-order and second-order methods is particularly apparent when going from a coarse source grid to a finer destination grid (see Section 3.6). Another difference is that the second-order method does not guarantee that after regridding the range of values in the destination field is within the range of values in the source field. For example, if the minimum value in the source field is 0.0, it is possible that after regridding the destination field contains negative values.

SCRIP applies gradients calculated in the longitudinal and latitudinal directions.

YAC, ESMF, and XIOS implement the second-order conservative algorithm based on [21]. For all four libraries, in cases where the gradient computation fails (for example due to a lack of neighbours, which can occur at land-sea mask borders), the algorithm automatically assumes a zero gradient, which is essentially a fall back to a first-order conservative remapping.

For ESMF, the same options used for the first-order conservative remapping (i.e., *–ignore_unmapped –ignore_degenerate*, and *–norm_type fracarea*) were activated.

2.1.4. Benchmark Metrics

The benchmark implements the calculation of regridding metrics proposed by the CANGA project. With the following definitions:

- Ψ^s: the analytical function on the source grid;
- Ψ^t: the analytical function on the target grid;
- $R\Psi^s$: the source analytical function regridded on the target grid;
- I_s: the integral on the source grid;
- I_t: the integral on the target grid;

The CANGA metrics are defined as:

- mean misfit: mean ($|R\Psi^s - \Psi^t| / |\Psi^t|$);
- maximum misfit: max ($|R\Psi^s - \Psi^t| / |\Psi^t|$);
- RMS (root mean square) misfit: RMS ($|R\Psi^s - \Psi^t| / |\Psi^t|$);
- L_{min}: (min Ψ^t − min $R\Psi^s$)/max ($|\Psi^t|$) (A positive L_{min} detects an overestimate of the function minimum (i.e., it reinforces the minimum) while a negative L_{min} detects some smoothing of the function minimum);
- L_{max}: (max $R\Psi^s$ − max Ψ^t)/max ($|\Psi^t|$) (A positive L_{max} detects an overestimate of the function maximum (i.e., it reinforces the maximum) while a negative L_{max} detects some smoothing of the function maximum);
- Source global conservation: $|I_t (R\Psi^s) - I_s (\Psi^s)| / I_s (\Psi^s)$;
- Target global conservation: $|I_t (R\Psi^s) - I_t (\Psi^t)| / I_t (\Psi^t)$.

We calculated these metrics for all libraries for all pairs of grids for the 4 functions for all algorithms except when the library did not support the algorithm.

2.2. Implementation of the Regridding Benchmark for SCRIP, YAC, ESMF and XIOS

The steps to realize in order to calculate the benchmark metrics for each regridding library is, of course to download the library sources, compile them, and develop a scripting environment to generate regridding weights activating the different regridding algorithms for the different pairs of grids. We went through these steps for YAC, ESMF and XIOS. For completeness, we also describe the environment used to generate the weights with the SCRIP library, as the benchmark metrics were also calculated for the SCRIP for comparison. These calculations were realized by different developers on different platforms, using the intel 18.0.1.163 compiler and associated intel mpi 2018.1.163. The current benchmark results, evaluating the quality of the regriddings, are not sensible to the platform used, while a benchmark evaluating the numerical performance of the libraries would be.

- SCRIP

The OASIS3-MCT, and therefore SCRIP, sources used for the regridding benchmark correspond to the trunk of the OASIS3-MCT git developer repository dated 05/05/2021. The environment used to calculate regridding weights with the SCRIP library in OASIS3-MCT is available on Zenodo (see the Data Availability section below). The benchmark tests were run on LENOVO cluster nemo at CERFACS (288 bi-socket nodes with 12 Intel cores E5-2680-v3 2.5 Ghz with 64 GB of memory).

- ESMF

The sources used for the results presented in Section 3 correspond to the branch ESMF_8_2_0_beta_snapshot_08. An environment developed to generate regridding weights with ESMF is available on Zenodo. As for SCRIP, the benchmark tests were run on LENOVO cluster nemo at CERFACS.

- YAC

YAC sources used for the regridding benchmark corresponds to a pre-release state of YAC v2.0.0 that was provided by the developers. All developments used in this version are now included in the official release YAC v2.3.0. The environment to calculate regridding weights with YAC is available on Zenodo. All regridding weight calculations were done on a PC Dell Precision M7720 with 6 cores Intel Xeon E-2186M, 64 Gb RAM.

- XIOS

The sources used for the results presented in Section 3 correspond to SVN revision 2134 dated 2021-04-29. The environment developed to generate regridding weights with XIOS is available on Zenodo. As for YAC, all regridding weight calculations were done on a PC Dell Precision M7720 with 6 cores Intel Xeon E-2186M, 64 Gb RAM.

Once the regridding weights had been generated, the benchmark metrics were calculated for the four libraries using different analytical functions using a specific scripting environment based on Python 3.7.7 available on Zenodo.

3. Benchmark Results

All benchmark metrics were calculated for:
- the four analytical functions: *sinusoid, harmonic, vortex, gulfstream* (see Section 2.1.2);
- the six pairs of relatively low-resolution grids matching an ocean grid with an atmospheric grid: *torc-bggd, torc-icos, torc-sse7, nogt-bggd, nogt-icos, nogt-sse7* in both directions (see Section 2.1.1); for the conservative remapping, we also analyse the regridding of the *vortex* function for *icos-icoh* and *nogt-icoh* in order to test the impact of that regridding on cases with large resolution difference (see Section 3.6);
- for the four regridding libraries: SCRIP (+SCRIP-L, i.e., with Lambert projection for conservative regridding), YAC, ESMF and XIOS;
- for all algorithms: nearest neighbour, 1st and 2nd order non-conservative, 1st and 2nd order conservative, except when the regridding library does not support the algorithm, such as, e.g., nearest neighbour for XIOS (see Section 2.1.3).

Results of all metric values and plots are available on Zenodo. The lists of the individual files containing metric values and plots are detailed in Appendices B and C, respectively.

We analysed all metrics obtained but we cannot of course discuss them all here. In the next paragraphs, we present specific cases, either to illustrate the main conclusions of our analysis or to highlight the specific problems observed. We note here that we show metric results for the SCRIP library, as a basis for comparison. However, if specific problems are revealed by the benchmark for the SCRIP, we do not further investigate them as the current objective is to evaluate other regridding libraries.

3.1. Nearest Neighbour Regridding

Figure 3 shows the mean, rms and maximum misfit for the different pairs of grids for the *harmonic* function for the nearest neighbour regridding. The three regridding libraries produce almost exactly the same, and very reasonable, results: the curves are superimposed and not distinguishable. This is also true for the other analytical functions (not shown).

We observed that the function used to define the coupling field has a strong impact on the maximum misfit, as illustrated on Figure 4, which shows the maximum misfit for the different pairs of grids for the four functions. The maximum misfit is directly linked to the gradient of the function, being much higher for example for the *gulfstream* function than for the slowly varying *sinusoid* function, as is expected for a nearest neighbour algorithm.

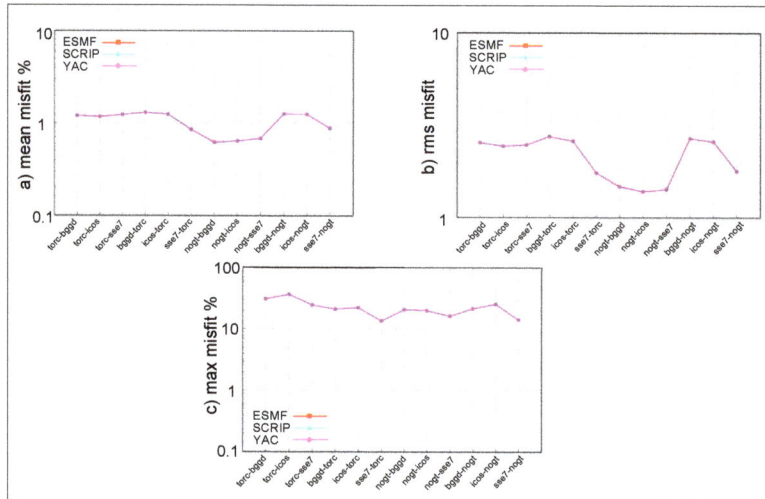

Figure 3. (**a**) mean, (**b**) rms and (**c**) maximum misfit for the different pairs of grids for the *harmonic* function for nearest neighbour algorithm for ESMF, SCRIP and YAC.

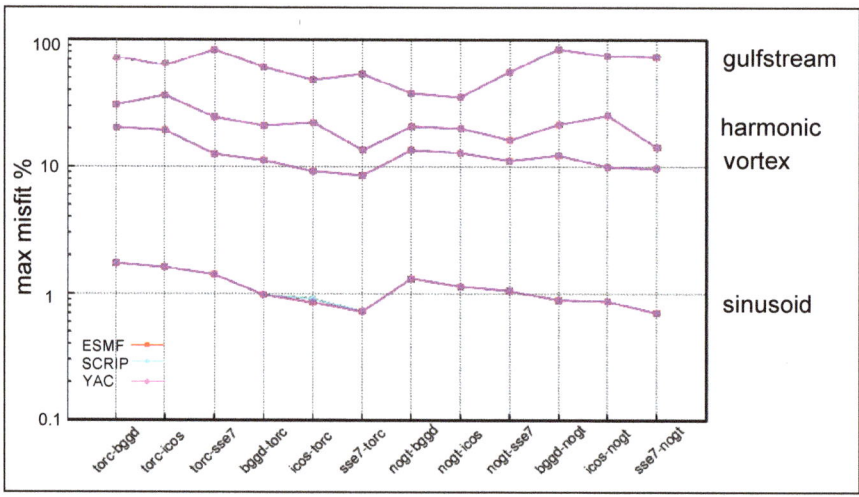

Figure 4. Maximum misfit for the different pairs of grids for the different functions *sinusoid, vortex, harmonic, gulfstream* for the nearest neighbour algorithm for ESMF, SCRIP and YAC.

3.2. 1st Order Non-Conservative Regridding

Figure 5 shows the mean, rms, and maximum misfit for the different pairs of grids for SCRIP, ESMF and YAC, for the *vortex* function for the first-order non-conservative regriddings described in Section 2.1.3. The algorithm in YAC is less accurate on average, i.e., the mean misfit is higher on average. This was also observed for the other analytical functions (not shown).

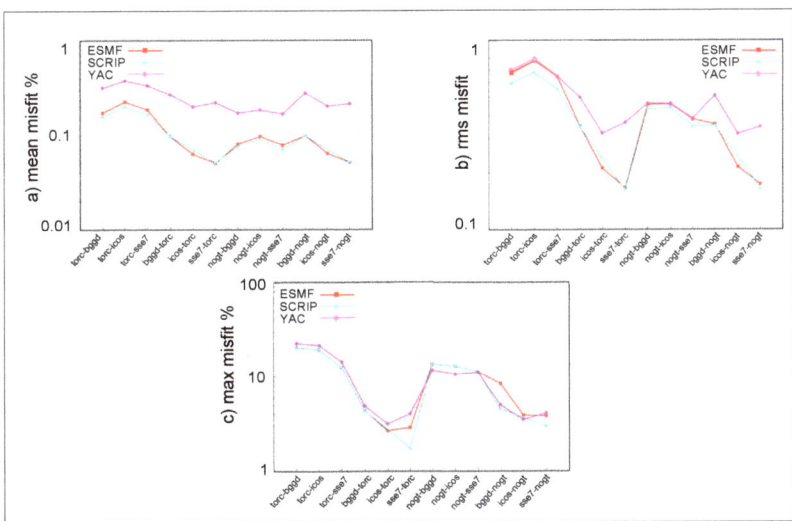

Figure 5. (**a**) mean, (**b**) rms, and (**c**) maximum misfit for the different pairs of grids for the *vortex* function for first-order non-conservative regridding for ESMF, SCRIP and YAC.

3.3. Second-Order Non-Conservative Regridding

Second-order non-conservative algorithms are available in SCRIP, ESMF and YAC (see details in Section 2.1.3). Figure 6 shows the mean misfit, rms misfit, maximum misfit, and L_{max} for the different pairs of grids for these three regridding libraries for the *gulfstream* function.

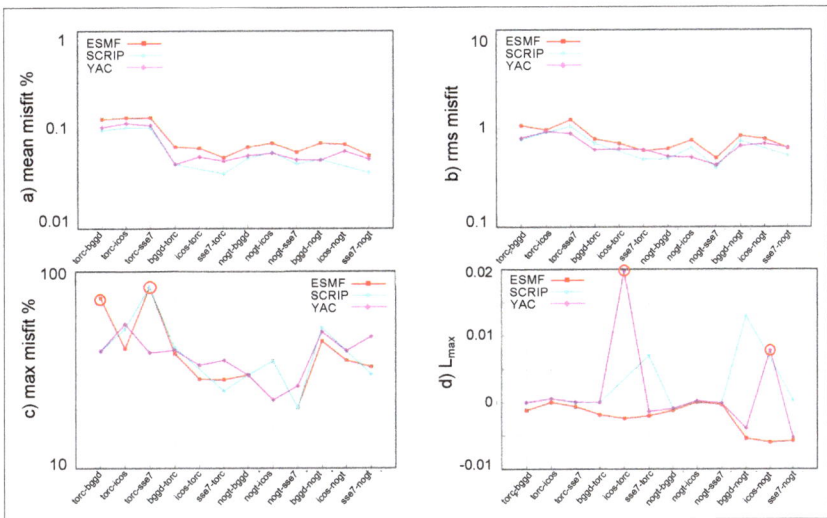

Figure 6. (**a**) mean, (**b**) rms, and (**c**) maximum misfit and (**d**) L_{max} for the different pairs of grids for the *gulfstream* function for second-order non-conservative algorithms for ESMF, SCRIP and YAC. The red circles identify anomalous regriddings detailed in the text.

On average, the SCRIP bicubic algorithm gives slightly better results for certain pairs of grids and the ESMF patch algorithm gives slightly less accurate results (Figure 6a). The averaging present in the ESMF patch algorithm smooths the regridded field and prevents overshoots or undershoots, as can be seen by the more negative values for L_{max}.

In Figure 6c, we note some high maximum misfit for ESMF for *torc-bggd* and *torc-sse7*, not present for the other functions (not shown). These anomalous points also appear for the bilinear regridding for the *gulfstream* function only (not shown). This led us to look for anomalous regridded values in the gulf stream region. The 2D plots of the misfit in that region for the *gulfstream* function for the *torc-bggd* regridding are shown at Figure 7. One anomalous value near the coast (in yellow) is indeed easy to identify for ESMF patch algorithm. The same anomalous point appears for the *torc-sse7* regridding (not shown). At the time of writing this paper, this particular case was under investigation with ESMF developers.

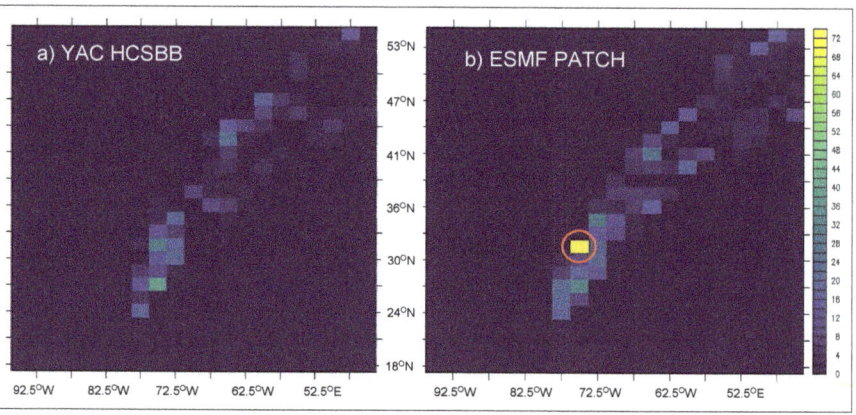

Figure 7. Misfit (%) for the *gulfstream* function in the gulf stream region for *torc-bggd* regridded with the second-order non-conservative algorithm for (**a**) YAC HCSBB and (**b**) ESMF PATCH. The red circle identifies the anomalous value near the coast for the ESMF patch algorithm discussed in the text.

Figure 6d also shows high values of L_{max} for *icos-torc* and *icos-nogt* for the *gulfstream* function that do not appear for the other functions (not shown). Figure 8 shows 2D plots of the regridded field in the gulf stream region for ESMF and YAC. Indeed, it confirms that, compared to ESMF, which tends to smooth the local maximum with its patch averaging algorithm, YAC gives higher, but a priori non-anomalous, values in the centre of the gulf stream.

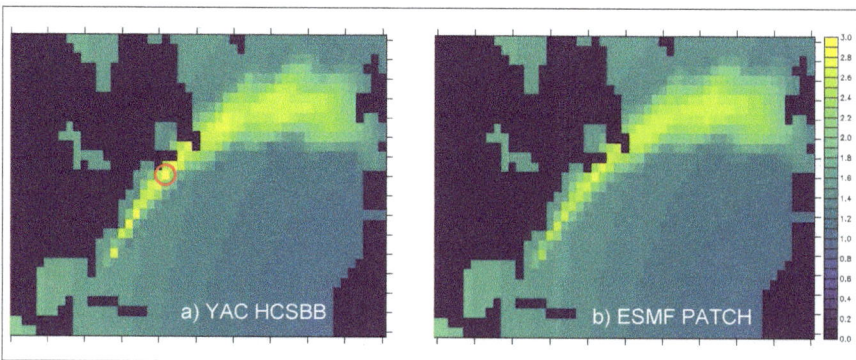

Figure 8. Misfit (%) for the *gulfstream* function in the gulf stream region for *icos-torc* regridded with the second-order non-conservative algorithm for (**a**) YAC HCSBB and (**b**) ESMF PATCH. The red circle identifies the highest, but a priori non-anomalous, value in the centre of the gulf stream for YAC discussed in the text.

3.4. First-Order Conservative Remapping with DESTAREA Normalisation

To evaluate the quality of the first-order conservative regridding, we started by looking at the results obtained with the DESTAREA normalisation option, which usually reveals problems that the FRACAREA option would hide, sometimes involving a cancellation of errors. Figure 9 shows the mean and the maximum misfits for the *harmonic* function for the four libraries. Here, for ESMF, *nogt* and *torc* are described with the *unstructured* grid format (see Section 2.1.3). It confirms the extremely wrong values obtained using the SCRIP library either activating (SCRIP-L) or not activating (SCRIP) the Lambert azimuthal projection, as mentioned in the introduction (see also [4,5]). The other libraries ESMF, YAC and XIOS produced practically the same and satisfactory results, with a mean misfit between 0.1% and 1% and a maximum misfit between 1% and 10% for all pairs of grids.

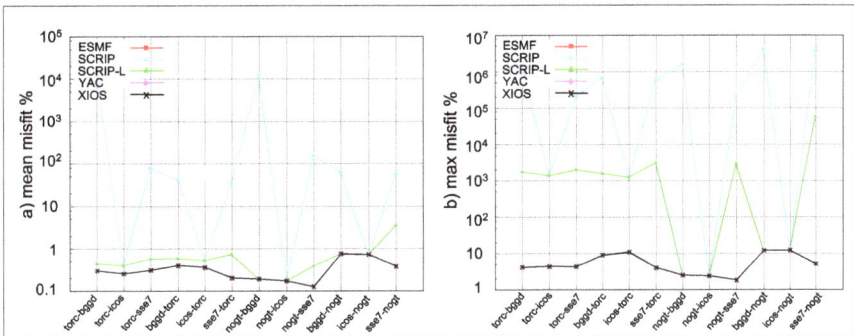

Figure 9. (**a**) mean and (**b**) maximum misfit for ESMF, SCRIP, SCRIP-L, YAC, and XIOS for the first-order conservative remapping with DESTAREA normalisation for the different pairs of grids for the *harmonic* function. For ESMF, *nogt* and *torc* are described with the *unstructured* format.

Figure 10 shows the source global conservation metric for the 4 functions for all regridding libraries for the different pairs of grids. Again, it is very clear that the SCRIP/SCRIP-L library presents some important problems with the first-order conservative remapping. On the contrary, ESMF, YAC, and XIOS show similar and very reasonable results, this metric being at maximum of the order of 1%.

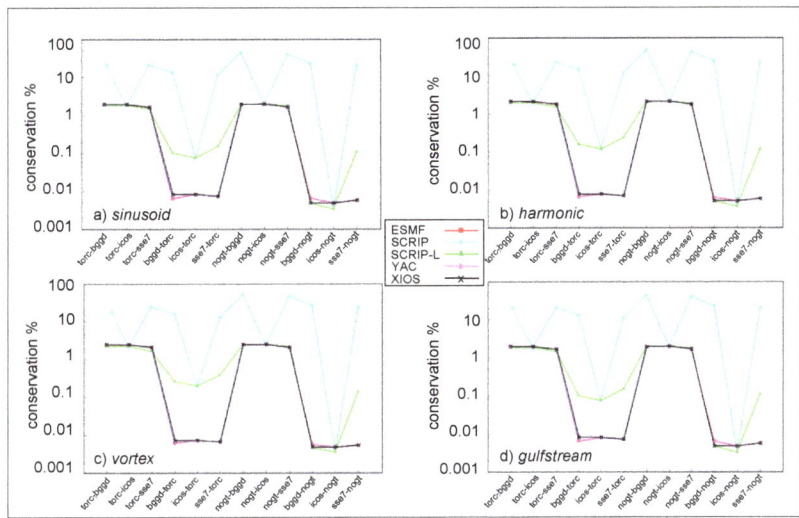

Figure 10. Source global conservation (%) for ESMF, SCRIP, SCRIP-L, YAC and XIOS for the 1st order conservative remapping with DESTAREA normalisation for the different pairs of grids for the 4 functions: (**a**) *sinusoid*, (**b**) *harmonic*, (**c**) *vortex*, and (**d**) *gulfstream*.

We then analysed the impact of the grid description format in ESMF. As explained in Section 2.1.3 two formats are supported to describe the grids with ESMF, either the so-called *SCRIP* format or the *unstructured* format. The results above were produced describing the ocean NEMO grids *nogt* and *torc* with the *unstructured* format. However, the *nogt* and *torc* grids are structured, and it is possible to describe them using the *SCRIP* format. As such, we repeated the first-order conservative regriddings for ESMF using the *SCRIP* format to describe the *nogt* and *torc* grids. Figure 11 shows the mean and the maximum misfit for the *harmonic* function in that case. The results are the same as on Figure 9, except that ESMF now presents anomalous mean and maximum misfits when *nogt* is the source grid. The same anomalies are observed for the other functions (not shown).

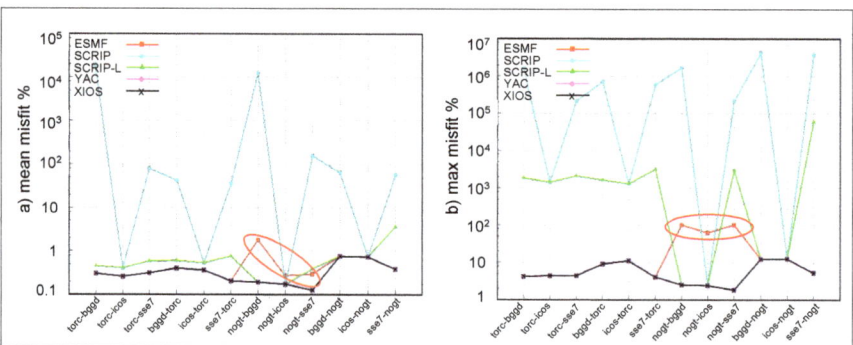

Figure 11. (**a**) Mean and (**b**) maximum misfit for ESMF, SCRIP, SCRIP-L, YAC, and XIOS for the first-order conservative remapping with DESTAREA normalisation for the different pairs of grids for the *harmonic* function. For ESMF, *nogt* and *torc* are described with the *SCRIP* structured format. The red oval shapes identify ESMF regriddings showing anomalous mean and maximum misfits when *nogt* is the source grid. These regriddings are discussed in the text.

Figure 12 shows the 2D plot of the misfit of the *harmonic* regridded function for *nogt-bggd* with *nogt* described (a) with the *SCRIP* format and (b) with the *unstructured* format. The problem, clearly linked to the north fold of the NEMO, disappears when *nogt* is described with the *unstructured* format.

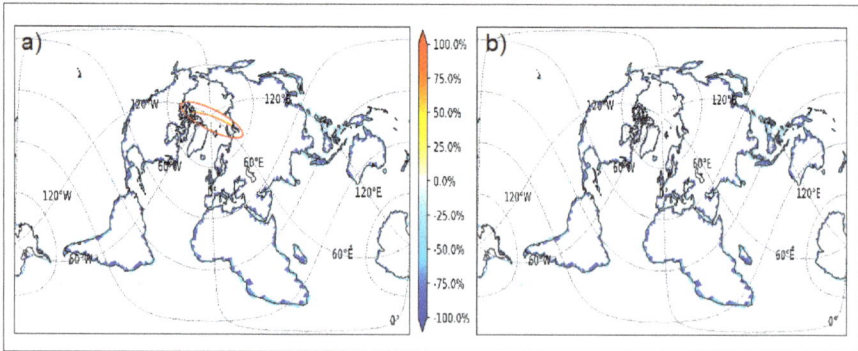

Figure 12. Two-dimensional plot of the misfit of the *harmonic* function regridded with ESMF first-order conservative remapping for *nogt-bggd* with *nogt* described (**a**) with the *SCRIP* format and (**b**) with the *unstructured* format. The red oval shape identifies in (**a**) the grid points linked to the north fold of the NEMO grid showing anomalous misfit when the *nogt* grid is described in the SCRIP format.

It is interesting to note that the regridding does not show any specific problem when *torc* is the source grid. This is certainly linked to the type of grid in the north fold. For *torc*, the north fold is such that in the (i,j) space the third-to-last row folds on the last row and the penultimate row folds on to itself. For *nogt*, the penultimate row folds on the last row. As for the anomaly identified for the patch regridding for the *gulfstream* function (see Section 3.3), this problem is, at the time of writing, under investigation with ESMF developers.

3.5. 1st Order Conservative Remapping with FRACAREA Normalisation

Figure 13 shows the maximum misfit for the first-order conservative regridding with FRACAREA normalisation for the four functions for all pairs of grids. For ESMF, the *nogt* grid is described with the *unstructured* format to avoid specific problems linked to the north fold (see Section 3.4). All regridding libraries have the same maximum misfit, except SCRIP and SCRIP-L, which we will not further discuss here. As expected, the maximum misfit is higher for the functions with sharper gradients. For example, the maximum misfit is higher for the *harmonic* function than for the *sinusoid* function for all pairs of grids. For the *gulfstream* function (Figure 13d), the maximum misfit for *torc-sse7* is particularly high. As this is the case for all regridding libraries and not for the other functions, this is probably linked to the sharp gradients of the *gulfstream* function.

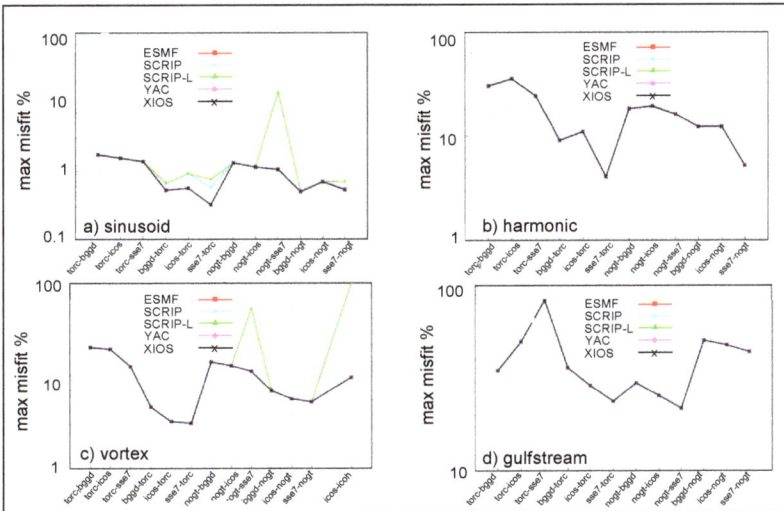

Figure 13. Maximum misfit for ESMF, SCRIP, SCRIP-L, YAC, and XIOS for the first-order conservative remapping with FRACAREA normalisation for the different pairs of grids for the 4 functions: (**a**) *sinusoid*, (**b**) *harmonic*, (**c**) *vortex*, (**d**) *gulfstream*.

For the source global conservation metric (not shown), ESMF, YAC, and XIOS show similar and very good results, this metric being less than 0.01% in all cases. The source global conservation for the *icos-icoh* pair of grids for the *vortex* function, also calculated for that regridding, is remarkably small, being of the order of 10^{-9}.

3.6. Second-Order Conservative Remapping with FRACAREA Normalisation

Figure 14 shows the mean, maximum, rms misfits, and the source global conservation for the second-order conservative remapping with the FRACAREA normalisation for the different pairs of grids for all regridding libraries for the *harmonic* function. Besides SCRIP and SCRIP-L, which we will not further analyse here, we see that all regridding libraires show more or less the same behaviour with good global conservation. This is not surprising, as they all implement the same algorithm based on [21].

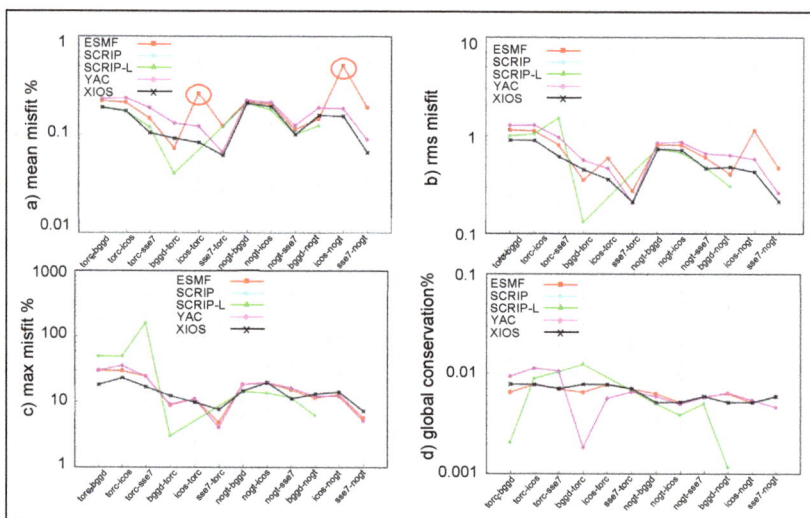

Figure 14. (**a**) Mean, (**b**) rms, (**c**) maximum misfit, and (**d**) source global conservation for the different pairs of grids for the *harmonic* function for second-order conservative remapping with FRACAREA normalisation. The red circles identify anomalous regriddings for ESMF when the source grid is the icosahedral one (*icos*) detailed in the text.

The only particularity seems to be for ESMF, when the source grid is the icosahedral one (*icos*), which shows a relatively high mean misfit. To better qualify this anomaly, we zoomed in on the 2D representation of the misfit for the *icos-nogt* case, as shown on Figure 15. The misfit shows an alternating positive and negative pattern which causes the relatively high mean misfit for ESMF. Work is underway with ESMF developers to solve this issue.

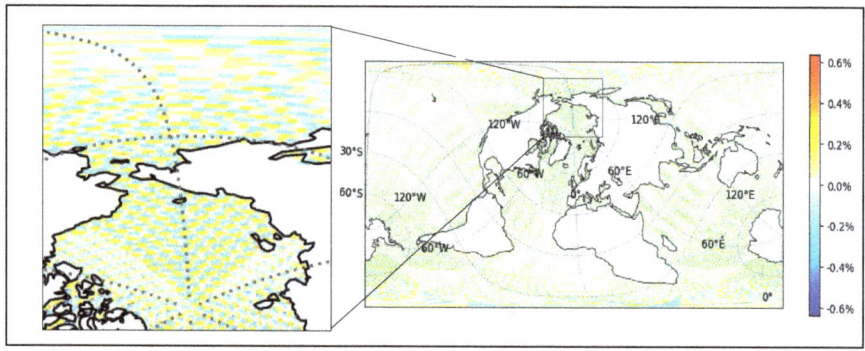

Figure 15. Misfit (%) on the target grid *nogt* for the *icos-nogt* second-order conservative remapping with FRACAREA normalisation for ESMF, with a zoom on the left.

Figure 16 shows L_{min} and L_{max} for the second-order conservative remapping with FRACAREA normalisation for the *gulfstream* function, which presents some outstanding values (the other functions do not present such outstanding values). XIOS shows a strong undershoot for *torc-icos*, as shown by L_{min}, and a strong overestimate for *bggd-nogt* as shown by L_{max}.

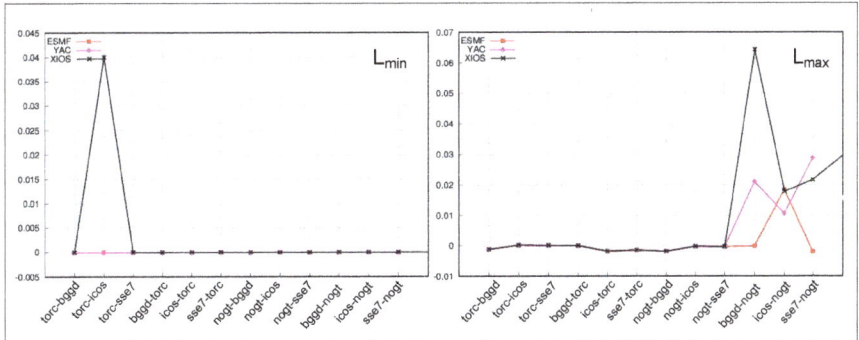

Figure 16. L_{min} and L_{max} for the different pairs of grids for the *gulfstream* function for 2nd order conservative remapping with FRACAREA normalisation.

To understand XIOS's undershoot of *torc-icos*, we looked at the 2D misfit in the gulf stream region for XIOS, ESMF and YAC (Figure 17). We observed one clearly outstanding point near the coast for XIOS. ESMF also shows some negative misfit at this point, but it is much smaller than XIOS. YAC does not show any important misfit at this point. This difference between the three regridding libraries has to be investigated in more detail. As they are based on the same algorithm, it must be linked to some implementation differences in the way the libraries calculate the gradients and eventually switch to a 1st order conservative remapping when the gradient cannot be calculated, e.g., near the coast.

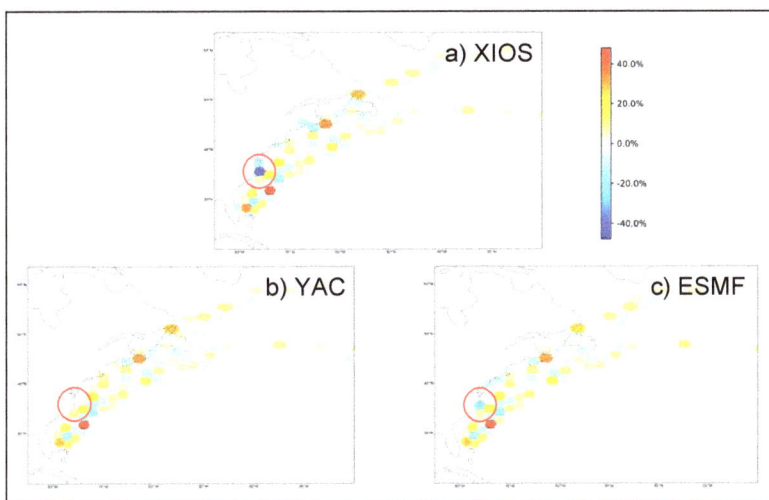

Figure 17. Misfit in the gulf stream region for the second-order conservative remapping of the *gulfstream* function for *torc-icos* for XIOS, YAC and ESMF. The red circles identify the grid point near the coast showing an outstanding value for XIOS.

To understand XIOS's overshoot of *bggd-nogt*, we looked at the 2D regridded *gulfstream* function in the gulf stream region for XIOS, ESMF and YAC (not shown). We observed that XIOS shows higher values near the centre of the gulf stream. As for YAC HCSBB (see Figure 8), this behaviour, which explains the overshoot, is most probably linked to some specificities in the algorithm but not to a bug in the implementation.

3.7. Comparison of Regridding Algorithms

It is also interesting to compare the results of the different algorithms for each specific library. Figure 18 shows 2D plots of the relative misfit for the remapping of the *vortex* function from the low-resolution icosahedral grid *icos* to the high-resolution icosahedral grid *icoh* with YAC for the (a) first-order conservative remapping and (b) the second-order conservative remapping (both with FRACAREA normalisation). We see the clear benefit of the second order compared to the first order, especially when this remapping involves two grids with very different resolutions. XIOS shows very similar results but not ESMF, probably because of the problem identified above for the second-order conservative remapping for *icos-nogt*, which also exists for *icos-icoh* (alternating positive and negative pattern in the misfit, see Figure 15).

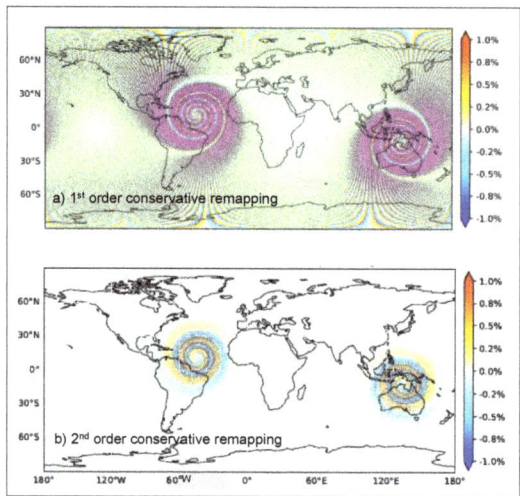

Figure 18. Misfit (%) for the (**a**) first-order and (**b**) second-order conservative remapping (both with FRACAREA normalisation) of the *vortex* function from the low-resolution icosahedral grid *icos* to the high-resolution icosahedral grid *icoh* with YAC.

Figure 19 shows the mean misfit and the source global conservation for the different regridding algorithms for ESMF and for YAC. We do not show the equivalent graphs for XIOS as this library supports only conservative regridding, which makes the comparison somewhat limited.

For both ESMF and YAC, Figure 19a,c, respectively, show that the mean misfit for the first-order conservative remapping is always higher than for the second-order remapping. This is expected and fully coherent with the 2D results shown above in Figure 18 for the *icos-icoh* pair of grids.

The comparison of the mean misfit between conservative and non-conservative algorithms does not lead to such clear-cut conclusions. We would expect non-conservative algorithms to show less error at the price of being non-conservative. For ESMF (Figure 19a), we see that this is the case for bilinear and patch when compared with first-order conservative remapping (the green and red curves are under the blue curve for all grid pairs) but their mean misfits are of about the same magnitude as that of the second-order conservative remapping (black curve). For YAC (see Figure 19c), this expectation is basically fulfilled for the HCSBB algorithm (red curve), which shows an error smaller than all of the other algorithms except for the second-order conservative remapping for *torc-icos*. However, the non-conservative first-order regridding (green curve) is the one showing in general the highest error.

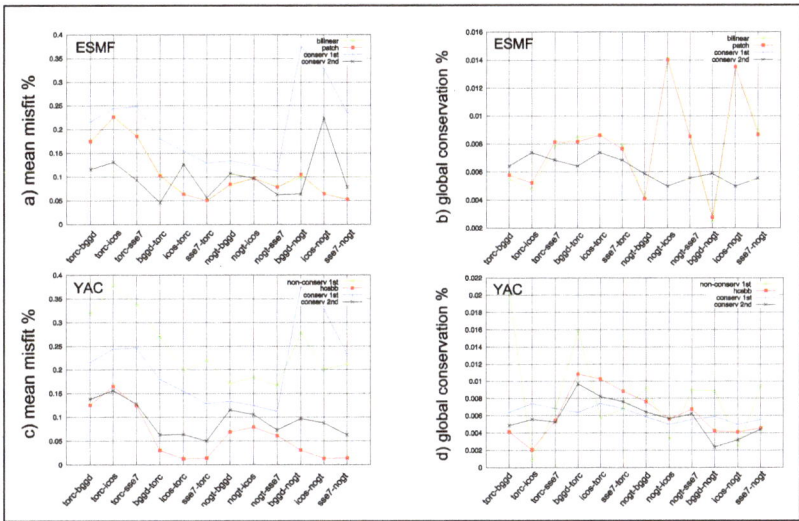

Figure 19. Mean misfit (%) for the different regridding algorithms in (**a**) ESMF and (**c**) YAC; source global conservation (%) for the different regridding algorithms in (**b**) ESMF and (**d**) YAC.

Regarding the global conservation (Figure 19b,d), we can observe that the non-conservative remappings (green and red curves) show much more variability with respect to the grid pairs than the first- or second-order remappings. This is reassuring, as it means that the conservative remapping guarantees a certain level of conservation. Still, we observe that for a few pairs of grids, e.g., for *torc-icos* for ESMF, the global conservation is better for non-conservative regriddings than for conservative remappings, which was unexpected a priori.

4. Discussion

This paper presents work done to benchmark the quality of four regridding libraries: SCRIP, YAC, ESMF, and XIOS, each evaluating five algorithms (see Section 2.1.3) with four different analytical functions (see Section 2.1.2) for six grids used in real ocean or atmosphere models (see Section 2.1.1).

This benchmark calculates some of the metrics proposed by the CANGA project and we can state that it provides a strong basis to analyse the quality of regridding libraries as it evaluates:

- their sensitivity, as we perform the metric calculation for six pairs of grids in both directions and, in addition, for the *icos-icoh* and *nogt-icoh* pairs for the *vortex* function for second-order conservative FRACAREA remapping;
- their global conservation, as we provide the source and target global conservation metrics.

As we provide and analyse L_{min} and L_{max} metrics, our benchmark also allows a first analysis of the regridding library monotonicity and dissipation (or smoothing). We also started to evaluate the performances of the libraries with a first scalability analysis, not shown here but in [5], that will be completed in the coming months. However, we do not address the regridding library consistency, i.e., the preservation of discretization order and accuracy.

The details of our analysis are the following (note that XIOS offers only first- and second-order conservative remapping):

- ESMF and YAC nearest neighbour regriddings produce almost exactly the same and very reasonable results than the SCRIP (Figure 3). The analytical function defining the field to be regridded has a strong impact on the maximum of the misfit, which is directly linked to the field gradient (Figure 4).
- For first-order non-conservative regridding, YAC, using an inverse distance weighting of the vertex values of the source polygon enclosing the target point, is less accurate

on average than the SCRIP or ESMF bilinear schemes (Figure 5). For second-order non-conservative regridding, the ESMF patch algorithm gives slightly less accurate results than the SCRIP bicubic or the YAC spherical Bernstein–Bézier polynomial algorithms (Figure 6). For first- and second-order non-conservative regridding, all results for ESMF and YAC are reasonable, except for ESMF in the case of the *gulfstream* function for *torc-bggd* and *torc-sse7* grid pairs, which show one anomalous value near the coast (Figure 7). For second-order non-conservative regridding in the case of *icos-torc* and *icos-nogt* for the *gulfstream* function, YAC also shows some higher, but a priori non-anomalous values, in the centre of the gulf stream (Figure 8).

- First-order conservative remapping with DESTAREA normalisation in YAC, ESMF, and XIOS show very similar and good results (Figures 9 and 10), except for ESMF when *nogt* is the source grid if this grid is described with the *SCRIP* (structured) format (Figures 11 and 12). For first-order conservative remapping with FRACAREA normalisation, YAC, ESMF, and XIOS show very similar and good results for all functions (Figure 13); this regridding raises no specific issues for any regridding library.
- YAC, ESMF, and XIOS show approximately the same behaviour with good global conservation for second-order conservative remapping with FRACAREA normalisation, implemented following [21] in the 3 libraries (Figure 14). One issue, however, is in ESMF when the source grid is the icosahedral one, *icos*, which shows a relatively high mean misfit for all functions, with an alternating positive and negative pattern (Figure 15). Another issue is present for XIOS, which shows a strong undershoot for the *gulfstream* function for *torc-icos*, with one clearly outstanding point near the coast.
- The second-order remapping always shows a lower mean misfit than the first-order remapping (Figure 19a,c), and the gain is very evident when going from a low-resolution to a high-resolution grid (Figure 18).
- Unexpectedly, conservative algorithms do not always offer better global conservation than non-conservative ones (Figure 19b,d).

This benchmark leads us to conclude that YAC, ESMF, and XIOS can all three be considered as high-quality regridding libraries, even if some details for few specific cases still need to be fixed. Interactions are currently going on with the library developers to address the very few problems observed.

Benchmarking libraries is always a delicate task as the environment has to be designed to not favour any library a priori. Benchmarking is more than a way to compare libraries and should be taken as a great opportunity for the users to interact with the developers, as we did during the exercise presented here, and for the developers to have their library tested in depth by expert users.

Author Contributions: Conceptualization, S.V. and A.P.; data curation, A.P. and G.J.; formal analysis, S.V. and G.J.; funding acquisition, S.V.; investigation, A.P. and G.J.; methodology, S.V. and A.P.; project administration, S.V.; software, A.P. and G.J.; supervision, S.V.; validation, S.V. and G.J.; visualization, S.V., A.P. and G.J.; writing—original draft, S.V. All authors have read and agreed to the published version of the manuscript.

Funding: This research was funded by the European Union's Horizon 2020 research and innovation programme in the framework of the IS-ENES3 project under grant agreement No 824084.

Data Availability Statement: The OASIS3-MCT sources correspond to the trunk of the OASIS git developer repository dated 5 May 2021. They are available on Zenodo at https://doi.org/10.5281/zenodo.5872502 (accessed on 18 January 2022). The environment used to calculate regridding weights with the SCRIP library in OASIS3-MCT is available in the tar file test_hybrid.tar on Zenodo at https://doi.org/10.5281/zenodo.5342548 (accessed on 31 August 2021). YAC sources correspond to a pre-release state of YAC v2.0.0 that was provided by the developers. All developments used in this version are now included in the official release YAC v2.3.0, available at https://www.dkrz.de/en/services/software-development (accessed on 18 January 2022) and on Zenodo at https://doi.org/10.5281/zenodo.5871066 (accessed on 18 January 2022). The environment to calculate regridding weights with YAC is registered on the CERFACS git developer repository at https://nitrox.cerfacs.fr/globc/OASIS3-MCT/oasis3

-mct_other/tree/master/generate_weights/YAC (accessed on 9 September 2021) and is available on Zenodo at https://doi.org/10.5281/zenodo.5872627 (accessed on 18 January 2022). ESMF sources correspond to the branch ESMF_8_2_0_beta_snapshot_08, which can be obtained with the git command "git clone –branchESMF_8_2_0_beta_snapshot_08–depth 1". They are available on Zenodo at https://doi.org/10.5281/zenodo.5871823 (accessed on 18 January 2022). An environment developed to generate regridding weights with ESMF is available in the tar file generate_weights_ESMF.tar on Zenodo at https://doi.org/10.5281/zenodo.5343048 (accessed on 31 August 2021). XIOS sources correspond to SVN revision 2134 dated 2021-04-29 that can be extracted with the SVN command "svn co -r 2134 http://forge.ipsl.jussieu.fr/ioserver/svn/XIOS/trunk XIOS". They are available on Zenodo at https://doi.org/10.5281/zenodo.5872716 (accessed on 18 January 2022). The environment developed to generate regridding weights with XIOS is available in the tar file generate_weights_XIOS.tar on Zenodo at https://doi.org/10.5281/zenodo.5342491 (accessed on 31 August 2021). The environment used to calculate the benchmark metrics for the four libraries, once the regridding weights were generated for each of them, is available in the tar file compare_interpolation.tar on Zenodo at https://doi.org/10.5281/zenodo.5342778 (accessed on 31 August 2021). The tar file Regridding_Benchmark_metrics.tar, gathering the CSV files containing the benchmark metric values, is available on Zenodo at https://doi.org/10.5281/zenodo.5343166 (accessed on 31 August 2021). The tar file Regridding_Benchmark_metrics_plots.tar, gathering the plots of the regridding benchmark metric, is available on Zenodo at https://doi.org/10.5281/zenodo.5347696 (accessed on 31 August 2021).

Acknowledgments: We would like to warmly thank the regridding library developers for their availability and willingness to interact with us during this benchmarking exercise: Moritz Hanke from DKRZ for YAC, Robert Oehmke from NCAR for ESMF and Yann Meurdesoif from CEA/IPSL for XIOS. Interacting with them transformed this benchmarking exercise from a purely analytical work into a community effort, improving regridding libraries used in Earth System Modelling.

Conflicts of Interest: The authors declare no conflict of interest. The funders had no role in the design of the study; in the production, analyses, or interpretation of data; in the writing of the manuscript, or in the decision to publish the results.

Appendix A. Analytical Functions

This appendix contains the definition of the four analytical functions, expressed in Fortran 90, used to define the coupling fields, i.e., *sinusoid* (see Figure A1), *harmonic* (see Figure A2), *vortex* (see Figure A3), *gulfstream* (see Figure A4).

(A) *sinusoid*

```fortran
!! ANALYTICAL SINUSOID
SUBROUTINE function_sinusoid(ni, nj, xcoor, ycoor, fnc_ana)
!
IMPLICIT NONE
!
INTEGER, PARAMETER :: wp = SELECTED_REAL_KIND(12,307) ! double
!
INTEGER, INTENT(IN) :: ni, nj
REAL(kind=wp), DIMENSION(ni,nj), INTENT(IN)  :: xcoor, ycoor
REAL(kind=wp), DIMENSION(ni,nj), INTENT(OUT) :: fnc_ana
!
REAL (kind=wp), PARAMETER    :: dp_pi=3.14159265359
REAL (kind=wp), PARAMETER    :: dp_conv = dp_pi/180.
REAL(kind=wp) :: dp_length, coef, coefmult
INTEGER       :: i,j
!
DO j=1,nj
  DO i=1,ni

    dp_length = 1.2*dp_pi
    coef = 2.
    coefmult = 1.
    fnc_ana(i,j) = coefmult*(coef - COS( dp_pi*(ACOS( COS(xcoor(i,j)*dp_conv)*COS(ycoor(i,j)*dp_conv) )/dp_length)) )

  ENDDO
ENDDO
END SUBROUTINE function_sinusoid
```

Figure A1. Fortran 90 code defining the *sinusoid* analytical function.

(B) harmonic

```fortran
!
!!! HARMONIC FROM TEMPEST-REMAP
SUBROUTINE function_harmonic(ni, nj, xcoor, ycoor, fnc_ana)
  !
  IMPLICIT NONE
  !
  INTEGER, PARAMETER :: wp = SELECTED_REAL_KIND(12,307) ! double
  !
  INTEGER, INTENT(IN) :: ni, nj
  REAL(kind=wp), DIMENSION(ni,nj), INTENT(IN)  :: xcoor, ycoor
  REAL(kind=wp), DIMENSION(ni,nj), INTENT(OUT) :: fnc_ana
  !
  REAL (kind=wp), PARAMETER    :: dp_pi = 3.14159265359
  REAL (kind=wp), PARAMETER    :: dp_conv = dp_pi/180.
  !
  INTEGER         :: i,j
  !
  DO j=1,nj
    DO i=1,ni
      fnc_ana(i,j) = 2.0 + SIN( 2.0 * ycoor(i,j)*dp_conv ) ** 16.0 * COS( 16.0 * xcoor(i,j)*dp_conv )
    ENDDO
  ENDDO
END SUBROUTINE function_harmonic
```

Figure A2. Fortran 90 code defining the *harmonic* analytical function.

(C) vortex

```fortran
!
!!! VORTEX FROM TEMPEST-REMAP (as in XIOS)
SUBROUTINE function_vortex(ni, nj, xcoor, ycoor, fnc_ana)
  !
  IMPLICIT NONE
  !
  INTEGER, PARAMETER :: wp = SELECTED_REAL_KIND(12,307) ! double
  !
  INTEGER, INTENT(IN) :: ni, nj
  REAL(kind=wp), DIMENSION(ni,nj), INTENT(IN)  :: xcoor, ycoor
  REAL(kind=wp), DIMENSION(ni,nj), INTENT(OUT) :: fnc_ana
  !
  REAL(kind=wp), PARAMETER :: dp_pi = 3.14159265359
  REAL(kind=wp), PARAMETER :: dLon0  = 5.5
  REAL(kind=wp), PARAMETER :: dLat0  = 0.2
  REAL(kind=wp), PARAMETER :: dR0    = 3.0
  REAL(kind=wp), PARAMETER :: dD     = 5.0
  REAL(kind=wp), PARAMETER :: dT     = 6.0
  REAL(kind=wp) :: dp_length, dp_conv

  REAL(kind=wp) :: dSinC, dCosC, dCosT, dSinT
  REAL(kind=wp) :: dTrm, dX, dY, dZ
  REAL(kind=wp) :: dlon, dlat
  REAL(kind=wp) :: dRho, dVt, dOmega

  INTEGER          :: i,j
  CHARACTER(LEN=7) :: cl_anaftype="vortex"

  dp_conv = dp_pi/180.
  dSinC = SIN( dLat0 )
  dCosC = COS( dLat0 )

  DO j=1,nj
    DO i=1,ni
      ! Find the rotated longitude and latitude of a point on a sphere
      !          with pole at (dLon0, dLat0).
      dCosT = COS( ycoor(i,j)*dp_conv )
      dSinT = SIN( ycoor(i,j)*dp_conv )

      dTrm = dCosT * COS( xcoor(i,j)*dp_conv - dLon0 )
      dX   = dSinC * dTrm - dCosC * dSinT
      dY   = dCosT * SIN( xcoor(i,j)*dp_conv - dLon0 )
      dZ   = dSinC * dSinT + dCosC * dTrm

      dlon = ATAN2( dY, dX )
      IF( dlon < 0.0 ) dlon = dlon + 2.0 * dp_pi
      dlat = ASIN( dZ )

      dRho = dR0 * COS(dlat)
      dVt  = 3.0 * SQRT(3.0)/2.0/COSH(dRho)/COSH(dRho)*TANH(dRho)
      IF (dRho == 0.0) THEN
        dOmega = 0.0
      ELSE
        dOmega = dVt / dRho
      END IF

      fnc_ana(i,j) = 2.0 * ( 1.0 + TANH( dRho / dD * SIN( dLon - dOmega * dT ) ) )
    END DO
  END DO
!
END SUBROUTINE function_vortex
```

Figure A3. Fortran 90 code defining the *vortex* analytical function.

(D) *gulfstream*

```fortran
!
!! ANALYTICAL GULF STREAM
SUBROUTINE function_gulfstream(ni, nj, lon, lat, fnc_ana)
!!*************************************************
!
  INTEGER, PARAMETER :: wp = SELECTED_REAL_KIND(12,307) ! double

  INTEGER, INTENT(IN) :: ni, nj
  REAL (kind=wp), DIMENSION(ni,nj), INTENT(IN)  :: lon, lat
  REAL(kind=wp), DIMENSION(ni,nj), INTENT(OUT) :: fnc_ana

  REAL (kind=wp), PARAMETER :: coef=2., dp_pi=3.14159265359
  REAL (kind=wp) :: dp_length, dp_conv
  INTEGER :: i, j

  ! Analytical Gulf Stream
  REAL (kind=wp) :: gf_coef, gf_ori_lon, gf_ori_lat, &
                 & gf_end_lon, gf_end_lat, gf_dmp_lon, gf_dmp_lat
  REAL (kind=wp) :: gf_per_lon
  REAL (kind=wp) :: dx, dy, dr, dth, dc, dr0, dr1

  ! Parameters for analytical function
  dp_length = 1.2*dp_pi
  dp_conv = dp_pi/180.
  gf_coef = 1.0 ! Coefficient for Gulf Stream term (0.0 = no Gulf Stream)
  gf_ori_lon = -80.0 ! Origin of the Gulf Stream (longitude in deg)
  gf_ori_lat =  25.0 ! Origin of the Gulf Stream (latitude in deg)
  gf_end_lon =  -1.8 ! End point of the Gulf Stream (longitude in deg)
  gf_end_lat =  50.0 ! End point of the Gulf Stream (latitude in deg)
  gf_dmp_lon = -25.5 ! Point of the Gulf Stream decrease (longitude in deg)
  gf_dmp_lat =  55.5 ! Point of the Gulf Stream decrease (latitude in deg)

  dr0 = SQRT(((gf_end_lon - gf_ori_lon)*dp_conv)**2 + &
      & ((gf_end_lat - gf_ori_lat)*dp_conv)**2)
  dr1 = SQRT(((gf_dmp_lon - gf_ori_lon)*dp_conv)**2 + &
      & ((gf_dmp_lat - gf_ori_lat)*dp_conv)**2)

  DO j=1,nj
    DO i=1,ni

      ! Original OASIS fcos analytical test function
      fnc_ana(i,j)=(coef-COS(dp_pi*(ACOS(COS(lat(i,j)*dp_conv)*&
                 & COS(lon(i,j)*dp_conv)/dp_length)))
      gf_per_lon = lon(i,j)
      IF (gf_per_lon >  180.0) gf_per_lon = gf_per_lon - 360.0
      IF (gf_per_lon < -180.0) gf_per_lon = gf_per_lon + 360.0
      dx = (gf_per_lon - gf_ori_lon)*dp_conv
      dy = (lat(i,j) - gf_ori_lat)*dp_conv
      dr = SQRT(dx*dx + dy*dy)
      dth = ATAN2(dy, dx)
      dc = 1.3*gf_coef
      IF (dr > dr0) dc = 0.0
      IF (dr > dr1) dc = dc * COS(dp_pi*0.5*(dr-dr1)/(dr0-dr1))
      fnc_ana(i,j) = fnc_ana(i,j) + &
                   & (MAX(1000.0*SIN(0.4*(0.5*dr+dth) + &
                   & 0.007*COS(50.0*dth) + 0.37*dp_pi),999.0) - 999.0) * dc

    ENDDO
  ENDDO

END SUBROUTINE function_gulfstream
```

Figure A4. Fortran 90 code defining the *gulfstream* analytical function.

Appendix B. List of CSV Files Containing Metrics Values

The files included in the tar file Regridding_Benchmark_metrics.tar available on Zenodo at https://doi.org/10.5281/zenodo.5343166 (accessed on 31 August 2021) are listed here below in Table A1. This tar file contains the regridding benchmark metrics calculated for all pairs of grids (see Section 2.1.1) and all functions (see Section 2.1.2) for all regridding libraries SCRIP (+SCRIP-L, i.e., with Lambert projection for conservative regridding), YAC, ESMF, and XIOS, and for all algorithms (see Section 2.1.3), except in cases where the regridding library does not support the algorithm (e.g., nearest neighbour for XIOS).

The name of the file is given as R_A_f.csv, where R is the regridding library, A is the algorithm, and f is the function (here *classic* is equivalent to *sinusoid*). The algorithm A can be:

- "DISTWGT_1" for nearest neighbour
- "BILINEAR" for first-order non conservative
- "BICUBIC" for second-order non-conservative
- "CONSERV" for first-order conservative
- "CONS2ND" for second-order conservative

For conservative remapping, A also contains the normalisation option "FRACAREA" or "DESTAREA".

For XIOS, there are therefore no files for nearest neighbour, first- and second-order non-conservative algorithms as they are not supported in XIOS. For the first- and second-order conservative remapping for SCRIP, there are two files: one with (SCRIP-L) and one without (SCRIP) the Lambert azimuthal projection. For ESMF, all results are provided for the version tagged ESMF_8_2_0_beta_snapshot_08 (ESMF-820bs08). For ESMF, for first- and second-order conservative algorithms, the *nogt* grid was described as *unstructured*, as it correctly supports the north fold of the NEMO grid (see Section 3.4), in that case, R = "ESMF-820bs08-U". For first-order conservative with DESTAREA normalisation, results are also provided that describe the *nogt* grid as with the *SCRIP* format for comparison, in that case, R = "ESMF-820bs08".

Table A1. Regridding benchmark files included in the tar file Regridding_Benchmark_metrics.tar available on Zenodo at https://doi.org/10.5281/zenodo.5343166 (accessed on 31 August 2021).

ESMF-820bs08-U_CONS2ND_FRACAREA_classic.csv	SCRIP_CONS2ND_FRACAREA_harmonic.csv
ESMF-820bs08-U_CONS2ND_FRACAREA_gulfstream.csv	SCRIP_CONS2ND_FRACAREA_vortex.csv
ESMF-820bs08-U_CONS2ND_FRACAREA_harmonic.csv	SCRIP_CONSERV_DESTAREA_classic.csv
ESMF-820bs08-U_CONS2ND_FRACAREA_vortex.csv	SCRIP_CONSERV_DESTAREA_gulfstream.csv
ESMF-820bs08-U_CONSERV_DESTAREA_classic.csv	SCRIP_CONSERV_DESTAREA_harmonic.csv
ESMF-820bs08-U_CONSERV_DESTAREA_gulfstream.csv	SCRIP_CONSERV_DESTAREA_vortex.csv
ESMF-820bs08-U_CONSERV_DESTAREA_harmonic.csv	SCRIP_CONSERV_FRACAREA_classic.csv
ESMF-820bs08-U_CONSERV_DESTAREA_vortex.csv	SCRIP_CONSERV_FRACAREA_gulfstream.csv
ESMF-820bs08-U_CONSERV_FRACAREA_classic.csv	SCRIP_CONSERV_FRACAREA_harmonic.csv
ESMF-820bs08-U_CONSERV_FRACAREA_gulfstream.csv	SCRIP_CONSERV_FRACAREA_vortex.csv
ESMF-820bs08-U_CONSERV_FRACAREA_harmonic.csv	SCRIP_DISTWGT_1_classic.csv
ESMF-820bs08-U_CONSERV_FRACAREA_vortex.csv	SCRIP_DISTWGT_1_gulfstream.csv
ESMF-820bs08_BICUBIC_classic.csv	SCRIP_DISTWGT_1_harmonic.csv
ESMF-820bs08_BICUBIC_gulfstream.csv	SCRIP_DISTWGT_1_vortex.csv
ESMF-820bs08_BICUBIC_harmonic.csv	XIOS_CONS2ND_FRACAREA_classic.csv
ESMF-820bs08_BICUBIC_vortex.csv	XIOS_CONS2ND_FRACAREA_gulfstream.csv
ESMF-820bs08_BILINEAR_classic.csv	XIOS_CONS2ND_FRACAREA_harmonic.csv
ESMF-820bs08_BILINEAR_gulfstream.csv	XIOS_CONS2ND_FRACAREA_vortex.csv
ESMF-820bs08_BILINEAR_harmonic.csv	XIOS_CONSERV_DESTAREA_classic.csv
ESMF-820bs08_BILINEAR_vortex.csv	XIOS_CONSERV_DESTAREA_gulfstream.csv
ESMF-820bs08_CONSERV_DESTAREA_classic.csv	XIOS_CONSERV_DESTAREA_harmonic.csv
ESMF-820bs08_CONSERV_DESTAREA_gulfstream.csv	XIOS_CONSERV_DESTAREA_vortex.csv
ESMF-820bs08_CONSERV_DESTAREA_harmonic.csv	XIOS_CONSERV_FRACAREA_classic.csv
ESMF-820bs08_CONSERV_DESTAREA_vortex.csv	XIOS_CONSERV_FRACAREA_gulfstream.csv
ESMF-820bs08_DISTWGT_1_classic.csv	XIOS_CONSERV_FRACAREA_harmonic.csv
ESMF-820bs08_DISTWGT_1_gulfstream.csv	XIOS_CONSERV_FRACAREA_vortex.csv
ESMF-820bs08_DISTWGT_1_harmonic.csv	YAC_BICUBIC_classic.csv
ESMF-820bs08_DISTWGT_1_vortex.csv	YAC_BICUBIC_gulfstream.csv
SCRIP-L_CONS2ND_FRACAREA_classic.csv	YAC_BICUBIC_harmonic.csv
SCRIP-L_CONS2ND_FRACAREA_gulfstream.csv	YAC_BICUBIC_vortex.csv
SCRIP-L_CONS2ND_FRACAREA_harmonic.csv	YAC_BILINEAR_classic.csv
SCRIP-L_CONS2ND_FRACAREA_vortex.csv	YAC_BILINEAR_gulfstream.csv
SCRIP-L_CONSERV_DESTAREA_classic.csv	YAC_BILINEAR_harmonic.csv
SCRIP-L_CONSERV_DESTAREA_gulfstream.csv	YAC_BILINEAR_vortex.csv
SCRIP-L_CONSERV_DESTAREA_harmonic.csv	YAC_CONS2ND_FRACAREA_classic.csv

Table A1. *Cont.*

SCRIP-L_CONSERV_DESTAREA_vortex.csv	YAC_CONS2ND_FRACAREA_gulfstream.csv
SCRIP-L_CONSERV_FRACAREA_classic.csv	YAC_CONS2ND_FRACAREA_harmonic.csv
SCRIP-L_CONSERV_FRACAREA_gulfstream.csv	YAC_CONS2ND_FRACAREA_vortex.csv
SCRIP-L_CONSERV_FRACAREA_harmonic.csv	YAC_CONSERV_DESTAREA_classic.csv
SCRIP-L_CONSERV_FRACAREA_vortex.csv	YAC_CONSERV_DESTAREA_gulfstream.csv
SCRIP_BICUBIC_classic.csv	YAC_CONSERV_DESTAREA_harmonic.csv
SCRIP_BICUBIC_gulfstream.csv	YAC_CONSERV_DESTAREA_vortex.csv
SCRIP_BICUBIC_harmonic.csv	YAC_CONSERV_FRACAREA_classic.csv
SCRIP_BICUBIC_vortex.csv	YAC_CONSERV_FRACAREA_gulfstream.csv
SCRIP_BILINEAR_classic.csv	YAC_CONSERV_FRACAREA_harmonic.csv
SCRIP_BILINEAR_gulfstream.csv	YAC_CONSERV_FRACAREA_vortex.csv
SCRIP_BILINEAR_harmonic.csv	YAC_DISTWGT_1_classic.csv
SCRIP_BILINEAR_vortex.csv	YAC_DISTWGT_1_gulfstream.csv
SCRIP_CONS2ND_FRACAREA_classic.csv	YAC_DISTWGT_1_harmonic.csv
SCRIP_CONS2ND_FRACAREA_gulfstream.csv	YAC_DISTWGT_1_vortex.csv

Appendix C. List of Metric Plots

The plots included in the tar file Regridding_Benchmark_metrics_plots.tar available on Zenodo at https://doi.org/10.5281/zenodo.5347696 (accessed on 31 August 2021) are listed here below in Table A2.

This tar file contains the regridding benchmark metric plots calculated for all pairs of grids (see Section 2.1.1), for all functions (see Section 2.1.2), for all regridding libraries (SCRIP, SCRIP-L, i.e., with Lambert projection for conservative regridding, YAC, ESMF, and XIOS) and for all algorithms (see Section 2.1.3), except when the regridding library does not support the algorithm (e.g., bilinear for XIOS).

There is one plot for each algorithm, for each function, for all metrics, for all pairs of grids, and for all regridding libraries. The name of the file is given as plot_remap_metrics_A_f.pdf, where A is the algorithm (see Appendix B) and f is the function (here *classic* is equivalent to *sinusoid*).

For XIOS, there is no plot for nearest neighbour, first- and second-order non-conservative algorithms as they are not supported by XIOS.

For ESMF, all plots are provided for the version tagged ESMF_8_2_0_beta_snapshot_08 with *nogt* described with the *unstructured* format, as it correctly supports the north fold of the NEMO grid (see Section 3.4).

Table A2. Regridding benchmark plots included in the tar file Regridding_Benchmark_metrics_plots.tar available on Zenodo at https://doi.org/10.5281/zenodo.5347696 (accessed on 31 August 2021).

plot_remap_metrics_BICUBIC_classic.pdf
plot_remap_metrics_BICUBIC_gulfstream.pdf
plot_remap_metrics_BICUBIC_harmonic.pdf
plot_remap_metrics_BICUBIC_vortex.pdf
plot_remap_metrics_BILINEAR_classic.pdf
plot_remap_metrics_BILINEAR_gulfstream.pdf
plot_remap_metrics_BILINEAR_harmonic.pdf
plot_remap_metrics_BILINEAR_vortex.pdf
plot_remap_metrics_CONS2ND_FRACAREA_classic.pdf
plot_remap_metrics_CONS2ND_FRACAREA_gulfstream.pdf
plot_remap_metrics_CONS2ND_FRACAREA_harmonic.pdf
plot_remap_metrics_CONS2ND_FRACAREA_vortex.pdf
plot_remap_metrics_CONSERV_DESTAREA_classic.pdf
plot_remap_metrics_CONSERV_DESTAREA_gulfstream.pdf
plot_remap_metrics_CONSERV_DESTAREA_harmonic.pdf
plot_remap_metrics_CONSERV_DESTAREA_vortex.pdf
plot_remap_metrics_CONSERV_FRACAREA_classic.pdf
plot_remap_metrics_CONSERV_FRACAREA_classic_sansconservationMTR.pdf

Table A2. *Cont.*

plot_remap_metrics_CONSERV_FRACAREA_gulfstream.pdf
plot_remap_metrics_CONSERV_FRACAREA_harmonic.pdf
plot_remap_metrics_CONSERV_FRACAREA_vortex.pdf
plot_remap_metrics_DISTWGT_1_classic.pdf
plot_remap_metrics_DISTWGT_1_gulfstream.pdf
plot_remap_metrics_DISTWGT_1_harmonic.pdf
plot_remap_metrics_DISTWGT_1_vortex.pdf

References

1. Craig, A.; Valcke, S.; Coquart, L. Development and performance of a new version of the OASIS coupler, OASIS3-MCT_3.0. *Geosci. Moaccessed on 18 January 2022del Dev.* **2017**, *10*, 3297–3308. [CrossRef]
2. Jones, P. Conservative remapping: First-and second-order conservative remapping. *Mon. Weather Rev.* **1999**, *127*, 2204–2210. [CrossRef]
3. Valcke, S.; Piacentini, A. Analysis of SCRIP Conservative Remapping in OASIS3-MCT—Part A, Technical Report TR/CMGC/19-129, CERFACS, France. 2019. Available online: https://oasis.cerfacs.fr/wp-content/uploads/sites/114/2021/08/GLOBC_TR_Valcke-SCRIP_CONSERV_TRNORM_partA_2019.pdf (accessed on 18 January 2022).
4. Jonville, G.; Valcke, S. Analysis of SCRIP Conservative Remapping in OASIS3-MCT—Part B, Technical Report TR/CMGC/19-155, CERFACS, France. 2019. Available online: https://oasis.cerfacs.fr/wp-content/uploads/sites/114/2021/08/GLOBC_TR_Jonville-SCRIP_CONSERV_TRNORM_partB_2019.pdf (accessed on 18 January 2022).
5. Valcke, S.; Piacentini, A.; Jonville, G. Benchmarking of Regridding Libraries Used in Earth System Modelling: SCRIP, YAC, ESMF and XIOS, Technical Report TR/CMGC/21-145, CERFACS, France. 2021. Available online: https://oasis.cerfacs.fr/wp-content/uploads/sites/114/2021/11/GLOBC-TR_Valcke_Report_regridding_analysis_final_2021.pdf (accessed on 18 January 2022).
6. Deconinck, W.; Bauer, P.; Diamantakis, M.; Hamrud, M.; Kühnlein, C.; Maciel, P.; Mengaldo, G.; Quintino, T.; Raoult, B.; Smolarkiewicz, P.K.; et al. Atlas: A library for numerical weather prediction and climate modelling. *Comput. Phys. Commun.* **2017**, *220*, 188–204. [CrossRef]
7. Mahadevan, V.S.; Grindeanu, I.; Jacob, R.; Sarich, J. Improving climate model coupling through a complete mesh representation: A case study with E3SM (v1) and MOAB (v5.x). *Geosci. Model Dev.* **2020**, *13*, 2355–2377. [CrossRef]
8. Energy Exascale Earth System Model. Available online: https://e3sm.org (accessed on 18 January 2022).
9. Hanke, M.; Redler, R.; Holfeld, T.; Yastremsky, M. YAC 1.2.0: New aspects for coupling software in Earth system modelling. *Geosci. Model Dev.* **2016**, *9*, 2755–2769. [CrossRef]
10. Hanke, M.; Redler, R. *New Features with YAC 1.5.0*; DWD Report 2019; DWD: Hamburg, Germany, 2019. [CrossRef]
11. Alfeld, P.; Neamtu, M.; Schumaker, L.L. Bernstein-Bezier polynomials on spheres and sphere-like surfaces. *Comput. Aided Geom. Des.* **1996**, *13*, 333–349. [CrossRef]
12. Collins, N.; Theurich, G.; DeLuca, C.; Suarez, M.; Trayanov, A.; Balaji, V.; Li, P.; Yang, W.; Hill, C.; da Silva, A. Design and Implementation of Components in the Earth System Modeling Framework. *Int. J. High Perfor. Comput. Apps.* **2005**, *19*, 341–350. [CrossRef]
13. Earth System Modelling Framework. Available online: https://earthsystemmodeling.org (accessed on 18 January 2022).
14. The XML-IO-Server. Available online: http://forge.ipsl.jussieu.fr/ioserver (accessed on 18 January 2022).
15. Piacentini, A. Evaluation of Atlas 0.21 Interpolation Capability, Technical Report TR/CMGC/20-105, CERFACS, France. 2020. Available online: https://oasis.cerfacs.fr/wp-content/uploads/sites/114/2021/11/GLOBC-TR-Piacentini-atlas-0.21-analysis_2020.pdf (accessed on 18 January 2022).
16. CANGA/MIRA. Available online: https://github.com/CANGA/Remapping-Intercomparison (accessed on 18 January 2022).
17. NEMO Community Ocean Model. Available online: https://www.nemo-ocean.eu (accessed on 18 January 2022).
18. The LMDZ model. Available online: https://forge.ipsl.jussieu.fr/igcmg_doc/wiki/Doc/Models/LMDZ (accessed on 18 January 2022).
19. ARPEGE-Climate. Available online: https://www.umr-cnrm.fr/spip.php?article124 (accessed on 18 January 2022).
20. DYNAMICO. Available online: https://www.lmd.polytechnique.fr/~{}dubos/DYNAMICO/ (accessed on 18 January 2022).
21. Kritsikis, E.; Aechtner, M.; Meurdesoif, Y.; Dubos, T. Conservative interpolation between general spherical meshes. *Geosci. Model Dev.* **2017**, *10*, 425–431. [CrossRef]

Article

Coupled Neural–Glial Dynamics and the Role of Astrocytes in Alzheimer's Disease

Swadesh Pal [1] and Roderick Melnik [1,2,*]

[1] MS2Discovery Interdisciplinary Research Institute, Wilfrid Laurier University, Waterloo, ON N2L 3C5, Canada; spal@wlu.ca
[2] BCAM—Basque Centre for Applied Mathematics, 48009 Bilbao, Spain
* Correspondence: rmelnik@wlu.ca

Abstract: Neurodegenerative diseases such as Alzheimer's (AD) are associated with the propagation and aggregation of toxic proteins. In the case of AD, it was Alzheimer himself who showed the importance of both amyloid beta ($A\beta$) plaques and tau protein neurofibrillary tangles (NFTs) in what he called the "disease of forgetfulness". The amyloid beta forms extracellular aggregates and plaques, whereas tau proteins are intracellular proteins that stabilize axons by cross-linking microtubules that can form largely messy tangles. On the other hand, astrocytes and microglial cells constantly clear these plaques and NFTs from the brain. Astrocytes transport nutrients from the blood to neurons. Activated astrocytes produce monocyte chemoattractant protein-1 (MCP-1), which attracts anti-inflammatory macrophages and clears $A\beta$. At the same time, the microglia cells are poorly phagocytic for $A\beta$ compared to proinflammatory and anti-inflammatory macrophages. In addition to such distinctive neuropathological features of AD as amyloid beta and tau proteins, neuroinflammation has to be brought into the picture as well. Taking advantage of a coupled mathematical modelling framework, we formulate a network model, accounting for the coupling between neurons and astroglia and integrating all three main neuropathological features with the brain connectome data. We provide details on the coupled dynamics involving cytokines, astrocytes, and microglia. Further, we apply the tumour necrosis factor alpha (TNF-α) inhibitor and anti-$A\beta$ drug and analyze their influence on the brain cells, suggesting conditions under which the drug can prevent cell damage. The important role of astrocytes and TNF-α inhibitors in AD pathophysiology is emphasized, along with potentially promising pathways for developing new AD therapies.

Keywords: astrocytes; neural–glial coupled dynamics; Alzheimer's disease; multiple scales; data assimilation; data-driven dynamic environments; biologic TNF-α inhibitors; neuroinflammation; AD drug development; biomarkers

1. Introduction

Alzheimer's disease (AD) is one of the most common late-life dementias, with colossal social and economic impacts. The study by the Institute for Health Metrics and Evaluation published this year in [1] predicts 153 million people will be living with Alzheimer's disease by 2050. While there are various medical products that help manage the symptoms of AD, as of today, there is only one drug officially approved by the FDA that was designed to treat a possible cause of this form of dementia, rather than the symptoms. Yet, the cost and controversy are limiting the use of this drug, known as aducanuman and marketed as aduhelm. Much of this controversy is related to whether or not the build-up of a protein called amyloid β in the brain can be used as a biomarker. According to the AD "amyloid cascade hypothesis", this build-up causes neurodegeneration, but the link between clearance of amyloid β from the brain and deceleration of memory loss and cognitive decline requires further clarification. With over 99% of drugs developed for AD having failed in clinical trials, in addition to more traditional targets related to $A\beta$ and tau proteins, there is an increasing interest in the potential of TNF-α inhibition to prevent AD and improve cognitive function [2].

With underlying interconnections between the processes and factors mentioned above, computational experiments, based on mathematical modelling and computer simulations, can effectively supplement in vivo and in vitro research. In this paper, we present a multiscale model for the onset and evolution of AD that accounts for the diffusion and agglomeration of amyloid beta ($A\beta$) peptide (amyloid cascade hypothesis) and the spreading of the disease through neuron-to-neuron transmission (prionoid hypothesis). Indeed, to cover such diverse facets of AD in a single model, different spatial and temporal scales must be taken into account: microscopic spatial scales to describe the role of the neurons, macroscopic spatial and short temporal (minutes, hours) scales for the description of relevant diffusion processes in the brain, and large temporal scales (years, decades) for the description of the global development of AD. The way in which we combine distinct scales in a single model with brain connectome data assimilation forms the core and major novelty of the paper. Following closely the biomedical literature on AD, we briefly describe the processes that we shall include in our model. In the neurons and their interconnections, several microscopic phenomena take place. We know that $A\beta$ monomers are present in healthy individuals, and therefore, they are unlikely to be toxic. Furthermore, the τ monomer is non-toxic [3]. On the other hand, $A\beta$ oligomers are highly toxic, playing an important role in the process of cerebral damage, as postulated by the already mentioned amyloid cascade hypothesis.

In the analysis of neurodegenerative diseases, AD in particular, it is important to account for the coupling between neuronal and glial dynamics. Furthermore, given the importance of astrocytes (collectively known as astroglia) in amyloid production [4], several coupled models have been recently developed in this direction, describing Alzheimer's $A\beta$ accumulation based on calcium-dependent exosome release from astrocytes [5]. It represents a shift from a more traditional view, considering astrocytes as non-excitable brain cells, to a deeper investigation of reactive functions of astrocytes (e.g., increasing the calcium concentration level in response to neurotransmitters and neuromodulators) and their synaptic communication with neurons and other brain cells via what is sometimes labelled as astrocytic networks. Hence, with the ready availability of the data from the brain connectome, derived from various AD mouse models and obtained with the help of transcriptomics and other technologies [6], it is enlightening to go beyond single astrocyte's consideration and to develop network models allowing such data assimilation (see, e.g., [7–10] and references therein). This idea is pursued further in this paper.

By now, it is well known that the neuronal and astroglial networks of the brain are innately interwoven, with astrocytes carrying out a multitude of functions in various brain processes, including homoeostasis and neurogenesis, with both positive and negative effects reported [11]. In particular, they are critical in defining the normal operation of the nervous system, but they could also actively contribute to the pathogenesis of AD and other neurodegenerative disorders [8]. As observed in experiments on astroglial atrophy at earlier stages of such neurodegenerative diseases such as Alzheimer's, Parkinson's, and various forms of dementia, they play a major role in them, leading to disruptions in synaptic connectivity, disbalance in neurotransmitter homoeostasis, and neuronal death through increased excitotoxicity [12,13]. They maintain their importance in the progression of these diseases at the later stages as well, in particular through their activation and contribution to the neuroinflammatory component of grey matter in pathological neurodegeneration. Given the significance of the contribution of neuroinflammation to Alzheimer's disease (AD) progression [14], our better understanding and ultimately controlling of these coupled neuronal–astroglial networks become increasingly important, opening the door to developing future therapies. For this to happen, increasing attention is being paid to a relationship between the astrocytes' effects in the brain and such fundamental processes as synaptic transmission, cognition, and myelination [11]. At the same time, conclusive experimental studies of the role of astrocytes remain extremely challenging, given that the multiple functionalities of these cells are dependent on numerous (and sometimes contradictory) factors during the disease progression. Researchers have shown that both microglia and astrocytes are very heterogeneous in their functions in the diseased brain [15–17]. Indeed, on the one hand, they can contribute to the clearance of $A\beta$ and limit the growing inflammation in the brain, while, on the other hand, they may neglect their metabolic role and release neurotoxins, contributing in this way to AD neurodegenerative processes. This leads to a situation where mathematical and computational models developed in a data-driven environment may very efficiently complement the progress made in the experimental domain.

While we briefly touch on other aspects, in this paper, our focus is mainly on the role of astrocytes in Alzheimer's disease via their dynamic interactions with agglomerations of $A\beta$ peptides. Not only AD is typified by such agglomerations, along with activated glial cells, but also because $A\beta$ plaques trigger intracellular NFT formation, neuronal cell death, neuroinflammation, and gliosis, whereas reactive astrocytes in AD, surrounding these plaques, may additionally contribute to the overall amyloid burden in the brain by secreting $A\beta$ [4]. Indeed, today, we know that a reactive character of astrocytes in AD is usually expressed by intermediate filament proteins and cellular hypertrophy, as well as that these star-shaped glial cells can regulate synaptic communication and modulate brain network functions [7].

The rest of the paper is organised as follows. We develop a network mathematical model for brain connectome data assimilation in Section 2. With the help of brain connectome data, in Section 3, we provide details on two groups of computational experiments elucidating the role of cytokines and astrocytes in AD and giving further details on AD TNF-α inhibitor drugs, quantifying their influence on the reduction of neuronal damage. All numerical results, reported in this section, were obtained with our new network model. Several possible extensions of this work are discussed in Section 4, with concluding remarks given in Section 5.

2. AD Network Model for Brain Connectome Data Assimilation

In this section, we develop a network model based on the consideration originally presented in [18], where a PDE model on AD was discussed. Before going to the full network model in the brain connectome, we first define the diffusion and chemoattraction terms in a network [19,20]. Suppose the network graph $\mathcal{G}$ has V number of nodes and E number of edges. For $j,k = 1,2,3,\ldots,V$, the elements of the adjacency matrix $\mathbf{W}^l$ corresponding to the graph $\mathcal{G}$ are

$$W^l_{jk} = \frac{n_{jk}}{l^2_{jk}}, \qquad (1)$$

where n_{jk} is the mean fiber number and l^2_{jk} is the mean length squared between the nodes j and k. We define a matrix $\mathbf{L}^l$ with entries

$$L^l_{jk} = (D^l_{jj} - W^l_{jk}), \ j,k = 1,2,3,\ldots,V, \qquad (2)$$

where $D^l_{jj} = \sum_{k=1}^{V} W^l_{jk}$. Therefore, at each node j, we take the contribution of the diffusion term for a dummy variable denoted below as u in the following form

$$(\Delta u)_j = -\sum_{k=1}^{V} L^l_{jk} u_k. \qquad (3)$$

Similarly, at each node j, we find the chemoattraction term as

$$(\nabla \cdot (v \nabla u))_j = \left(\sum_{k=1}^{V} L^c_{jk} v_k\right)\left(\sum_{k=1}^{V} L^c_{jk} u_k\right) - v_j \sum_{k=1}^{V} L^l_{jk} u_k, \qquad (4)$$

where $L^c_{jk} = (D^c_{jj} - W^c_{jk})$ with $W^c_{jk} = n_{jk}/l_{jk}$ and $D^c_{jj} = \sum_{k=1}^{V} W^c_{jk}$.

Now, at the node j in the brain connectome, we are ready to define a network model for Alzheimer's disease incorporating the astrocytes' dynamics [18]. We use j as the node index in each of the upcoming equations. Suppose N_j is the density of the living neurons and N_0 is the reference density of the neurons in brain cells. Inside the neurons, amyloid beta A_β^i is constitutively produced from APP at a rate λ_β^i and degraded at a rate $d_{A_\beta^i}$. In the early stage of disease progression, A_β^i is overproduced by reactive oxygen species (ROS) factor R. Therefore, the dynamics of A_β^i is given by

$$\frac{dA_{\beta j}^i}{dt} = \left(\lambda_\beta^i(1+R) - d_{A_\beta^i} A_{\beta j}^i\right)\frac{N_j}{N_0}. \tag{5}$$

The density of extracellular amyloid beta peptides (A_β^o), depends on different factors, such as neuronal death, microglias, astrocytes, etc. The equation for A_β^o is given by

$$\begin{aligned}\frac{dA_{\beta j}^o}{dt} =& A_{\beta j}^i \left|\frac{dN_j}{dt}\right| + \lambda_N \frac{N_j}{N_0} + \lambda_A \frac{A_j}{A_0} \\ & - \left(d_{A_\beta^o \widehat{M}}(\widehat{M}_{1j} + \theta \widehat{M}_{2j}) + d_{A_\beta^o M}(M_{1j} + \theta M_{2j})\right)\frac{A_{\beta j}^o}{A_{\beta j}^o + \overline{K}_{A_\beta^o}},\end{aligned} \tag{6}$$

where A_0 is the reference astrocyte cell density and $\overline{K}_{A_\beta^o}$ is a Michaelis–Menten coefficient [21]. The first term on the right-hand side of (6) is the contribution due to neuronal death. The second and third terms of (6) are the growths released from amyloid precursor protein (APP) [22] and astrocytes [23], respectively. The last multiplying factor is the clearance of A_β^o by peripheral macrophages $\widehat{M}_1$ and $\widehat{M}_2$ and the activated microglias M_1 and M_2. Here, $0 \le \theta < 1$ as $\widehat{M}_1$ and M_1 are more effective in clearing the extracellular $A\beta$ compared to $\widehat{M}_2$ and M_2. APP on live neurons shed $A\beta$ peptides in both the intracellular and extracellular space [18]. We assumed that most of the A_β^o is produced from dead neurons, so the production from the live neurons is neglected.

The second most critical factor in AD is the tau protein. Suppose that the tau protein is constitutively produced, and the degradation rates are $\lambda_{\tau 0}$ and d_τ, respectively. Due to the abnormal concentrations of $A\beta$, i.e., when the production of A_β^i exceeds a threshold, say $A_{\beta c}^i$, glycogen synthase kinase-type 3 (GSK-3) becomes activated, and it mediates the hyperphosphorylation of tau proteins. Suppose d_τ is the degradation rate of tau proteins due to ROSs. Then, the rate of change of tau protein is given by

$$\frac{d\tau_j}{dt} = \left(\lambda_{\tau 0} + \lambda_\tau R - d_\tau \tau_j\right)\frac{N_j}{N_0}. \tag{7}$$

Inside the neurons, NFTs form from the hyperphosphorylation of tau proteins and are released into the extracellular space after the death of the neurons [24–27]. The equations for NFTs inside the neurons and the extracellular space are given by

$$\frac{dF_{ij}}{dt} = \left(\lambda_F \tau_j - d_{F_i} F_{ij}\right)\frac{N_j}{N_0}, \tag{8}$$

$$\frac{dF_{oj}}{dt} = F_{ij}\left|\frac{dN_j}{dt}\right| - d_{F_o} F_{oj}, \tag{9}$$

respectively.

Due to NFTs' formation in the brain cell, microtubules are depolymerised and destructed, leading to neuron death [24–27]. Not only NFTs, proinflammatory and anti-inflammatory cytokines are also responsible for neuronal death in the brain. Including these factors in the dynamics for N, we obtain

$$\frac{dN_j}{dt} = -d_{NF}\frac{F_{ij}}{F_{ij}+K_{F_i}}N_j - d_{NT}\frac{T_{\alpha j}}{T_{\alpha j}+K_{T_\alpha}}\frac{1}{1+\gamma I_{10j}K_{I_{10}}}N_j, \quad (10)$$

where T_α and I_{10} denote the proinflammatory and anti-inflammatory cytokines, respectively.

Astrocytes are primarily activated by the proinflammatory cytokines T_α [28], but they are also activated by the extracellular amyloid beta A_β^o [23]. Therefore, the equation for astrocytes is given by

$$\frac{dA_j}{dt} = \lambda_{AA_\beta^o}A_{\beta j}^o + \lambda_{AT_\alpha}T_{\alpha j} - d_A A_j. \quad (11)$$

Microglias and peripheral macrophages clear the NFTs in the extracellular space and keep neurons healthy. Therefore, the dynamics of neuronal death is given by

$$\begin{aligned}\frac{dN_{dj}}{dt} =& d_{NF}\frac{F_{ij}}{F_{ij}+K_{F_i}}N_j + d_{NT}\frac{T_{\alpha j}}{T_{\alpha j}+K_{T_\alpha}}\frac{1}{1+\gamma I_{10j}K_{I_{10}}}N_j \\ &- d_{N_dM}(M_{1j}+M_{2j})\frac{N_{dj}}{N_{dj}+\overline{K}_{N_d}} - d_{N_d\widehat{M}}(\widehat{M}_{1j}+\widehat{M}_{2j})\frac{N_{dj}}{N_{dj}+\overline{K}_{N_d}}.\end{aligned} \quad (12)$$

Amyloid beta oligomers are soluble, and they diffuse in the brain tissue [29–31]. Incorporating the diffusion of the oligomers in the network model along with its production (from A_β^o) and degradation, we obtain

$$\frac{dA_{oj}}{dt} = -D_{A_o}\sum_{k=1}^{V}L_{jk}^l A_{ok} + \lambda_{A_o}A_{\beta j}^o - d_{A_o}A_{oj}, \quad (13)$$

where D_{A_o} is the diffusion coefficient.

In the AD-affected brain, dying neurons produce nonhistone chromatin-associated protein (HMGB-1), and it diffuses in the brain cells [32,33]. The reaction–diffusion equation for the PDE-based model of [18] is simplified into the ODE in the network as follows:

$$\frac{dH_j}{dt} = -D_H\sum_{k=1}^{V}L_{jk}^l H_k + \lambda_H N_{dj} - d_H H_j. \quad (14)$$

Microglias travel in the brain cell [34]. Activated microglias are chemoattracted to the cytokines' high mobility group box 1 (HMGB-1). Furthermore, microglias are activated by the extracellular NFTs and soluble oligomers. The M_1 and M_2 phenotypes are characterised by the proinflammatory and anti-inflammatory signals from T_α and I_{10}, respectively. These two types of microglias satisfy the following equations

$$\begin{aligned}\frac{dM_{1j}}{dt} =& M_{1j}\sum_{k=1}^{V}L_{jk}^l H_k - \left(\sum_{k=1}^{V}L_{jk}^c M_{1k}\right)\left(\sum_{k=1}^{V}L_{jk}^c H_k\right) - \lambda_{M_1 T_\beta}\frac{T_{\beta j}}{T_{\beta j}+K_{T_\beta}}M_{1j} \\ &- d_{M_1}M_{1j} + M_G^0\left(\lambda_{MF}\frac{F_{oj}}{F_{oj}+K_{F_o}} + \lambda_{MA}\frac{A_{oj}}{A_{oj}+K_{A_o}}\right)\frac{\beta\epsilon_1}{\beta\epsilon_1+\epsilon_2},\end{aligned} \quad (15)$$

$$\begin{aligned}\frac{dM_{2j}}{dt} =& M_{2j}\sum_{k=1}^{V}L_{jk}^l H_k - \left(\sum_{k=1}^{V}L_{jk}^c M_{2k}\right)\left(\sum_{k=1}^{V}L_{jk}^c H_k\right) + \lambda_{M_1 T_\beta}\frac{T_{\beta j}}{T_{\beta j}+K_{T_\beta}}M_{1j} \\ &- d_{M_2}M_{2j} + M_G^0\left(\lambda_{MF}\frac{F_{oj}}{F_{oj}+K_{F_o}} + \lambda_{MA}\frac{A_{oj}}{A_{oj}+K_{A_o}}\right)\frac{\epsilon_2}{\beta\epsilon_1+\epsilon_2},\end{aligned} \quad (16)$$

where $\epsilon_1 = T_{\alpha j}/(T_{\alpha j} + K_{T_\alpha})$, $\epsilon_2 = I_{10j}/(I_{10j} + K_{I_{10}})$. The parameter β is the ratio of the proinflammatory and anti-inflammatory environment, and it determines the relative strengths of T_α and I_{10}. Here, the ratios $\beta\epsilon_1/(\beta\epsilon_1 + \epsilon_2)$ and $\epsilon_2/(\beta\epsilon_1 + \epsilon_2)$ in the right-hand sides measure the activated microglias becoming M_1 and M_2 macrophages, respectively.

Depending on the relative concentrations of T_α and I_{10}, the incoming macrophages are divided into two phenotypes $\widehat{M}_1$ and $\widehat{M}_2$ [35]. Furthermore, the phenotype of macrophages $\widehat{M}_1$ change to the macrophages $\widehat{M}_2$ under the signal T_β. Therefore, the peripheral macrophages satisfy the following equations:

$$\frac{d\widehat{M}_{1j}}{dt} = \widehat{M}_{1j} \sum_{k=1}^{V} L^l_{jk} A_{ok} - \left(\sum_{k=1}^{V} L^c_{jk} \widehat{M}_{1k}\right)\left(\sum_{k=1}^{V} L^c_{jk} A_{ok}\right) - \lambda_{\widehat{M}_1 T_\beta} \frac{T_{\beta j}}{T_{\beta j} + K_{T_\beta}} \widehat{M}_{1j} \quad (17)$$
$$- d_{\widehat{M}_1} \widehat{M}_{1j} + \alpha(P_j)(M_0 - \widehat{M}_j) \frac{\beta\epsilon_1}{\beta\epsilon_1 + \epsilon_2},$$

$$\frac{d\widehat{M}_{2j}}{dt} = \widehat{M}_{2j} \sum_{k=1}^{V} L^l_{jk} A_{ok} - \left(\sum_{k=1}^{V} L^c_{jk} \widehat{M}_{2k}\right)\left(\sum_{k=1}^{V} L^c_{jk} A_{ok}\right) + \lambda_{\widehat{M}_1 T_\beta} \frac{T_{\beta j}}{T_{\beta j} + K_{T_\beta}} \widehat{M}_{1j} \quad (18)$$
$$- d_{\widehat{M}_2} \widehat{M}_{2j} + \alpha(P_j)(M_0 - \widehat{M}_j) \frac{\epsilon_2}{\beta\epsilon_1 + \epsilon_2},$$

where $\widehat{M}_j = \widehat{M}_{1j} + \widehat{M}_{2j}$ and $\alpha(P_j) = \alpha P_j/(P_j + K_P)$.

T_α is produced by proinflammatory macrophages M_1 and $\widehat{M}_1$. T_β and I_{10} are produced by M_2 and $\widehat{M}_2$. Therefore, the equations for T_α, T_β and I_{10} are in the form

$$\frac{dT_{\alpha j}}{dt} = -D_{T_\alpha} \sum_{k=1}^{V} L^l_{jk} T_{\alpha k} + \lambda_{T_\alpha M_1} M_{1j} + \lambda_{T_\alpha \widehat{M}_1} \widehat{M}_{1j} - d_{T_\alpha} T_{\alpha j}, \quad (19)$$

$$\frac{dT_{\beta j}}{dt} = -D_{T_\beta} \sum_{k=1}^{V} L^l_{jk} T_{\beta k} + \lambda_{T_\beta M} M_{2j} + \lambda_{T_\beta \widehat{M}} \widehat{M}_{2j} - d_{T_\beta} T_{\beta j}, \quad (20)$$

$$\frac{dI_{10j}}{dt} = -D_{I_{10}} \sum_{k=1}^{V} L^l_{jk} I_{10k} + \lambda_{I_{10} M} M_{2j} + \lambda_{I_{10} \widehat{M}} \widehat{M}_{2j} - d_{I_{10}} I_{10j}. \quad (21)$$

Activated astrocytes and microglias produce monocyte chemoattractant protein-1 (MCP-1) [36,37], and it is assumed to be of the M_2 phenotype. Hence,

$$\frac{dP_j}{dt} = -D_P \sum_{k=1}^{V} L^l_{jk} P_k + \lambda_{PA} A_j + \lambda_{PM_2} M_{2j} - d_P P_j. \quad (22)$$

We have used the same estimated parameter values developed by Hao and Friedman [18] for the network model (5)–(22). Further details are provided in Tables 1 and 2 with $\lambda_{AA^o_\beta} = 1.793$ and $\lambda_{AT_\alpha} = 1.54$. We have assumed that these parameter values are uniform for all the nodes in the brain connectome.

Table 1. Parameter values.

Parameter	Value	Parameter	Value	Parameter	Value
$d_{A^i_\beta}$	9.51	$d_{A^i_\beta}$	9.51	K_{M_1}	0.03
d_{F_i}	2.77×10^{-3}	d_{F_o}	2.77×10^{-4}	K_{M_2}	0.017
d_T	0.277	d_N	1.9×10^{-4}	$K_{\widehat{M}_1}$	0.04
d_{NF}	3.4×10^{-4}	d_{NT}	1.7×10^{-4}	$K_{\widehat{M}_2}$	0.007
$d_{N_d M}$	0.06	$d_{N_d \widehat{M}}$	0.02	K_{F_i}	3.36×10^{-10}
d_A	1.2×10^{-3}	d_{A_o}	0.951	K_{F_o}	2.58×10^{-11}

Table 1. Cont.

Parameter	Value	Parameter	Value	Parameter	Value
d_{M_1}	0.015	d_{M_2}	0.015	K_M	0.47
$d_{\widehat{M}_1}$	0.015	$d_{\widehat{M}_2}$	0.015	$K_{I_{10}}$	2.5×10^{-6}
d_H	58.71	$d_{I_{10}}$	16.64	K_{T_β}	2.5×10^{-7}
d_{T_α}	55.45	d_{T_β}	333	K_{T_α}	4×10^{-5}
d_P	1.73	A_0	0.14	$K_{\widehat{M}}$	0.47
R_0	6	M_0	0.05	K_{A_o}	10^{-7}
M_G^0	0.47	N_0	0.14	K_P	6×10^{-9}
$\overline{K}_{A_\beta^o}$	7×10^{-3}	$\overline{K}_{N_d}$	10^{-3}		

Table 2. Parameter values.

Parameter	Value	Parameter	Value	Parameter	Value
D_{A_o}	4.32×10^{-2}	D_H	8.11×10^{-2}	D_P	2×10^{-1}
D_{T_α}	6.55×10^{-2}	D_{T_β}	6.55×10^{-2}	$D_{I_{10}}$	6.04×10^{-2}
λ_β^i	9.51×10^{-6}	λ_N	8×10^{-9}	λ_A	8×10^{-10}
λ_{τ_0}	8.1×10^{-11}	λ_τ	1.35×10^{-11}	λ_F	1.662×10^{-3}
λ_{PA}	6.6×10^{-8}	λ_{PM_2}	1.32×10^{-7}	λ_{A_o}	5×10^{-2}
λ_H	3×10^{-5}	λ_{MF}	2×10^{-2}	λ_{MA}	2.3×10^{-3}
$\lambda_{M_1 T_\beta}$	6×10^{-3}	$\lambda_{\widehat{M}_1 t_\beta}$	6×10^{-4}	$\lambda_{T_\beta M}$	1.5×10^{-2}
$\lambda_{T_\beta \widehat{M}}$	1.5×10^{-2}	$\lambda_{T_\alpha M_1}$	3×10^{-2}	$\lambda_{T_\alpha \widehat{M}_1}$	3×10^{-2}
$\lambda_{I_{10} M_2}$	6.67×10^{-3}	$\lambda_{I_{10} \widehat{M}_2}$	6.67×10^{-3}	θ	0.9
α	5	β	10	γ	1

3. Numerical Results Based on the Network Model

In this section, we report two groups of computational experiments, focusing on (a) the role of cytokines and astrocytes and (b) the importance of TNF-α inhibitors in reducing neuronal damage in AD. We considered a high-resolution brain connectome structure consisting of $V = 1015$ vertices and $E = 70{,}892$ edges; the data source is available for the patients' connectome data at https://braingraph.org (accessed on 20 April 2022) [38]. The network model developed in Section 2 was implemented by using the C programming language and Matlab. We simulated the network model (5)–(22) for each node $j = 1, 2, \ldots, V$ with uniform initial conditions for all the nodes [18,39]: $A_\beta^i = 10^{-6}, A_\beta^o = 10^{-8}, \tau = 1.37 \times 10^{-10}, F_i = 3.36 \times 10^{-10}, F_o = 3.36 \times 10^{-11}, N = 0.14, A = 0.14, M_1 = M_2 = 0.02, \widehat{M}_1 = \widehat{M}_2 = N_d = 0, H = 1.3 \times 10^{-11}, T_\beta = 10^{-6}, T_\alpha = 2 \times 10^{-5}, I_{10} = 10^{-5}, P = 5 \times 10^{-9}$. Table 3 lists all the variables used in the model, and the units of all these variables are given in g/mL. We used the value of R as

$$R = \begin{cases} R_0 t/100 & 0 \leq t \leq 100, \\ R_0 & t > 100. \end{cases}$$

Table 3. The variables of the model and their functions.

Variable	Function	Variable	Function
A_β^i	Amyloid beta inside neurons	A_β^o	Amyloid beta outside neurons
τ	hyperphosphorylated tau protein	F_i	Neurofibrillary tangle inside neurons
F_o	Neurofibrillary tangle outside neurons	N	Live neurons
A	Astrocytes	N_d	Dead neurons
A_o	Amyloid beta oligomer	H	High mobility group box 1
M_1	Proinflammatory microglias	M_2	Anti-inflammatory microglias
$\widehat{M}_1$	Peripheral proinflammatory macrophages	$\widehat{M}_2$	Peripheral anti-inflammatory macrophages
T_α	Tumour necrosis factor alpha	T_β	Transforming growth factor beta
I_{10}	Interleukin 10	P	Monocyte chemoattractant protein-1

3.1. Computational Experiments on the Role of Cytokines and Astrocytes in AD

After integrating the brain connectome data, we computed all the components involved in Equations (1)–(4). We plot the average densities of twelve variables for the network model in Figure 1. These variables have an influence on the model, but we mainly focused on the astrocyte and microglia variables and the other variables directly associated with these two. Extracellular $A\beta$ and the pro-inflammatory cytokines T_α play a crucial role in the growth of astrocytes in the brain (see Equation (11)).

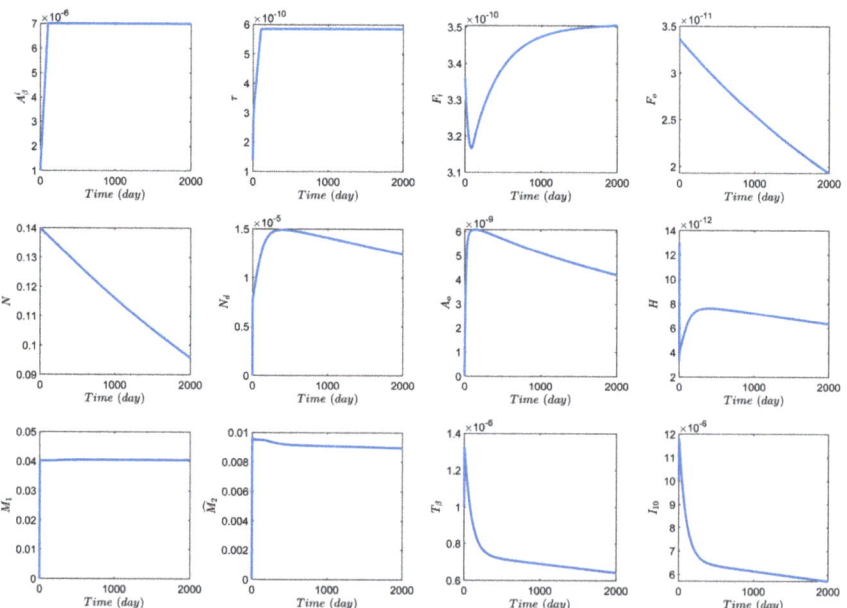

Figure 1. Average concentration of the variables with respect to time.

The density change in the astrocytes or the microglia causes a change in the MCP-1's density, following from Equation (22). Figure 2 shows the density change in astrocytes and MCP-1 by changing the growth parameters in astrocytes, while the other concentrations do not change as such. As additional factors have been taken into account, here we obtained a different result as compared to [18]. As time progresses, the densities of extracellular $A\beta$, $A\beta$-oligomers, astrocytes, and MCP-1 decrease.

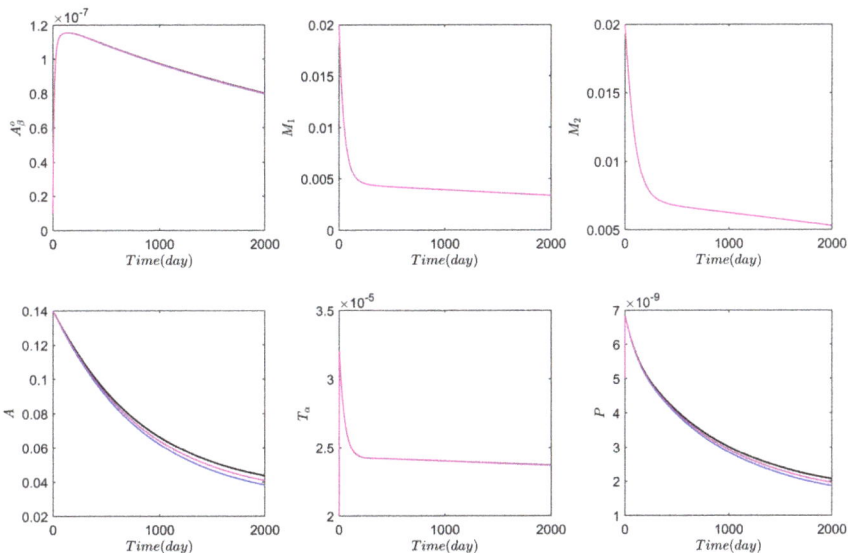

Figure 2. Average solutions of the six variables associated with astrocytes and microglias. We take all the parameter values from Tables 1 and 2. Blue curve: $\lambda_{AA_\beta^o} = 1.793$ and $\lambda_{AT_\alpha} = 1.4$; black curve: $\lambda_{AA_\beta^o} = 1.793$ and $\lambda_{AT_\alpha} = 1.7$; magenta curve: $\lambda_{AA_\beta^o} = 1.65$ and $\lambda_{AT_\alpha} = 1.54$.

3.2. Computational Experiments on AD Drugs

Next, our attention is drawn to soluble inflammatory cytokines. It is well known that soluble cytokine receptors regulate inflammatory and immune events by functioning as antagonists of cytokine signalling. Among various biologic medical products that, by interfering with tumour necrosis factor (TNF), are used to treat autoimmune diseases, caused by an overactive immune response, etanercept (marketed as enbrel) is quite popular. In its essence, it is an amalgamation protein, produced by recombinant DNA, that fuses the TNF receptor to the constant end of the immunoglobulin G1 antibody. Etanercept acts as a TNF-α inhibitor, where TNF-α is considered to be the key regulator of the inflammatory/immune response in many organ systems. We used the amount of etanercept as a parameter in a series of computational experiments that will be discussed here.

Before going to the numerical results involving etanercept, we provide our motivation for focusing on TNF-α and further insight into its leading role in AD pathophysiology. First, we recall that three distinctive neuropathological features of AD are: (a) extracellular deposits of $A\beta$ peptides assembled in plaques, (b) intraneuronal accumulation of hyperphosphorylated tau proteins forming tangles, and (c) chronic inflammation [40]. While (a) and (b) we addressed earlier, here, we note that as far as (c) is concerned, the pro-inflammatory cytokine TNF-α plays a critical role. Moreover, with existing evidence, indicating that TNF-α signalling frequently makes pathologies related to (a) and (b) worse, a growing interest is seen in modulating this signalling and developing anti-TNF-α AD therapies, allowing improved cognitive performance. Compared to other approaches in developing AD treatments [41], TNF-α inhibitors have been consistently rated favourably. While the full range of pathogenetic mechanisms underlying neuronal death and dysfunction in AD remain unclear, most recent analyses convincingly imply that TNF-mediated neuroinflammation is linked to AD neuronal necroptosis [42]. Furthermore, given the pathophysiological importance of the entire TNF-TNFR1/2 system, more and more attention is currently paid to its other components as well, in particular to tumour necrosis factor receptor 2 (TNFR2) of the cytokines, which promotes neuronal survival downstream. While both TNFR1 and TNFR2 can induce pro-inflammatory activities, it is TNFR2 that can also elicit strong anti-inflammatory activities and has protective effects. Recent studies (e.g., [43]) indicate that the TNF pathway can contribute to resilience in AD. The latter concept is important in understanding heterogeneity in cognitive and behavioural phenotypes of AD, which requires involvement not only of $A\beta$ and tau proteins, but other molecular factors as well. This leads, among other things, to the investigation of genetic variants of the TNFR2 pathway as a marker of resilience and the TNFR2 pathway itself as a target for developing new AD therapies [43].

As we already mentioned earlier, debates over AD drugs continue today, with the first drug able to remove amyloid approved only in 2021 (it is also the first new AD drug approved since 2003). The controversy around this new drug, aducanumab, is effectively centred on AD biomarkers and whether the extent of amyloid plaques can be considered as one of them because some scientists believe that they are more like a side-effect of the disease process. In the meantime, this controversy has generated a burst of new research activities and the development of another drug, known as donanemab, which is currently in late-stage clinical trials. Considered to be an important advance in amyloid pathology, it is expected to be able to treat early symptoms of AD. In the meantime, scientists are in agreement that new treatments, drugs, and therapies are urgently needed [44], and mathematical modelling and computational experiments will be playing an increasingly important role in these new developments.

In what follows, we account for the fact that many clinical trials of drugs aimed at preventing or clearing the $A\beta$ and tau protein pathology have failed to demonstrate efficacy and that one of the possible treatments could be based on TNF-α inhibitors (suggested also in [18]). For the treatment, we first ran the model for what corresponds to 300 days in order to ensure that AD has been diagnosed, and then, we applied continuous treatment by the drug from Day 300 until the end of 10 years. In this case, we can replace Equation (19) with

$$\frac{dT_{\alpha j}}{dt} = -D_{T_\alpha} \sum_{k=1}^{V} L_{jk}^l T_{\alpha k} + \lambda_{T_\alpha M_1} M_{1j} + \lambda_{T_\alpha \widehat{M}_1} \widehat{M}_{1j} - d_{T_\alpha} T_{\alpha j} - f T_{\alpha j}, \qquad (23)$$

where f is proportional to the amount of etanercept. We simulated this equation with the full network model for the brain connectome and plotted the result in Figure 3. For this set of computations, we took $f = 10 d_{T_\alpha}$ along with $\lambda_{AA_\beta^o} = 1.793$ and $\lambda_{AT_\alpha} = 1.4$. After applying the drug, we observed reduced neuronal damage in the brain (see the middle-top sub-figure in Figure 3).

The drug aducanumab is considered to be one of the most effective in clearing $A\beta$. In this case, we replaced Equation (6) with

$$\frac{dA_{\beta j}^o}{dt} = A_{\beta j}^i \left|\frac{dN_j}{dt}\right| + \lambda_N \frac{N_j}{N_0} + \lambda_A \frac{A_j}{A_0}$$
$$- \left(d_{A_\beta^o \widehat{M}}(\widehat{M}_{1j} + \theta \widehat{M}_{2j}) + d_{A_\beta^o M}(M_{1j} + \theta M_{2j})(1+g)\right) \frac{A_{\beta j}^o}{A_{\beta j}^o + \overline{K}_{A_\beta^o}}, \qquad (24)$$

where g is proportional to the amount of the dosing level of the drug aducanumab. We simulated the entire network model (5)–(22) with (24) with $g = 10$, $\lambda_{AA_\beta^o} = 1.793$, and $\lambda_{AT_\alpha} = 1.54$. Figure 4 depicts the effect of the drug in $A\beta$ aggregation in the advancement of time. It has a pronounced effect only on the extracellular $A\beta$ concentrations, not on neuronal death. However, the TNF-α inhibitors have a strong effect of reducing neuronal death. These results agree with those obtained in [18].

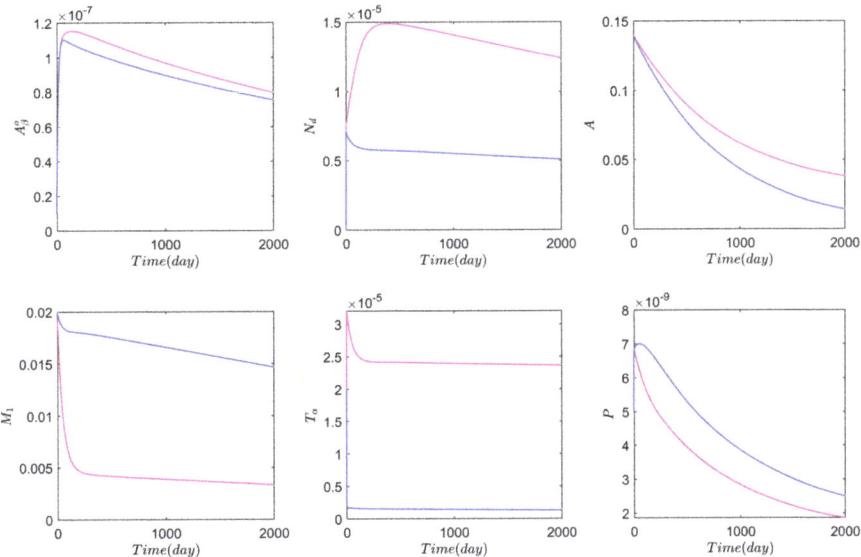

Figure 3. Average solutions of the six variables associated with astrocytes and microglias. We take all the parameter values from Tables 1 and 2 with $\lambda_{AA_\beta^o} = 1.793$ and $\lambda_{AT_\alpha} = 1.4$. Magenta and blue curve curves correspond to the absence and presence of TNF-α inhibitor, respectively.

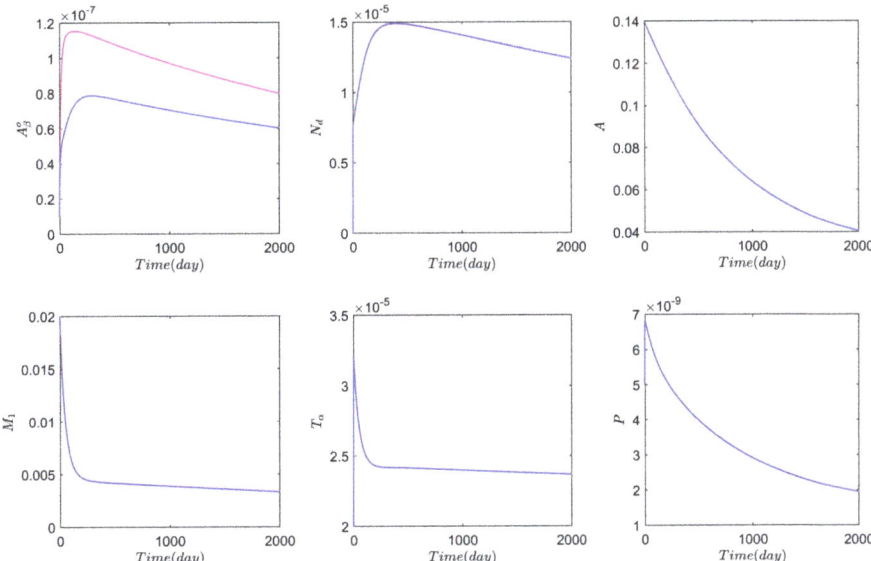

Figure 4. Average solutions of the six variables associated with astrocytes and microglias. We take all the parameter values from Tables 1 and 2 with $\lambda_{AA_\beta^o} = 1.793$ and $\lambda_{AT_\alpha} = 1.54$. Magenta and blue curve curves correspond to the absence and presence of anti-$A\beta$ drug, respectively.

4. Discussion and Future Directions

There are several recently developed models, dealing with astrocytes, that consider averaged characteristics such as the collective exosomal release rate in astrocytes. Such models have an advantage of their generalisations to account for temperature effects, which is an important consideration, given the recent discovery of an intrinsic connection between the temperature dependence of exosome release and $A\beta$ neurotoxicity [5]. Among other things, such models can describe the synapse and astrocyte couplings and allow replicating typical calcium oscillations in astrocytes under the influence of $A\beta$. Therefore, it would be instructive to extend the network model proposed here to account for thermal effects.

Tau proteins play a more prominent role than the amyloid hypothesis suggests. The τPs are usually considered as secondary agents in the disease even though: (i) other τP-related diseases (tauopathies), such as frontotemporal lobar degeneration, are mostly dominated by tau spreading; (ii) brain atrophy in AD is directly correlated with large concentrations of NFT; (iii) the τP distribution determines disease staging, and lowering τP levels prevent neuronal loss; (iv) τP reduces neural activity and is the main factor associated with cognitive decline. This motivates another possible extension of the developed model by combining the tau protein dynamics more precisely [19,45], rather than considering only a linear contribution of the tau protein. In addition, the dynamics of the variables in different regions in the brain connectome would give a better understanding of the disease progression [45]. Following this recent study, we note that although we chose here uniform parameter values all over the brain connectome, this can be extended to different parameter values in respective regions according to the clinical data.

Regarding the route connected with TNF inhibitors, in addition to the possible extensions already mentioned in the previous section, we will also mention that one of the existing difficulties lies with the fact that classical biologic TNF-α inhibitor macromolecules cannot cross the blood–brain barrier [46,47]. This requires the development of blood–brain-barrier-penetrating TNF-α inhibitors, and from a modelling point of view, further extensions of the models developed here may be needed to qualitatively estimate this factor.

5. Conclusions

We constructed a network model to study neurodegenerative disorders in the brain connectome, focusing on Alzheimer's disease. The developed model can capture the concentrations of the variables in different regions of the brain connectome, which could not be identified by earlier-developed simple PDE-based models. All three distinctive neuropathological features of AD, including amyloid beta and tau proteins, as well as neuroinflammation were considered in the network model for brain connectome data assimilation. Special attention was given to the role of cytokines and astrocytes, as well as to the influence of anti-$A\beta$ and TNF-α inhibitor drugs in AD pathophysiology. We showed that etanercept has good efficacy in most of the aspects, including neuronal death, while aducanumab has a good efficacy only in reducing the aggregation of extracellular amyloid beta. Among other applications, one may choose the developed methodology to address the diffusion and chemoattraction challenges by evaluating the corresponding term's contributions in the network model. Finally, potentially promising pathways for developing new AD therapies were also discussed.

Author Contributions: Conceptualisation, S.P. and R.M.; initial formal analysis and visualisation, S.P.; writing, S.P. and R.M.; initial draft preparation, S.P.; supervision of the study and review, R.M. All authors have read and agreed to the published version of the manuscript.

Funding: This research was funded by the Natural Sciences and Engineering Research Council (NSERC) of Canada and Canada Research Chairs (CRC) Program.

Acknowledgments: The authors are grateful to the Natural Sciences and Engineering Research Council (NSERC) of Canada and Canada Research Chairs (CRC) Program, and R.M. also acknowledges the support of the BERC 2018-2021 program, the Spanish Ministry of Science, Innovation, and

Universities through the Agencia Estatal de Investigacion (AEI) BCAM Severo Ochoa excellence accreditation SEV-2017-0718, and the Basque Government fund AI in BCAM EXP. 2019/00432.

Conflicts of Interest: The authors declare no conflict of interest.

References

1. Nichols, E.; Steinmetz, J.D.; Vollset, S.E.; Fukutaki, K.; Chalek, J.; Abd-Allah, F.; Abdoli, A.; Abualhasan, A.; Abu-Gharbieh, E.; Akram, T.T.; et al. Estimation of the global prevalence of dementia in 2019 and forecasted prevalence in 2050: An analysis for the Global Burden of Disease Study 2019. *Lancet* **2022**, *7*, e105–e125. [CrossRef]
2. Torres-Acosta, N.; O'Keefe, J.H.; O'Keefe, E.L.; Isaacson, R.; Small, G. Therapeutic potential of TNF-α inhibition for Alzheimer's disease prevention. *J. Alzheimer's Dis.* **2020**, *78*, 619–626. [CrossRef] [PubMed]
3. Mroczko, B.; Groblewska, M.; Litman-Zawadzka, A. The Role of Protein Misfolding and Tau Oligomers (TauOs) in Alzheimer's Disease (AD). *Int. J. Mol. Sci.* **2019**, *20*, 4661. [CrossRef]
4. Frost, G.R.; Li, Y.M. The role of astrocytes in amyloid production and Alzheimer's disease. *Open Biol.* **2017**, *7*, 170228. [CrossRef] [PubMed]
5. Shaheen, H.; Singh, S.; Melnik, R. A neuron-glial model of exosomal release in the onset and progression of Alzheimer's disease. *Front. Comput. Neurosci.* **2021**, *15*, 653097. [CrossRef]
6. Lowe, R.; Shirley, N.; Bleackley, M.; Dolan, S.; Shafee, T. Transcriptomics technologies. *PLoS Comput. Biol.* **2017**, *13*, e1005457. [CrossRef] [PubMed]
7. Smit, T.; Deshayes, N.A.; Borchelt, D.R.; Kamphuis, W.; Middeldorp, J.; Hol, E.M. Reactive astrocytes as treatment targets in Alzheimer's disease—Systematic review of studies using the APPswePS1dE9 mouse model. *Glia* **2021**, *69*, 1852–1881. [CrossRef]
8. Preman, P.; Alfonso-Triguero, M.; Alberdi, E.; Verkhratsky, A.; Arranz, A.M. Astrocytes in Alzheimer's disease: Pathological significance and molecular pathways. *Cells* **2021**, *10*, 540. [CrossRef]
9. Raj, A.; Tora, V.; Gao, X.; Cho, H.; Choi, J.Y.; Ryu, Y.H.; Lyoo, C.H.; Franchi, B. Combined Model of Aggregation and Network Diffusion Recapitulates Alzheimer's Regional Tau-Positron Emission Tomography. *Brain Connect.* **2021**, *11*, 624–638. [CrossRef]
10. Bertsch, M.; Franchi, B.; Raj, A.; Tesi, M.C. Macroscopic modelling of Alzheimer's disease: Difficulties and challenges. *Brain Multiphysics* **2021**, *2*, 100040. [CrossRef]
11. Planas-Fontanez, T.M.; Sainato, D.M.; Sharma, I.; Dreyfus, C.F. Roles of astrocytes in response to aging, Alzheimer's disease and multiple sclerosis. *Brain Res.* **2021**, *1764*, 147464. [CrossRef]
12. Verkhratsky, A.; Olabarria, M.; Noristani, H.N.; Yeh, C.Y.; Rodriguez, J.J. Astrocytes in Alzheimer's disease. *Neurotherapeutics* **2010**, *7*, 399–412. [CrossRef] [PubMed]
13. Liddelow, S.A.; Guttenplan, K.A.; Clarke, L.E.; Bennett, F.C.; Bohlen, C.J.; Schirmer, L.; Bennett, M.L.; Münch, A.E.; Chung, W.S.; Peterson, T.C.; et al. Neurotoxic reactive astrocytes are induced by activated microglia. *Nature* **2017**, *541*, 481–487. [CrossRef] [PubMed]
14. Birch, A.M. The contribution of astrocytes to Alzheimer's disease. *Biochem. Soc. Trans.* **2014**, *42*, 1316–1320. [CrossRef] [PubMed]
15. Deczkowska, A.; Keren-Shaul, H.; Weiner, A.; Colonna, M.; Schwartz, M.; Amit, I. Disease-Associated Microglia: A Universal Immune Sensor of Neurodegeneration. *Cell* **2018**, *173*, 1073–1081. [CrossRef]
16. Habib, N.; Centini, G.; Lazzeri, L.; Amoruso, N.; El Khoury, L.; Zupi, E.; Afors, K. Bowel Endometriosis: Current Perspectives on Diagnosis and Treatment. *Int. J. Women's Health* **2020**, *12*, 35–47. [CrossRef] [PubMed]
17. Liddelow, S.A.; Barres, B.A. Reactive Astrocytes: Production, Function, and Therapeutic Potential. *Immunity* **2017**, *46*, 957–967. [CrossRef]
18. Hao, W.; Friedman, A. Mathematical model on Alzheimer's disease. *BMC Syst. Biol.* **2016**, *10*, 108. [CrossRef]
19. Thompson, T.B.; Chaggar, P.; Kuhl, E.; Goriely, A. Protein-protein interactions in neurodegenerative diseases: A conspiracy theory. *PLoS Comput. Biol.* **2020**, *16*, e1008267. [CrossRef]
20. Pal, S.; Melnik, R. Pathology dynamics in healthy-toxic protein interaction and the multiscale analysis of neurodegenerative diseases. In *Computational Science—ICCS 2021*; Paszynski, M., Kranzlmüller, D., Krzhizhanovskaya, V.V., Dongarra, J.J., Sloot, P.M., Eds.; Lecture Notes in Computer Science; Springer: Cham, Switzerland, 2021; Volume 12746, pp. 528–540.
21. Roskoski, R. Michaelis-Menten Kinetics. In *Reference Module in Biomedical Sciences*; Elsevier: Amsterdam, The Netherlands, 2015.
22. Seeman, P.; Seeman, N. Alzheimer's disease: Beta-amyloid plaque formation in human brain. *Synapse* **2011**, *65*, 1289–1297. [CrossRef]
23. Zhao, J.; O'Connor, T.; Vassar, R. The contribution of activated astrocytes to Aβ production: Implications for Alzheimer's disease pathogenesis. *J. Neuroinflamm.* **2011**, *8*, 150. [CrossRef]
24. Liu, Z.; Li, P.; Wu, J.; Wang, Y.; Li, P.; Hou, X.; Zhang, Q.; Wei, N.; Zhao, Z.; Liang, H.; et al. The Cascade of Oxidative Stress and Tau Protein Autophagic Dysfunction in Alzheimer's Disease. *Alzheimer's Dis. Challenges Future* **2015**, *2*, 48347.
25. Bloom, G.S. Amyloid-beta and tau: The trigger and bullet in Alzheimer disease pathogenesis. *JAMA Neurol.* **2014**, *71*, 505–508. [CrossRef] [PubMed]
26. Mondragon-Rodriguez, S.; Perry, G.; Zhu, X.; Boehm, J. Amyloid Beta and tau proteins as therapeutic targets for Alzheimer's disease treatment: Rethinking the current strategy. *Int. J. Alzheimers Dis.* **2012**, *2012*, 630182. [CrossRef] [PubMed]

27. Wray, S.; Noble, W. Linking amyloid and tau pathology in Alzheimer's disease: The role of membrane cholesterol in Abeta-mediated tau toxicity. *J. Neurosci.* **2009**, *29*, 9665–9667. [CrossRef]
28. Garwood, C.J.; Pooler, A.M.; Atherton, J.; Hanger, D.P.; Noble, W. Astrocytes are important mediators of Abeta-induced neurotoxicity and tau phosphorylation in primary culture. *Cell Death Dis.* **2011**, *2*, 167. [CrossRef]
29. Kevrekidis, P.G.; Thompson, T.B.; Goriely, A. Anisotropic diffusion and traveling waves of toxic proteins in neurodegenerative diseases. *Phys. Lett. A* **2020**, *384*, 12935 [CrossRef]
30. Bertsch, M.; Franchi, B.; Meschini, V.; Tesi, M.C.; Tosin, A. A sensitivity analysis of a mathematical model for the synergistic interplay of amyloid beta and tau on the dynamics of Alzheimer's disease. *Brain Multiphys.* **2021**, *2*, 1–13. [CrossRef]
31. Waters, J. The concentration of soluble extracellular amyloid beta protein in acute brain slices from CRND8 mice. *PLoS ONE* **2010**, *5*, e15709. [CrossRef]
32. Gao, H.M.; Zhou, H.; Zhang, F.; Wilson, B.C.; Kam, W.; Hong, J.S. HMGB1 acts on microglia Mac1 to mediate chronic neuroinflammation that drives progressive neurodegeneration. *J. Neurosci.* **2011**, *31*, 1081–1092. [CrossRef]
33. Zou, J.Y.; Crews, F.T. Release of neuronal HMGB1 by ethanol through decreased HDAC activity activates brain neuroimmune signalling. *PLoS ONE* **2014**, *9*, e87915.
34. Savchenko, V.L.; McKanna, J.A.; Nikonenko, I.R.; Skibo, G.G. Microglia and astrocytes in the adult rat brain: Comparative immunocytochemical analysis demonstrates the efficacy of lipocortin 1 immunoreactivity. *Neuroscience* **2000**, *96*, 195–203. [CrossRef]
35. Hao, W.; Crouser, E.D.; Friedman, A. Mathematical model of sarcoidosis. *Proc. Natl. Acad. Sci. USA* **2014**, *111*, 16065–16070. [CrossRef] [PubMed]
36. Hohsfield, L.A.; Humpel, C. Migration of blood cells to beta-amyloid plaques in Alzheimer's disease. *Exp. Gerontol.* **2015**, *65*, 8–15. [CrossRef] [PubMed]
37. Theriault, P.; ElAli, A.; Rivest, S. The dynamics of monocytes and microglia in Alzheimer's disease. *Alzheimers Res. Ther.* **2015**, *7*, 41. [CrossRef] [PubMed]
38. Szalkai, B.; Kerepesi, C.; Varga, B.; Grolmusz, V. High-Resolution Directed Human Connectomes and the Consensus Connectome Dynamics. *PLoS ONE* **2019**, *14*, e0215473. [CrossRef]
39. Hao, W.; Friedman, A. The LDL-HDL profile determines the risk of atherosclerosis: A mathematical model. *PLoS ONE* **2014**, *9*, e90497. [CrossRef]
40. Decourt, B.; Lahiri, D.K.; Sabbagh, M.N. Targeting tumour necrosis factor alpha for Alzheimer's disease. *Curr. Alzheimer. Res.* **2017**, *14*, 412–425. [CrossRef]
41. Kern, D.M.; Lovestone, S.; Cepeda, M.S. Treatment with TNF-α inhibitors versus methotrexate and the association with dementia and Alzheimer's disease. *Alzheimer's Dement.* **2021**, *7*, e12163. [CrossRef]
42. Jayaraman, A.; Htike, T.T.; James, R.; Picon, C.; Reynolds, R. TNF-mediated neuroinflammation is linked to neuronal necroptosis in Alzheimer's disease hippocampus. *Acta Neuropathol. Commun.* **2021**, *9*, 159. [CrossRef]
43. Pillai, J.A.; Bebek, G.; Khrestian, M.; Bena, J.; Bergmann, C.C.; Bush, W.S.; Leverenz, J.B.; Bekris, L.M. TNFRSF1B gene variants and related soluble TNFR2 levels impact resilience in Alzheimer's disease. *Front. Aging Neurosci.* **2021**, *13*, 638922. [CrossRef]
44. Cummings, J.; Lee, G.; Zhong, K.; Fonseca, J.; Taghva, K. Alzheimer's disease drug development pipeline: 2021. *Alzheimer's Dement.* **2021**, *7*, e12179. [CrossRef] [PubMed]
45. Pal, S.; Melnik, R. Nonlocal models in the analysis of brain neurodegenerative protein dynamics with application to Alzheimer's disease. *arXiv* **2021**, arXiv:2112.11681.
46. Chang, R.; Yee, K.L.; Sumbria, R.K. Tumour necrosis factor α inhibition for Alzheimer's disease. *J. Cent. Nerv. Syst. Dis.* **2017**, *9*, 1–5. [CrossRef] [PubMed]
47. Chang, R.; Knox, J.; Chang, J.; Derbedrossian, A.; Vasilevko, V.; Cribbs, D.; Boado, R.J.; Pardridge, W.M.; Sumbria, R.K. Blood–Brain Barrier Penetrating Biologic TNF-α Inhibitor for Alzheimer's Disease. *Mol. Pharm.* **2017**, *14*, 2340–2349. [CrossRef] [PubMed]

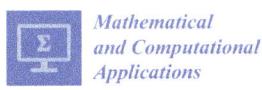

Article

Reduced Order Modeling Using Advection-Aware Autoencoders

Sourav Dutta [1,*], Peter Rivera-Casillas [2], Brent Styles [1,3] and Matthew W. Farthing [1]

1. Coastal and Hydraulics Laboratory, U.S. Army Engineer Research and Development Center, 3909 Halls Ferry Rd., Vicksburg, MS 39180, USA; brent.j.styles@usace.army.mil (B.S.); matthew.w.farthing@erdc.dren.mil (M.W.F.)
2. Information Technology Laboratory, U.S. Army Engineer Research and Development Center, 3909 Halls Ferry Rd., Vicksburg, MS 39180, USA; peter.g.rivera-casillas@erdc.dren.mil
3. Departmentof Electrical & Computer Engineering, University of Alabama in Huntsville, 301 Sparkman Drive, Huntsville, AL 35899, USA
* Correspondence: sourav.dutta@erdc.dren.mil

Abstract: Physical systems governed by advection-dominated partial differential equations (PDEs) are found in applications ranging from engineering design to weather forecasting. They are known to pose severe challenges to both projection-based and non-intrusive reduced order modeling, especially when linear subspace approximations are used. In this work, we develop an advection-aware (AA) autoencoder network that can address some of these limitations by learning efficient, physics-informed, nonlinear embeddings of the high-fidelity system snapshots. A fully non-intrusive reduced order model is developed by mapping the high-fidelity snapshots to a latent space defined by an AA autoencoder, followed by learning the latent space dynamics using a long-short-term memory (LSTM) network. This framework is also extended to parametric problems by explicitly incorporating parameter information into both the high-fidelity snapshots and the encoded latent space. Numerical results obtained with parametric linear and nonlinear advection problems indicate that the proposed framework can reproduce the dominant flow features even for unseen parameter values.

Keywords: deep autoencoder; advection-dominated flows; physics informed machine learning; LSTM; parametric model order reduction; non-intrusive reduced order modeling

1. Introduction

Modern scientific computing relies on efficient numerical simulation of complex physical systems, especially for applications that seek solutions at different time or parameter instances. For these types of applications, the relevant physical system is typically described by a set of parameterized nonlinear partial differential equations (PDEs). Numerical discretizations of such systems using a high-fidelity (finite element, finite volume, or finite difference type) computational solver can be prohibitively expensive as they generate high-dimensional representations of the solution in order to accurately resolve multiple time and space scales and underlying nonlinearities [1]. However, there is compelling scientific evidence to suggest that the underlying dynamics often exhibit low-dimensional structure [2]. Reduced order models (ROMs) can replace such expensive high-fidelity solvers by exploiting the intrinsic, low-rank structure of the simulation data in order to create more tractable models for the spatiotemporal evolution dynamics of the PDE system [3,4].

Among the many different classes of ROM techniques that have been developed over the years, projection-based ROMs occupy a prominent place across a wide range of applications [5]. Formally, this class of methods is based on the identification of a reduced set of basis functions (or modes) such that their linear superposition spans an optimal low-rank approximation of the solution manifold. The Proper Orthogonal Decomposition (POD) is one of the widely popular methods of this class that leverages the singular value decomposition (SVD) to determine an empirical basis of dominant, orthonormal modes that can help define the best possible linear subspace in which to project the PDE dynamics [6,7]. If the governing equations are known, Galerkin projection [8,9], or the Petrov–Galerkin projection [10,11], can be adopted to generate an interpretable ROM defined by the high-energy or dominant modes. For applications where the governing equations are not accessible, purely data-driven methods for non-intrusive ROM (NIROM) [12] have gained in popularity. In these methods, instead of a Galerkin-type projection, the expansion coefficients for the reduced solution are obtained via interpolation on the reduced basis space spanned by the set of dominant modes. However, since the reduced dynamics generally belong to nonlinear, matrix manifolds, a variety of interpolation and regression methods have been proposed, which are capable of enforcing the constraints characterizing those manifolds. Some notable examples in this class include dynamic mode decomposition (DMD) [13,14], radial basis function interpolation [15,16], and Gaussian process regression [17,18], to name a few. In addition, the emergence of modern machine learning (ML) methods has provided a transformative approach to effectively approximate and accelerate existing numerical models by leveraging the capabilities to incorporate multi-fidelity datastreams from diverse sources, seamlessly explore massive design spaces, and identify complex, multivariate correlations. A variety of data-driven, ML-based approximation frameworks have been proposed to model the propagation of system dynamics in the latent space. Some of the highly successful examples involve the use of deep neural networks (DNNs) [19], long-short-term memory (LSTM) networks [20,21], neural ordinary differential equations (NODE) [22,23], and temporal convolutional networks (TCNs) [24].

One fundamental assumption of linear reduced basis methods like POD is that any element in the solution manifold, $\mathcal{M}$ of the nonlinear PDE system can be accurately approximated using a linear combination of a small number of basis functions. Traditionally this concept is quantified by the Kolmogorov n-width, $\mathcal{D}_n$, which measures the error introduced by approximating any element f of $\mathcal{M}$ with an element g of a linear space E_n. A common heuristic approach to get a rough estimate of $\mathcal{D}_n$ for a particular discretized solution manifold is to examine the rate of decay of the singular values obtained by a SVD of the system snapshots. A fast rate of decay signifies a small $\mathcal{D}_n$, which indicates the existence of a low-dimensional space in which the high-fidelity nonlinear system can be approximated well. Many PDE systems of importance, however, exhibit transport-dominated behavior (e.g., advection-dominated flows, wave and shock propagation phenomena), which leads to a large Kolmogorov n-width. For instance, a stationary soliton wave solution can be perfectly captured by one spatial mode, as reflected in a rapid decay of the corresponding POD singular values, whereas a wave translating in time cannot be represented by a low-dimensional representation with POD/SVD. This is because the steep gradients and moving spatial discontinuities inherent to these problems often trigger temporal discontinuities. An accurate linear approximation of such temporal discontinuities requires a large number of basis functions, hindering the efficiency of a ROM [25,26], and often leads to an oscillatory approximation [27,28]. These inadequacies have inspired a growing number of works in recent years, which focus on constructing an efficient alternative—nonlinear ROMs for transport-dominated systems. A brief review will be provided in the following section. The focus of this work will be the study of ML-based, non-intrusive reduced order modeling strategies for transport-dominated systems.

2. Related Work

As the traditional approach for constructing a ROM for transport-dominated problems has proven to be ineffective due to the limitations of a linear subspace approximation, there has been significant interest in alternative modifications for improving the accuracy of ROM approximations in these applications. These can be roughly classified into two distinct approaches: transformations of the linear subspace to facilitate better mode extraction and improved stability of projection-based ROMs [28–34], and the construction of low-rank representations in terms of nonlinear manifolds [35–37]. Most of the previous work in the first class of methods has been focused on either (a) sparse sampling of nonlinear terms to enable efficient approximation in a reduced subspace like gappy POD [38], GNAT [27], DEIM [39], or (b) pre-processing the linear subspace to embed the dominant advective features of the solution [28–31,40]. A collection of methods in this class has also been based around the concept of adaptivity. For instance, offline adaptive methods either extend [41] or create a weighted [42] snapshot database during the construction of the reduced model. Online adaptive methods, on the other hand, either rely on precomputed quantities to update the reduced basis online using interpolation, localization, and dictionary approaches [43–45], or allow for the incorporation of new data online [46,47]. Almost all of the above techniques from the first class of methods require some kind of problem-specific prior knowledge of the physical or numerical properties of the underlying nonlinear system, thus imposing some limitations on their applicability to experimental data or systems with no access to governing equations and closed-form solutions.

An alternative perspective on the limitations of linear subspace approximation in ROM design is based upon the observation that many PDE systems of importance, especially in fluids, contain symmetries such as rotations, translations, and scaling, which play a foundational role in the dynamics. Traditional ROM approaches such as the SVD-based methods are unable to handle these symmetries, and are only truly effective for dynamical systems where time and space interactions can be essentially decoupled through separation of variables [7]. To overcome these limitations, the second class of methods based on nonlinear manifold learning have recently gained a lot of research interest. Some of the earliest examples of methods in this class include Iso-map [48], Locally linear embedding (LLE) [49], Laplacian eigenmaps [50], and t-SNE [51]. However, these methods fail to provide a mapping from the low-dimensional nonlinear representation to the high-dimensional input, which is a crucial tool for dimension reduction applications. Many other novel approaches have been proposed to overcome this gap, such as self-organizing maps [52], kernel PCA [53], diffeormorphic dimensionality reduction [54], and autoencoders [55] (see [36] for a survey). In recent years, due to the tremendous progress in the development of high-performance software tools for the construction of neural networks based models, different types of autoencoder models [56,57] have emerged, as some of the most popular and powerful techniques for nonlinear manifold-based dimension reduction of PDEs. These have been successfully applied to different types of ROM applications such as deep fully-connected autoencoders [58,59], deep convolutional autoencoders (CAEs) [36,60], time-lagged autoencoders [61], shallow masked autoencoders [37], variational autoencoders [62], and deep delay autoencoders [63].

In this work, we propose an advection-aware (AA) autoencoder design, in which a high-fidelity system snapshot is mapped through a shared latent space to an approximation of itself and simultaneously to another arbitrary snapshot. For advection-dominated problems, this arbitrary snapshot can be chosen in a physics-guided manner to primarily represent the advective features, thus allowing the latent space to more efficiently identify reduced-representations of high-fidelity solution fields. We then employ LSTM neural networks to non-intrusively model the temporal evolution of these compressed latent representations. Moreover, our approach enables exploration of parametric search spaces by training on a combined parametric dataset of offline simulations. In contrast, the studies outlined in Refs. [36,60] use a convolutional autoencoder architecture to nonlinearly embed high-dimensional states, and this may pose problems when the high-fidelity simulation data is available on unstructured computational meshes. In addition, Ref. [36] adopts an intrusive approach by solving the governing equations on the nonlinear manifold defined by the CAE model, whereas Ref. [60] employs a similar idea of modeling latent dynamics using recurrent neural networks like an LSTM. Another interesting approach is proposed by Ref. [64], where the authors introduce the idea of imposing Lyapunov stability-preserving priors to the autoencoder-based model in order to improve the generalization performance for fluid flow prediction. While this approach is similar to ours in being motivated by the idea of physics-informed learning, the ultimate objective of the proposed design is different. Our approach also differs from the framework proposed in [65] as the system parameters such as shape of the profile, flow speed, and viscosity are explicitly embedded in the input feature space and the latent space, thus allowing independent training of the AA autoencoder and the LSTM networks. In Ref. [66], a registration-based approach is proposed, which trains a diffeomorphic mapping between the physical space and a new parameter-varying, spatio-temporal grid on which the solution of the PDE can be expressed in the form of a low-rank linear decomposition. This low-rank time/parameter-varying grid or manifold is utilized as an autoencoder type layer for reducing the dimension of high-fidelity snapshots. This is an elegant approach, but involves solving optimization problems with nonlinear constraints and performing repeated 2D interpolation tasks, both of which may potentially lead to efficiency issues and introduce approximation errors for large-scale problems.

The rest of the article is organized as follows. In Section 3, we provide a high-level overview of undercomplete autoencoders, followed by details on the proposed AA autoencoder network design and training strategies. We also include a brief review of LSTM networks, which have been adopted in this work to model the system dynamics in the nonlinear latent space. In Section 4, we present numerical results obtained with the proposed AA autoencoder model on two different types of parametric problems characterized by advection-dominated flow features. Finally, in Section 5, we present some concluding remarks and discuss plans for future work.

3. Methodology

3.1. Autoencoders

An autoencoder is a type of neural network that is designed to learn an approximation of the identity mapping, $\chi : \mathbf{v} \mapsto \tilde{\mathbf{v}}$ such that $\tilde{\mathbf{v}} \approx \mathbf{v}$ and $\chi : \mathbb{R}^N \mapsto \mathbb{R}^N$. This is accomplished using a two-part architecture. Figure 1 shows an example of a fully connected autoencoder network with two distinct parts. The first part is called an encoder, χ_e, which maps a high-dimensional input vector $\mathbf{v}$ to a low-dimensional latent vector $\mathbf{z}$ as given by $\mathbf{z} = \chi_e(\mathbf{v}; \theta_e)$ where $\mathbf{z} \in \mathbb{R}^m$ ($m \ll N$).

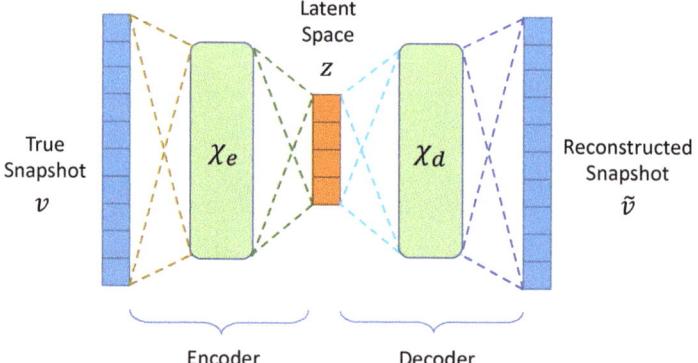

Figure 1. Fully connected autoencoder architecture.

The second part is called a decoder, χ_d, which maps the latent vector $\mathbf{z}$ to an approximation $\tilde{\mathbf{v}}$ of the high-dimensional input vector $\mathbf{v}$ and is defined as $\tilde{\mathbf{v}} = \chi_d(\mathbf{z}; \theta_d)$. The combination of these two parts yields an autoencoder of the form

$$\chi : \mathbf{v} \mapsto \chi_d \circ \chi_e(\mathbf{v}). \qquad (1)$$

This autoencoder network is trained by computing the optimal values of the parameters (θ_e^*, θ_d^*) that minimize the reconstruction error over all the training data

$$\theta_e^*, \theta_d^* = \underset{\theta_e, \theta_d}{\operatorname{argmin}} \mathcal{L}(\mathbf{v}, \widetilde{\mathbf{v}}), \tag{2}$$

where $\mathcal{L}(\mathbf{v}, \widetilde{\mathbf{v}})$ is a chosen measure of discrepancy between $\mathbf{v}$ and its approximation $\widetilde{\mathbf{v}}$. The restriction $\dim(\mathbf{z}) = m \ll N = \dim(\mathbf{v})$ forces the autoencoder model to learn the salient features of the input data via compression into a low-dimensional space and then reconstructing the input, instead of directly learning the identity function. Essentially, autoencoders can be thought of as a powerful generalization of the POD/SVD approach from learning a linear subspace to identifying an improved coordinate system on a nonlinear manifold. That is, with the choice of a linear, single-layer encoder of the form $\mathbf{z} = \mathbf{H}_E \mathbf{v}$, and a linear, single-layer decoder of the form $\widetilde{\mathbf{v}} = \mathbf{H}_D \mathbf{z}$, where $\mathbf{H}_E \in \mathbb{R}^{m \times N}$, $\mathbf{H}_D \in \mathbb{R}^{N \times m}$, and a squared reconstruction error as the loss function $\mathcal{L}(\widetilde{\mathbf{v}}, \mathbf{v}) = \|\mathbf{v} - \widetilde{\mathbf{v}}\|_2^2$, the autoencoder model has been shown to learn the same subspace as that spanned by the first m POD modes if $\mathbf{H} = \mathbf{H}_E = \mathbf{H}_D$. However, additional constraints are necessary to ensure that the columns of $\mathbf{H}$ form an orthonormal basis and follow an energy-based hierarchical ordering [67].

3.2. Advection-Aware Autoencoder Design

In this work, inspired by the registration-based nonlinear manifold learning idea [66] and the physics-informed autoencoder model design [64], we develop a new advection-aware autoencoder model that incorporates physical knowledge of the advection-dominated flow features into the autoencoder neural network via both an inductive bias as well as soft constraints. As shown in Figure 2, this AA autoencoder architecture is composed of three sub-networks. The first part, as usual, is called an encoder, χ_e, which maps a high-dimensional input snapshot $\mathbf{v}$ to a low-dimensional latent vector $\mathbf{z}$, and is defined by $\mathbf{z} = \chi_e(\mathbf{v}; \theta_e)$ where $\mathbf{z} \in \mathbb{R}^m$ ($m \ll N$).

Two independent decoder networks are also defined, which map the latent vector to (i) a transformed (or "shifted") version of the high-dimensional input snapshot, and (ii) back to the true high-dimensional input snapshot. The first of these two decoders is called a shift decoder, $\boldsymbol{\phi}_s$, that maps the latent vector $\mathbf{z}$ to an approximation $\widetilde{\mathbf{v}}_s$ of a suitably defined "shifted" snapshot $\mathbf{v}_s$ that encapsulates the dominant advective features of the flow problem, $\widetilde{\mathbf{v}}_s = \boldsymbol{\phi}_s(\mathbf{z}; \theta_s)$. This can be achieved by following a registration-type approach, where a high-fidelity snapshot at a randomly chosen time point in the simulation is selected to be the candidate output target for the shift decoder. The arbitrariness of the choice could be partially resolved if a known physical characteristic of the flow like dissipation or multiscale oscillations indicates that a particular time point such as the initial solution or a time point at the beginning of a period is a more preferable candidate for the output target of $\boldsymbol{\phi}_s$. This approach is flexible, by design, as it does not require any additional knowledge of the dominant advection patterns of the flow, such as speed of propagation, in order to train the shift decoder, and has been adopted for the numerical experiments in this study. Alternatively, if some partial knowledge of the dominant advective flow features are available, then a transported snapshot could be defined, following the ideas in [29,30]. In this approach, a set of time-varying coordinates are defined by transporting the high-dimensional computational grid using the dominant advection properties such as speed and direction of propagation. Then, the true simulation snapshots are mapped onto the time-varying grid using a suitable interpolation technique to produce the time-varying output target for the shift decoder. The primary advantage of this approach is that it preserves some of the structural properties of any secondary features of the flow problem such as frictional dissipation or the wake patterns trailing a moving vessel. In this way, this approach allows an improved isolation of the dominant advective features of the flow. However, the additional requirements of physical knowledge about the flow, and the potential approximation errors introduced by the interpolation technique are some of the primary issues that need to resolved.

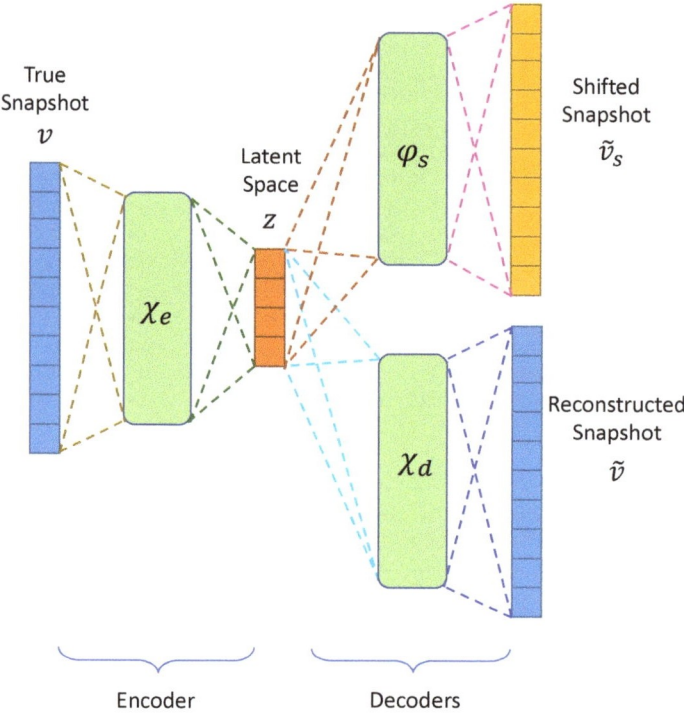

Figure 2. Advection-aware autoencoder architecture. An encoder network χ_e extracts the dominant features of $v \in \mathbb{R}^N$ into a compressed latent space $z \in \mathbb{R}^k$. One decoder network ϕ_s maps the latent vector to the shifted snapshot, $v_s \in \mathbb{R}^N$. The second decoder network χ_d maps the latent vector back to an approximation of itself, $\tilde{v} \in \mathbb{R}^N$.

The third and final sub-network is called a true decoder, χ_d, which maps the latent vector $\mathbf{z}$ to an approximation $\tilde{\mathbf{v}}$ of the high-dimensional input snapshot $\mathbf{v}$, and is defined as a traditional decoder network as $\tilde{\mathbf{v}} = \chi_d(\mathbf{z}; \theta_d)$. The combination of the encoder network χ_e and the true decoder network χ_d yields an autoencoder network given by (1).

Such a network enables us to express high-dimensional snapshots in terms of low-dimensional nonlinear manifolds, and can be employed in traditional non-intrusive reduced order modeling frameworks. On the other hand, the combination of the encoder network χ_e and the shift decoder network ϕ_s creates a nonlinear mapping between the true snapshots $\mathbf{v}$ and the "shifted" snapshots $\mathbf{v}_s$ in the high-dimensional physical space, such that this transformation map learns about the dominant advective features of the flow.

$$\hat{\chi} : \mathbf{v} \mapsto \phi_s \circ \chi_e(\mathbf{v}). \tag{3}$$

The primary contribution of this AA autoencoder design is that simultaneous training of these two partially-coupled autoencoder and transformation maps, χ and $\hat{\chi}$, respectively, endows the intermediate nonlinear, latent space manifold with the information about the dominant advection characteristics of the flow, as well. The simultaneous training can be performed by defining two separate loss functions. The shift loss, $\mathcal{L}_1$, is defined as a measure of discrepancy between the prediction of the shift decoder and the high-dimensional "shifted" snapshot, $\mathcal{L}_1 = \|\mathbf{v}_s - \phi_s \circ \chi_e(\mathbf{v})\|_V$, where $\|\ \|_V$ denotes a chosen error norm. The second loss function is called the reconstruction loss, $\mathcal{L}_2$, and is defined as the error between the prediction of the true decoder and the high-dimensional true snapshot, $\mathcal{L}_2 = \|\mathbf{v} - \chi_d \circ \chi_e(\mathbf{v})\|_V$, in the same error norm $\|\|_V$. The AA autoencoder is trained by computing the optimal values of the parameters $(\theta_e^*, \theta_d^*, \theta_s^*)$ that simultaneously minimize a weighted combination of these two loss components over all the training data

$$\theta_e^*, \theta_d^*, \theta_s^* = \underset{\theta_e, \theta_d, \theta_s}{\mathrm{argmin}} \left\{ w_1 \mathcal{L}_1(\mathbf{v}, \widetilde{\mathbf{v}}_s) + w_2 \mathcal{L}_2(\mathbf{v}, \widetilde{\mathbf{v}}) \right\}, \tag{4}$$

where w_1, w_2 are the weights of the linear combination that could either be fixed during training or be a part of the trainable hyperparameters. In this work, AA autoencoder networks are trained to produce an advection-informed latent space representation of the high-fidelity numerical solution of a parametric linear advection problem and a parametric viscous advecting shock problem. LSTM networks are trained to model the temporal dynamics of the latent space coefficients. The AA autoencoder and the LSTM dynamics model are combined to construct a fully non-intrusive, physics-aware reduced order model for the advection-dominated test problems.

3.3. Long-Short-Term Memory (LSTM) Network

An LSTM network is a special type of recurrent neural network (RNN) that is well-suited for performing classification and regression tasks based on time series data. The main difference between the traditional RNN and the LSTM architecture is the capability of an LSTM memory cell to retain information over time, and an internal gating mechanism that regulates the flow of information in and out of the memory cell [68]. A very concise overview of LSTM networks as applied in the context of model reduction can be found in Ref. [60].

The LSTM cell consists of three parts, also known as gates, that have specific functions. The first part, called the forget gate, chooses whether the information coming from the previous step in the sequence is to be remembered or can be forgotten. The second part, called the input gate, tries to learn new information from the current input to this cell. The third and final part, called the output gate, passes the updated information from the current step to the next step in the sequence. The basic LSTM equations for an arbitrary input vector $\mathbf{u}$ are

$$\begin{aligned}
\text{input gate:} & \quad \zeta_i = \alpha_S \circ \mathcal{F}_i(\mathbf{u}), \\
\text{forget gate:} & \quad \zeta_f = \alpha_S \circ \mathcal{F}_f(\mathbf{u}), \\
\text{cell state:} & \quad \mathbf{c}_t = \zeta_f \odot \mathbf{c}_{t-1} + \zeta_i \odot (\alpha_T \circ \mathcal{F}_a(\mathbf{u})), \\
\text{output gate:} & \quad \zeta_o = \alpha_S \circ \mathcal{F}_o(\mathbf{u}), \\
\text{output:} & \quad \mathbf{h}_t = \zeta_o \circ \alpha_T(\mathbf{c}_t).
\end{aligned} \tag{5}$$

Here, $\mathcal{F}$ refers to a linear transformation defined by a matrix multiplication and bias addition, that is, $\mathcal{F}(\mathbf{x}) = \mathbf{W}\mathbf{x} + \mathbf{b}$, where $\mathbf{W} \in \mathbb{R}^{n \times m}$ is a matrix of layer weights, $\mathbf{b} \in \mathbb{R}^n$ is a vector of bias values, and $\mathbf{x} \in \mathbb{R}^m$ is a vector of layer activations. Also, α_S and α_T denote sigmoid and hyperbolic tangent activation functions, which are usually the default choices in an LSTM network, and $\mathbf{x} \odot \mathbf{y}$ denotes a Hadamard product of two vectors $\mathbf{x}$ and $\mathbf{y}$. In the context of reduced order modeling, the vector $\mathbf{u}$ represents a linear or nonlinearly encoded snapshot vector, with which the LSTM network is trained to advance with time. The core concept of an LSTM network is the cell state $\mathbf{c}_t$, which behaves as the "memory" of the network. It can either allow greater preservation of past information, reducing the issues of short-term memory, or it can suppress the influence of the past depending on the actions of the various gates during the training process.

LSTM networks have proven to be an effective tool in the development of reduced order models for physical systems and have shown that they can outperform alternate classical methods such as DMD and POD-Galerkin, as well as other flavors of RNNs that often suffer from issues with vanishing gradients and the transmission of long-term information [20,69–71]. Different ML methods for time series modeling, such as the neural ordinary differential equations (NODE) [23,59], spatial transformer networks [72], echo state networks [73], and residual networks (ResNets) [74] have also been shown to be very accurate in various ROM applications for dynamical systems. Unfortunately, many of these newer approaches are not readily available as packages or modules inside well-known machine learning libraries such as TensorFlow and PyTorch. However, LSTM implementations are included as part of the core, highly-efficient, GPU-accelerated modules of all these libraries. Hence, owing to the ease of implementation and the well-known success stories of LSTM-based prediction models for dynamical systems, we have adopted it as our method of choice for modeling of latent space dynamics.

We train an LSTM network to independently learn the temporal evolution of the latent space coefficients generated by the encoder of a pre-trained AA autoencoder model, following a similar approach as in [60,69]. The decoupling of the AA autoencoder training for a nonlinear embedding and the LSTM training for latent space dynamics allows for greater flexibility in our non-intrusive ROM development. If alternate time series learning methods are available that better suit the needs of the problem in a future time, the nonlinear manifold defined by the pre-trained AA autoencoder will not need to be retrained. Moreover, an end-to-end, simultaneous training of an AA autoencoder and a time series learning method like LSTM requires the development of a carefully weighted loss function that appropriately penalizes both the reconstruction and the forecast accuracy. This can often lead to significant loss in both the training efficiency as well as in the overall robustness of the training algorithm.

4. Results

In this section, we demonstrate the capability of the advection-aware autoencoder architecture to generate a compressed representation for high-fidelity snapshots of two different advection-dominated problems. Furthermore, we present numerical results to illustrate the potential of training reduced order models for the system dynamics in the latent space generated by the pre-trained AA autoencoder models. In this study, LSTM architectures are chosen to build these dynamics models for the purposes of illustration. However, the methodology could be easily adapted to use any other approximation framework that might be more appropriate for a particular problem.

4.1. Linear Advection Problem

Consider the advection of a circular Gaussian pulse traveling in the positive y-direction through a rectangular domain, $\Omega = [-100, 100] \times [0, 500]$ at a constant speed, c. The analytical solution is given by

$$u(x,y) = \exp\left\{-\left(\frac{(x-x_0)^2}{2\sigma_x^2} + \frac{(y-y_0-ct)^2}{2\sigma_y^2}\right)\right\}, \tag{6}$$

where (x_0, y_0) is the initial location of the center of the pulse, σ_x and σ_y define the support of the pulse in the x and y directions, respectively. The domain is uniformly discretized into 201 grid points in the x-direction and 501 grid points in the y-direction using $\Delta x = 1$ and $\Delta y = 1$, respectively, and generating 100,701 computational nodes. A uniform time discretization of $\Delta t = 1$ is used to generate 460 high-fidelity time snapshots for a circular Gaussian pulse parametrized by different values of the size of the pulse profile $\sigma_x = \sigma_y \equiv \sigma = \{5, 8, 10, 16, 20\}$, and traveling at a constant speed $c = 1$ from an initial location $(x_0, y_0) = (0, 40)$. Figure 3 depicts the relative information content (RIC) for a different number of POD modes obtained by taking a SVD of the high-fidelity snapshots.

As the singular values computed by SVD are arranged in the descending order of relative importance, the RIC values of the leading r POD modes can be defined as

$$\text{RIC}(\%) = \frac{\sum_{k=1}^{r} \lambda_k^2}{\sum_{k=1}^{M} \lambda_k^2} \times 100, \tag{7}$$

where λ_k is the kth singular value and M denotes the total number of time snapshots. For the set of snapshots generated with a given value of σ, the dotted vertical line indicates the number of leading POD modes required to attain a RIC value of 99.9%. For instance, $\sigma = 20$ signifies a flatter pulse profile and 17 POD modes contain 99.9% RIC for the corresponding system of snapshots, whereas 63 POD modes are needed to capture 99.9% RIC for a sharper pulse profile given by $\sigma = 5$. This illustrates the phenomena of relatively large Kolmogorov n-widths for even simple, linear advection problems, which severely limits the efficiency of low-dimensional approximation using SVD-generated linear subspaces.

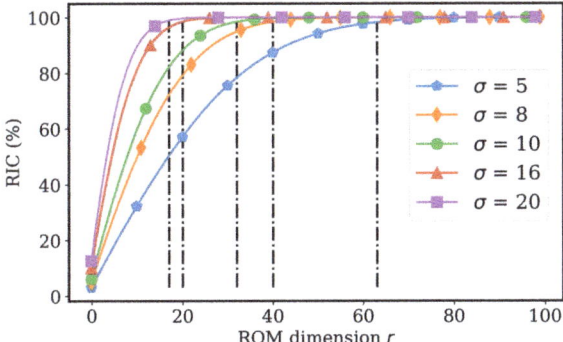

Figure 3. The relative information content for different number of retained POD modes. The singular values are computed by taking an SVD of the high-fidelity snapshots for an advecting circular Gaussian pulse (see Equation (6)) of varying width (σ) traveling at a constant speed $c = 1$.

In the first numerical example, the training dataset is constructed using 460 high-fidelity snapshots for each value of $\sigma_{train} = \{5, 10, 16\}$. The remaining snapshots corresponding to $\sigma_{test} = \{8, 20\}$ are used to create a test dataset. This creates a geometrically parameterized training and testing dataset.

The AA autoencoder network is trained on the parametric training set for 8000 epochs using the Adam optimizer with an initial learning rate of 1×10^{-4} that decays step-wise by 15% every 456 epochs. The training snapshots are all augmented by the value of the corresponding parameter. The training snapshots are divided into two sets—starting from the initial time point every alternate snapshot is used for training the AA autoencoder model, while the rest are reserved for validation during training. In this study, the losses computed on the validation points are solely used to monitor the extent of overfitting during training, and later to evaluate the accuracy of prediction on unseen time steps associated with a training parameter value. After exploring a large space of network design parameters, as described in Table 1, the results presented here are obtained with two of the most optimal AA autoencoder designs. In the first model (AA1), only the input feature is augmented by the parameter value; while in the second model (AA2), both the input feature and the output labels are augmented by the parameter value. The encoder network χ_e of both the models is constructed with three hidden layers composed of 629, 251 and 62 units that connect an input feature (i.e., augmented snapshot) of dimension $N = 100,702$ to an encoded latent vector representation of dimension k. For both the models, the decoder networks χ_d and ϕ_s are set up to be mirror images of the corresponding encoder network. The AA1 model uses the *selu* activation function for the hidden layers followed by a *linear* activation on the output layer, while the AA2 model uses the *swish* activation function for the hidden layers. The individual loss components $\mathcal{L}_1$ and $\mathcal{L}_2$ are defined as a weighted combination of the normalized mean square error (NMSE) loss and the pseudo-Huber loss, as defined below,

$$\text{NMSE Losses: } \mathcal{L}_1^{NMSE} = \frac{1}{N} \sum_{k=1}^{N} \frac{\|\mathbf{v}_{k,s} - \boldsymbol{\phi}_s \circ \chi_e(\mathbf{v}_k)\|_2^2}{\|\mathbf{v}_k\|_2^2},$$

$$\mathcal{L}_2^{NMSE} = \frac{1}{N} \sum_{k=1}^{N} \frac{\|\mathbf{v}_k - \chi_d \circ \chi_e(\mathbf{v}_k)\|_2^2}{\|\mathbf{v}_k\|_2^2},$$

$$\text{pseudo-Huber Losses: } \mathcal{L}_1^H = \frac{1}{N} \sum_{k=1}^{N} \delta^2 \left(\sqrt{1 + \left(\frac{\mathbf{a}_{k,1}}{\delta}\right)^2} - 1 \right), \quad (8)$$

$$\mathcal{L}_2^H = \frac{1}{N} \sum_{k=1}^{N} \delta^2 \left(\sqrt{1 + \left(\frac{\mathbf{a}_{k,2}}{\delta}\right)^2} - 1 \right),$$

where $\mathbf{a}_{k,1} = \mathbf{v}_{k,s} - \boldsymbol{\phi}_s \circ \chi_e(\mathbf{v}_k)$ and $\mathbf{a}_{k,2} = \mathbf{v}_k - \chi_d \circ \chi_e(\mathbf{v}_k)$.

Table 1. Hyperparameters explored to design the AA autoencoders for the linear advection example.

Hyperparameters	AA1	AA2
Input/Output	Augmented Input, non-augmented output	Augmented input and output
Hidden Units (50–1500)	629, 251, 62	629, 251, 62
Batch Size (8–128)	32	24
Latent Dimension (5–50)	15	15
Activation (ReLU, selu, linear, tanh, swish)	selu	swish

The pseudo-Huber loss is a smooth approximation of the Huber loss function that behaves as a L_2 squared loss by being strongly convex near the desired minimum and as a L_1 absolute loss with reduced steepness near the extreme values. The scale at which this transition happens and the steepness near the extreme values is controlled by the δ parameter.

A piece-wise segmented training approach is adopted for both the models in which only the $\mathcal{L}_2$ component of the total loss is minimized for the first 2500 epochs, followed by a weighted combination of both the loss components $w_1 \mathcal{L}_1 + w_2 \mathcal{L}_2$ for the rest of the training. The AA1 model is trained with mini batches of size 32 and the AA2 model is trained with mini batches of size 24, while both models generate a latent space of dimension 15.

The training trajectories for the AA1 and AA2 models are shown in Figure 4. Even though both models are trained for the same number of epochs, the lower training and validation loss values for the AA2 model (see the left panel of Figure 4) indicates a higher level of expressivity and overfitting due to the augmented dimension of the decoder outputs and the resultant higher number of network hyperparameters (weights and biases). As a result, the prediction errors for the test parameter values are found to be higher using the AA2 model than those obtained with the AA1 model. Less overfitting is usually an indication of better generalization performance, and hence the AA1 model is used to generate the field predictions for both the seen and unseen data in the rest of this example. On the other hand, when extrapolatory predictions or predictions for unseen data are not required, a slightly overfit model such as AA2 can be considered preferable. This is supported by the evolution of the losses corresponding to each decoder: $\mathcal{L}_1$ for the prediction of shifted snapshots and $\mathcal{L}_2$ for the reconstruction of the true solution (see the right panel of Figure 4). As the $\mathcal{L}_1$ loss is associated with the network's ability to map the true snapshot to a fixed snapshot, it is relatively easier to minimize and both models perform equally well in this task. However, due to the higher expressivity of the AA2 model, it is able to minimize the $\mathcal{L}_2$ loss much more than the AA1 model, thus leading to higher accuracy in the approximation of the true snapshots using training data.

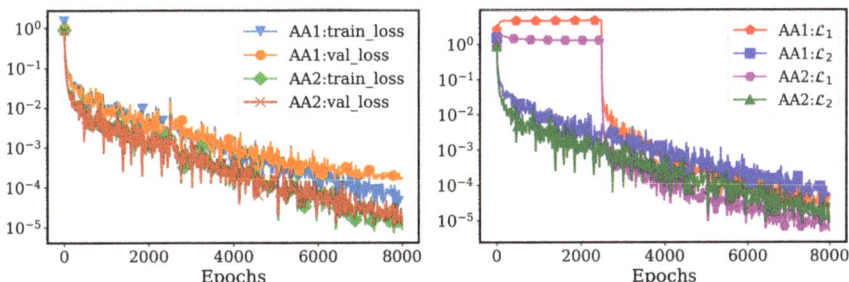

Figure 4. Training characteristics of two AA autoencoder networks trained using a parametric dataset of snapshots for a 1D advecting Gaussian pulse parameterized with varying support of the pulse profile, $\sigma = \{5, 10, 16\}$. AA1 denotes the model trained with the input features augmented by parameter values, while AA2 denotes the model where both input features and output labels are augmented. The left panel shows the decay of training and validation losses during training. The right panel shows the evolution of the loss components during training.

Figure 5 presents the predictions of the high-dimensional shifted and true snapshots obtained by the corresponding decoders ϕ_s and χ_d, respectively, of model AA1 as well as a comparison of the prediction performance for different training parameter values in terms of the spatial relative errors of the full-order predictions. The decoder predictions are evaluated for the two parameter values at the boundaries of the training range $\sigma = 5$ and $\sigma = 16$ as they present distinct challenges. Autoencoders are known to struggle with the extraction of discontinuities in the input feature space. The snapshot data for $\sigma = 5$ features a very steep gradient in the shape of the pulse profile which poses some of the same challenges as a discontinuous profile. Moreover, a single encoder network χ_e is being tasked, by design, to combine with two independent decoder networks χ_d and ϕ_s to map into both a stationary discontinuity as well as a moving discontinuity. Thus the spatial distribution of reconstruction error for the $\sigma = 5$ profile is more localized near the moving pulse, whereas that of the $\sigma = 16$ profile is more uniformly spread out across the spatial domain (see Figure 5b,d). Despite all of these minor differences in prediction performance, there is a high degree of agreement between the full order decoder predictions and the high-dimensional snapshots, with less than 4% relative error for all of the training parameter values (see Figure 5e).

Finally, prediction performance results of the AA1 model for high-fidelity snapshots generated with an unseen parameter value $\sigma = 8 \in \sigma_{test}$ are presented in Figure 6. Loss in accuracy with extrapolatory predictions for a geometrically parameterized dataset is one of the well-known challenges faced by both intrusive and non-intrusive reduced order modeling approaches, which requires particular attention to resolve representation issues posed by the topology of the parametric solution manifold [75]. Thus, as expected, there is a noticeable drop in accuracy for the prediction of full order solutions, with the errors being especially localized near the moving pulse profile (see Figure 6b). This effect is also reflected in the relative error plots for the two unseen parameter values in σ_{test}. However, the quality of predictions are still quite encouraging considering that these are purely extrapolatory predictions on an unseen parameter instance, without any special treatment of the solution manifold or modification of the training process.

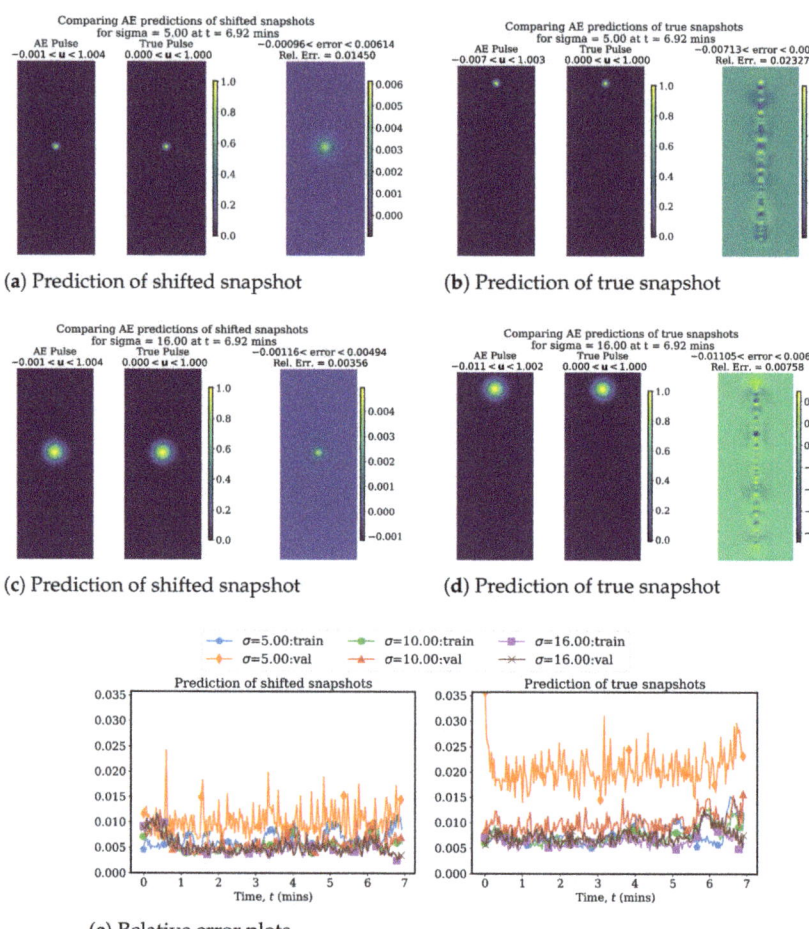

Figure 5. Prediction performance of ϕ_s and χ_d decoders on training data. (**a**,**b**) predictions of shifted and true snapshots, respectively, for pulse size $\sigma = 5$ and (**c**,**d**) predictions of shifted and true snapshots, respectively, for pulse size $\sigma = 16$ at an intermediate time $t = 6.92$ min using the AA1 model. (**e**) Relative errors for the decoder predictions using different values of the parameter from the training set.

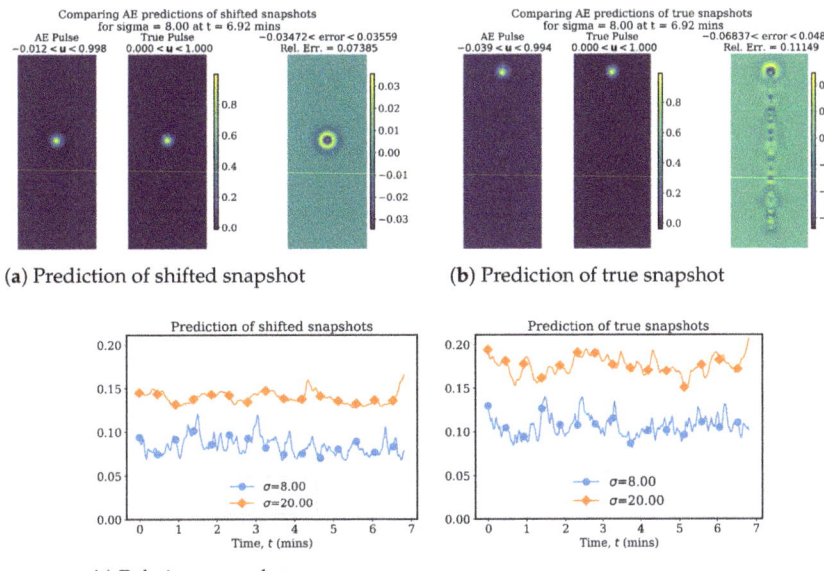

Figure 6. Prediction performance of ϕ_s and χ_d decoders on unseen data. (**a,b**) predictions of shifted and true snapshots, respectively, for unseen pulse size $\sigma = 8$ at time $t = 6.92$ min using the AA1 model. (**c**) Relative errors for the decoder predictions using two different values of the parameter from the unseen test set.

4.2. Advecting Viscous Shock Problem

The second numerical example is described by the one-dimensional viscous Burgers' equation (VBE) with Dirichlet boundary conditions [60] as given by

$$\dot{u} + u\frac{\partial u}{\partial x} = \nu \frac{\partial^2 u}{\partial x^2}, \quad (9)$$
$$u(x,0) = u_0, \quad x \in [0, L], \quad u(0,t) = u(L,t) = 0,$$

where we set $L = 1$ and the maximum time $t_{max} = 2$. The solution of the above equation is capable of generating shock discontinuities even with smooth initial conditions if the viscosity ν is sufficiently small, due to the advection-dominated behavior. We consider the initial condition

$$u(x,0) \equiv u_0 = \frac{x}{1 + \sqrt{\frac{1}{t_0}} \exp\left(Re \frac{x^2}{4}\right)}. \quad (10)$$

An analytical solution of this problem is given by

$$u(x,t) = \frac{\frac{x}{t+1}}{1 + \sqrt{\frac{t+1}{t_0}} \exp\left(Re \frac{x^2}{4t+4}\right)}, \quad (11)$$

where $t_0 = \exp(Re/8)$ and $Re = 1/\nu$. The high-fidelity snapshot data is generated by directly evaluating the analytical solution over a uniformly discretized spatial domain containing 200 grid points and for 500 uniform time steps. Figure 7 shows a visualization of the time evolution of the initial condition for three different values of $Re = 50, 300, 600$.

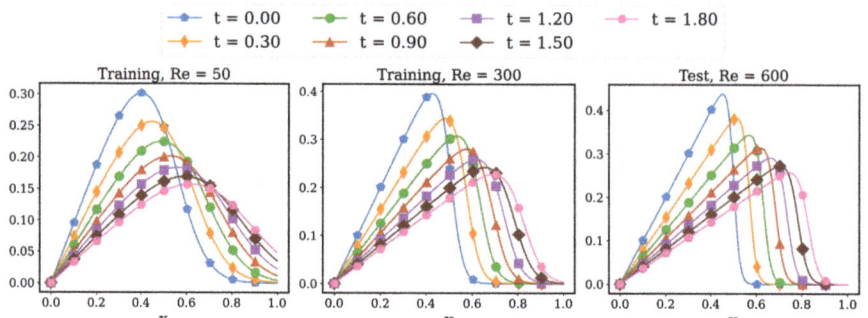

Figure 7. Time evolution of the high-fidelity snapshots for the advecting viscous shock problem (see Equation (9)), parameterized with variable Reynolds number, Re.

A parametric dataset is generated by collecting 500 high-fidelity snapshots for different values of the Reynolds number, $Re = \{50, 150, 300, 400, 500, 600\}$. The training dataset is constructed using snapshots for $Re_{train} = \{50, 150, 300, 500\}$. The remaining snapshots for $Re_{test} = \{400, 600\}$ constitute the test dataset. Figure 8 depicts the variation in RIC with the number of retained POD modes for snapshots corresponding to different Re values. The vertical dashed lines represent the number of POD modes required to attain 99.9% RIC for snapshots of a given Re value. The gradual rise in the number of POD modes required to attain 99.9% RIC with increasing values of Re clearly indicates how a growth of advection-dominated behavior raises the effective Kolmogorov n-width of the system.

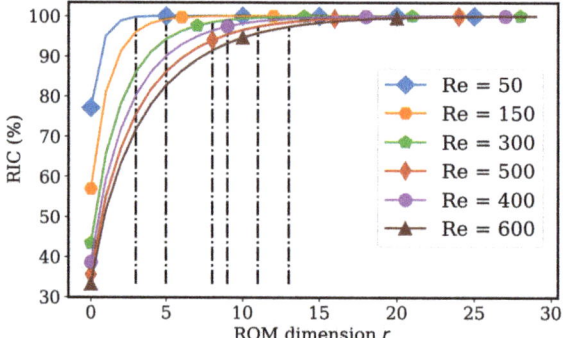

Figure 8. The relative information content for a different number of retained POD modes. The singular values are computed by taking an SVD of the high-fidelity snapshots for the advecting viscous shock problem (see Equation (9)), parameterized with variable Reynolds number, Re.

4.2.1. AA Autoencoder Models for Varying Advection Strength

In this section, we present the numerical results on the training of AA autoencoder networks for the viscous advecting shock problem parameterized with variable Re values, as discussed before. Following the idea of a registration-type approach, as discussed in Section 3.2, a high-fidelity simulation snapshot at roughly the midpoint of the simulation time period is chosen as the shifted snapshot for training the shift decoder. This choice is, however, arbitrary and any other high-fidelity snapshot could have been selected without affecting the effectiveness of the approach.

The results reported here are obtained with two different AA autoencoder models—AA3 and AA4. The primary objective of this comparison is to evaluate the ability of AA autoencoders not just to predict snapshots for unseen parameter values, but also to forecast solutions at time points not included in the time history of the training snapshots. With that objective in mind, the AA3 model is trained using all of the time snapshots available for each training parameter value, while the AA4 model is trained using the first 90% of the time snapshots, i.e., until $t = 1.80$, for each training parameter value. Similar to the previous numerical example, the available high-fidelity snapshots for each training parameter value are divided into two sets—starting from the initial time point every alternate snapshot is used for training the AA autoencoder model, while the rest are reserved for validation purposes. As in the previous example, the losses computed on the validation data points during training are solely used to monitor the extent of overfitting, and later to evaluate the accuracy of prediction on unseen data points corresponding to a training parameter value.

After a careful exploration of the design space, a set of optimal values for the hyperparameters were obtained to construct models AA3 and AA4 (see Table 2). Both the models are trained for 5000 epochs using minibatches of size 24 and employing the Adam optimizer. The AA3 model training is initialized with a learning rate of 5×10^{-4} that decays stepwise by 10% every 330 epochs, whereas the initial learning rate for the AA4 model is chosen to be 3×10^{-4} and it is allowed to decay by 10% every 309 epochs. Both the input features and the output labels are constructed by augmenting the training snapshots with the corresponding scaled parameter values. The encoder network χ_e for model AA3 is constructed with a single hidden layer of size 50 that connects an input feature (i.e., augmented snapshot) of dimension $N = 201$ to an encoded latent vector representation of dimension k. On the other hand, the AA4 model is defined with two hidden layers of sizes 100 and 50. For both the models, the decoder networks χ_d and ϕ_s are set up to be mirror images of the encoder network. From Figure 8 it can be seen that at least a minimum of 3 POD modes are required to attain 99% RIC for any of the chosen Re values. However, while exploring a range of possible latent space dimensions, $3 \leq k \leq 10$, it was observed that a latent space of dimension $k = 5$ was adequate in capturing the essential dynamical features of the entire parametric training dataset. Hence, $k = 5$ is selected as the optimal latent space dimension for both models AA3 and AA4. All hidden layers are endowed with the *swish* activation function, while the output layers are designed to have a *linear* activation. The individual loss components $\mathcal{L}_1$ and $\mathcal{L}_2$ are defined by a weighted combination of the NMSE loss and the pseudo-Huber loss, as discussed in the previous numerical example.

Table 2. Hyperparameters to design the AA autoencoders for the viscous advecting shock example.

Hyperparameters	AA3	AA4
Input/Output	Augmented input and output	Augmented input and output
Hidden Units (50–150)	50	100, 50
Batch Size (8–128)	24	24
Latent Dimension (3–10)	5	5
Activation (*selu, tanh, swish*)	*swish*	*swish*
Initial Learning Rate (1×10^{-3}–1×10^{-5})	5×10^{-4}	3×10^{-4}

Figure 9 shows the salient features of the training process for models AA3 and AA4. The left plot shows the decay of the training and validation losses during training, and the right plot shows the decay of the two loss components, $\mathcal{L}_1$ and $\mathcal{L}_2$, for both models. Due to its higher capacity (more hidden layers and more neurons) model AA4 is capable of attaining lower values of training and validation losses as compared to model AA3. This is possibly an indication that model AA4 is able to learn the essential features of the high-dimensional state space more effectively, thus enabling improved prediction over data points that lie within the bounds of the training time history, as will be shown in the later experiments. On the other hand, this also causes the $\mathcal{L}_1$ loss component of model AA4 to have a sharper decay than the $\mathcal{L}_2$ component, whereas model AA3 shows a more balanced decay of the two loss components. The latter trait is considered more preferable, as the effectiveness of any latent space dynamics model is dependent upon the accuracy of the true decoder χ_d, that is measured by the $\mathcal{L}_2$ loss component. Therefore, in situations when the entire time history is available for model training, a smaller capacity AA autoencoder model like AA3 is capable of achieving the desirable training outcomes. Hence, following the principle of parsimony, model AA3 is chosen to generate the latent space representations that are used to train LSTM dynamics models in the next two sections.

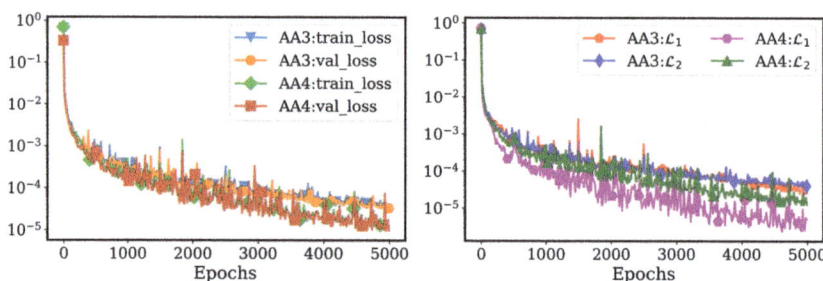

Figure 9. Training characteristics of two AA autoencoder networks trained using a parametric dataset of snapshots for the advecting viscous shock problem parameterized with variable Reynolds number, $Re = \{50, 150, 300, 500\}$. AA3 denotes the model trained with the entire time snapshot history for every parameter value, while AA4 denotes the model trained with the initial 90% of the time snapshot history for each parameter value. The left panel shows the decay of training and validation losses during training. The right panel shows the evolution of the loss components during training.

Figures 10 and 11 present a performance comparison of models AA3 and AA4 in predicting the true and shifted snapshots for two of the training parameter values, $Re = 50$ and $Re = 500$, respectively. Each individual plot shows the evolution of either a solution field or an error field in the $x - t$ space, where the x-axis represents time and the y-axis represents space. The plots in the left column depict the solution fields predicted by the AA3 and AA4 models, the plots in the middle column depict the high-dimensional solution fields, and the plots in the right column depict the pointwise error between the high-dimensional and the predicted solution fields. It is evident from the first two columns that models AA3 and AA4 are able to qualitatively capture both the true and shifted solution fields. A closer look at the right column reveals that the approximation error of model AA3 is randomly spread throughout the simulation time history (see Figures 10b and 11a,b), whereas the approximation error of model AA4 rises gradually as predictions are sought further away from the end point of the training time history, i.e., $t > 1.8$. This confirms the previously discussed observation that, due to the higher network capacity, model AA4 offers more accurate predictions than model AA3 for time points that lie within the bounds of the training time history that is common to both models, i.e., $0 < t \leq 1.8$. However, model AA4 gradually loses predictive capability over the unseen time points given by $t > 1.8$, whereas model AA3 still generates accurate predictions, despite its lower network capacity.

Figure 12 depicts the time trajectory of the spatial relative errors for the predictions obtained by models AA3 and AA4 over the parametric training dataset. The spatial relative errors are computed as the ratio of the spatial l^2-norm of the prediction error to the spatial l^2-norm of the high-dimensional solution at every computational time point. Figure 12a shows the spatial relative errors for the shift decoder predictions while Figure 12b shows the corresponding errors for the true decoder predictions, for every value in the parametric training dataset, $Re_{train} = \{50, 150, 300, 500\}$. The relative error plots not only validate the previously discussed observations about the prediction capabilities of models AA3 and AA4, but also highlight that even for $t > 1.8$, model AA4 offers encouraging extrapolatory predictions with a relative error of less than 4%.

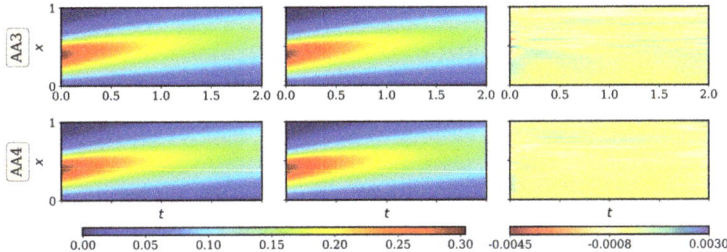

(**a**) Prediction of true snapshots for $Re = 50$ using models AA3 and AA4

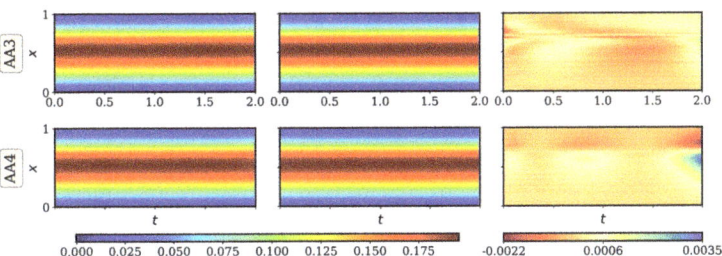

(**b**) Prediction of shifted snapshots for $Re = 50$ using models AA3 and AA4

Figure 10. Prediction performance of χ_d and ϕ_s decoders on training data. Predictions of (**a**) true and (**b**) shifted snapshots for a training parameter value, $Re = 50$, using models AA3 and AA4. The left column shows the predicted solutions, the center column shows the high-fidelity solutions, and the right column shows the error between the two.

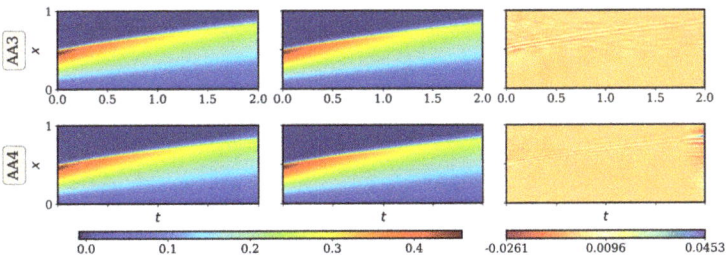

(**a**) Prediction of the true snapshots for $Re = 500$ using models AA3 and AA4

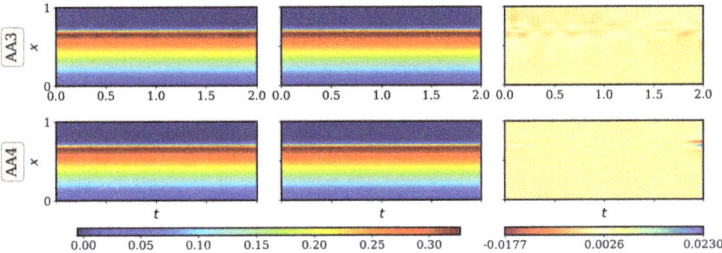

(**b**) Prediction of the shifted snapshots for $Re = 500$ using models AA3 and AA4

Figure 11. Prediction performance of χ_d and ϕ_s decoders on training data. Predictions of (**a**) true and (**b**) shifted snapshots for a training parameter value, $Re = 500$, using models AA3 and AA4. The left column shows the predicted solutions, the center column shows the high-fidelity solutions and the right column shows the error between the two.

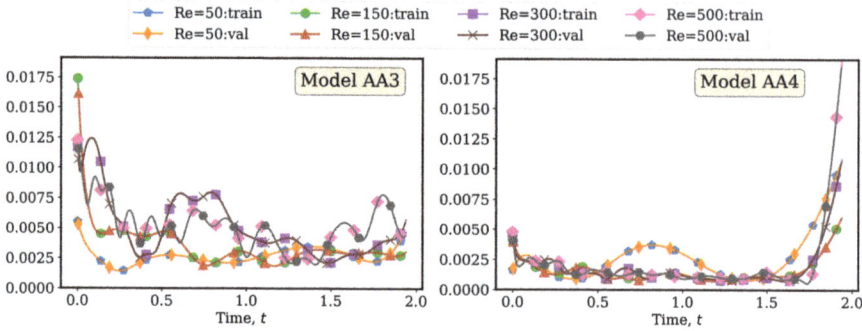

(**a**) Relative errors of the shift decoder predictions

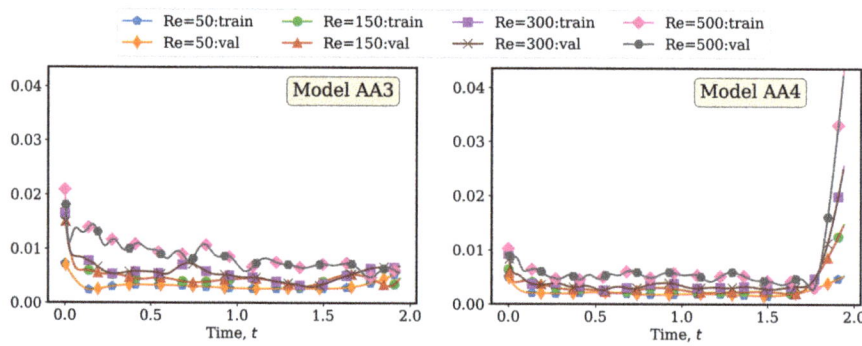

(**b**) Relative errors of the true decoder predictions

Figure 12. Relative errors of (**a**) ϕ_s and (**b**) χ_d predictions using the AA3 and AA4 models for snapshots generated with the training parameter values.

Figures 13 and 14 show the performance of models AA3 and AA4 while predicting the true and shifted snapshots using parameter values from the test dataset, $Re = 400$ and $Re = 600$, respectively. Figure 15 presents the spatial relative errors of the predictions made by models AA3 and AA4 for the two test parameter values. Even for the unseen test parameter values, the true and shifted solution fields computed by the AA autoencoder models are closely aligned with the high-dimensional solution fields. This is reflected in the error field plots as well as the spatial relative error plots, which are bounded below the 5% relative error. The results in this section demonstrate that even for a nonlinear advection-dominated problem, a trained AA autoencoder network can offer accurate extrapolatory predictions for unseen parameter instances as well as short-term extrapolatory predictions for unseen time.

4.2.2. LSTM Models for System Dynamics

In this section, numerical results are presented for the modeling of the temporal evolution of the latent space coefficients defined by a pre-trained AA autoencoder network for the advective viscous shock problem parametrized by variable Re. As the focus of this work is to demonstrate the efficiency and flexibility of the AA autoencoder architecture, hence for the sake of simplicity, the latent space dynamics are modeled in an autoregressive fashion using traditional LSTM networks.

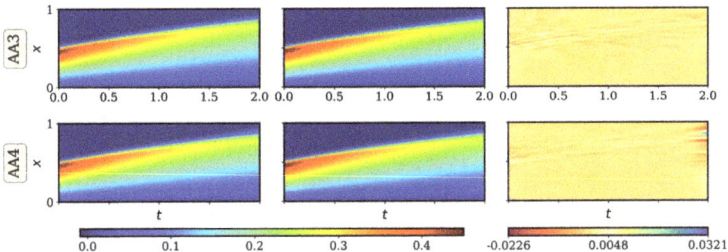

(a) Prediction of true snapshots for $Re = 400$ using models AA3 and AA4

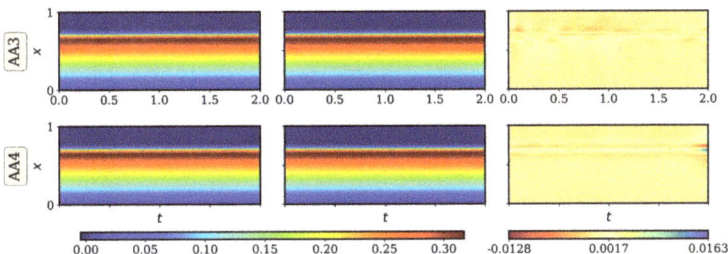

(b) Prediction of shifted snapshots for $Re = 400$ using models AA3 and AA4

Figure 13. Prediction performance of χ_d and ϕ_s decoders on unseen data. Predictions of (a) true and (b) shifted snapshots for a test parameter value, $Re = 400$, using models AA3 and AA4. The left column shows the predicted solutions, the center column shows the high-fidelity solutions and the right column shows the error between the two.

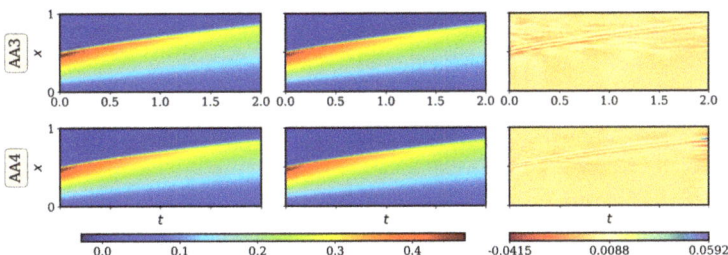

(a) Prediction of true snapshots for $Re = 600$ using models AA3 and AA4

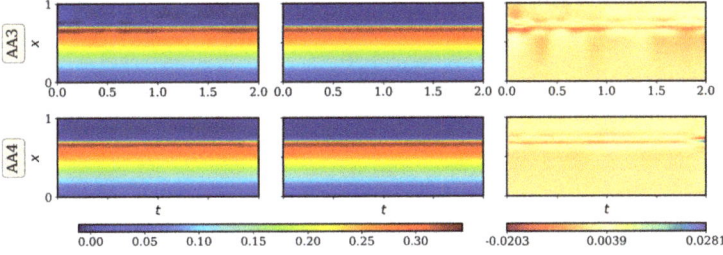

(b) Prediction of shifted snapshots for $Re = 600$ using models AA3 and AA4

Figure 14. Prediction performance of χ_d and ϕ_s decoders on unseen data. Predictions of (a) true and (b) shifted snapshots for a test parameter value, $Re = 600$, using models AA3 and AA4. The left column shows the predicted solutions, the center column shows the high-fidelity solutions, and the right column shows the error between the two.

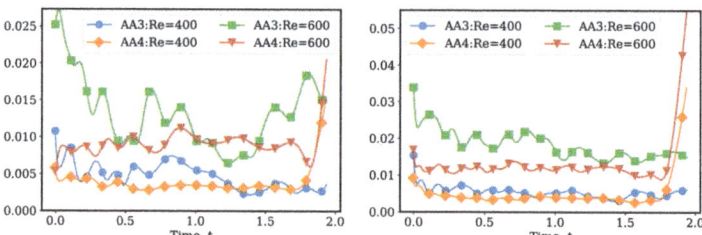

Figure 15. Relative errors of ϕ_s (left) and χ_d (right) predictions using the AA3 and AA4 models for snapshots generated with the unseen test parameter values.

Two different approaches are adopted to construct these dynamics models. In the first method, an independent LSTM network is trained for the encoded snapshots corresponding to each parameter value in $Re_{train} = \{50, 150, 300, 500\}$. The AA3 model is chosen to compute the encoded latent representations of the high-dimensional snapshots. For the datasets characterized by weaker advection, i.e., $Re = 50$ and $Re = 150$, a smaller capacity LSTM network is defined using two stacked LSTM cells with 32 hidden dimensions each and *swish* activation. For the datasets with higher values of $Re = 300$ and $Re = 500$, a higher capacity network consisting of two stacked LSTM cells with 150 hidden units was found to be necessary to accurately capture the dynamics. The network is trained to read an input consisting of 5 time steps and predict the next element in the time series. The first 90% time steps are used for training the LSTM model and the remaining 10% are used for testing the extrapolatory predictive capability of the trained LSTM model. Training is performed for 4000 epochs using the Adam optimizer with minibatches of size 24 and with an initial learning rate of 2×10^{-5} that decays by 10% every 304 epochs. For minimization of the network hyperparameters, losses in the latent space predictions are computed using the NMSE loss. Scaling the input features used for training the LSTM model was found to offer no additional benefits to the training process or improved prediction accuracy. Some preliminary testing with the use of dropout layers also yielded inconclusive evidence to support or recommend their use for further training.

In Figure 16, the latent space coefficients of the high-dimensional snapshots as defined by the AA3 autoencoder model are compared with the predictions generated by the individual LSTM models. As the AA3 model defines a latent space of dimension 5, hence each plot depicts 5 latent space modes. The plots in panels (**a**)–(**c**) are obtained with LSTM models that are trained on the latent space coefficients corresponding to snapshots in the parametric training dataset, i.e., $Re = 50, 300, 500$, respectively. The modal trajectories in panel (**d**) are obtained by evaluating the latent space coefficients for a test parameter value $Re = 400$ using the AA3 model and then training a LSTM model to learn the evolution of these coefficients. As mentioned before, the LSTM models are trained on the first 90% time steps of each timeseries and the boundary of the training data is marked by a dashed vertical line in each plot. The encoded true snapshots and the LSTM predictions for both the training and even the test parameter values display a high degree of agreement, especially for time steps within the LSTM training time window. It is also encouraging to observe that for a short length outside the training time window, even the extrapolatory predictions obtained from the trained LSTM models are in agreement with the encoded true snapshots. This behavior can also be seen in Figure 17 where the true decoder of the AA3 model is applied to the LSTM predictions and the results are compared with the high-dimensional snapshots. These plots are populated with the predicted and high-dimensional solution snapshots at four different intermediate times with the x-axis representing the spatial grid. The plotting time steps are distributed uniformly throughout the simulation time window and are chosen in such a way that the final time step $t = 1.90$ lies outside the LSTM training time window. It is clear that the trained autoencoder(AE)-LSTM model captures the viscous advecting shock-like feature fairly well, even for the extrapolatory time step. This demonstrates the ease of constructing dynamics models in a latent space defined by the parametric AA autoencoder model, even while adopting a standard implementation of a simple and lightweight LSTM network. While some discrepancies emerge with extrapolatory and longer-time prediction windows, this can be attributed to the well-known issues with the autoregressive modeling of time series data using standard LSTM networks [70]. However, for applications where time series predictions are desired over shorter time windows, the proposed AA autoencoder+LSTM approach shows the capacity for effective extrapolatory predictions.

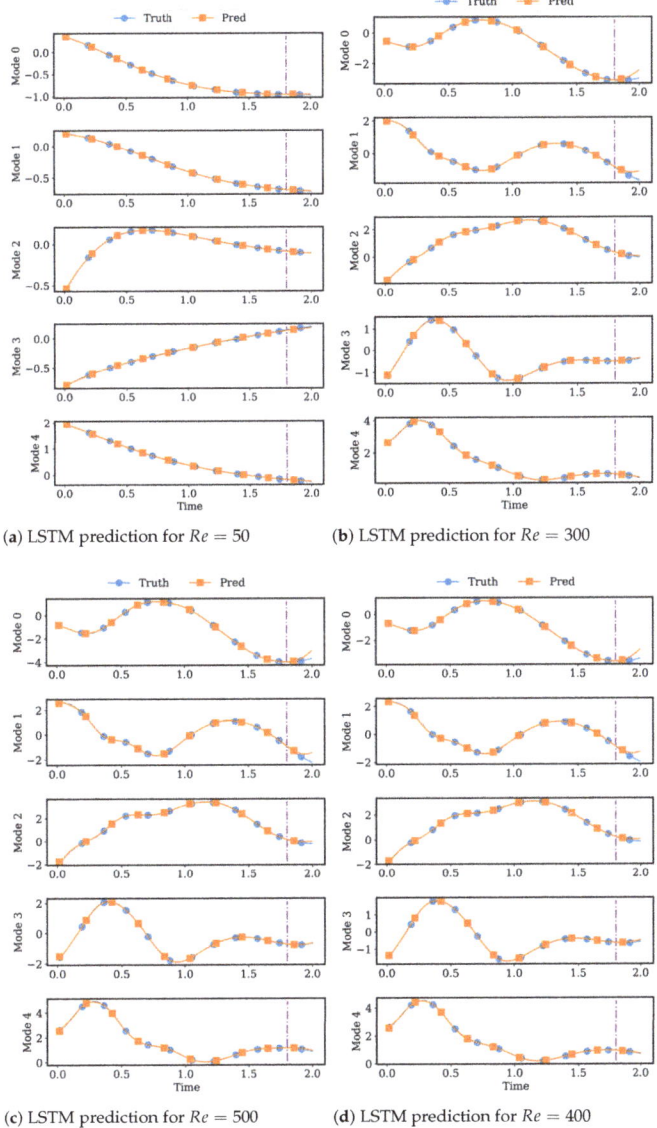

Figure 16. Comparing latent space predictions obtained using a parametric AA autoencoder and LSTM models. The LSTM models are trained separately for each training parameter value as presented in (**a**) $Re = 50$, (**b**) $Re = 300$, (**c**) $Re = 500$ and for a test parameter value shown in (**d**) $Re = 400$. All LSTM models are trained using the first 90% of the total time steps (as demarcated by the vertical lines in each figure), and the remaining time steps are used for evaluating extrapolatory predictions.

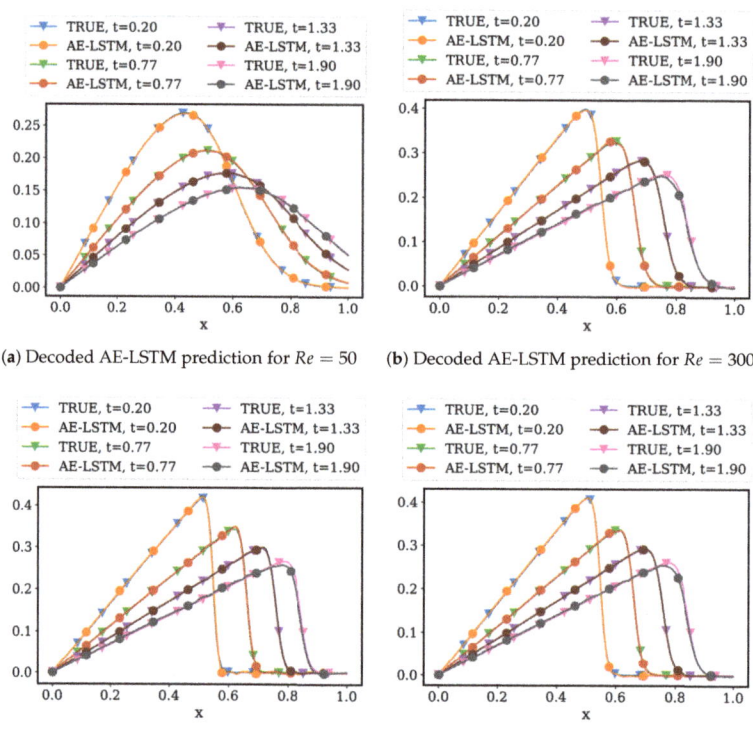

Figure 17. Comparing predictions of the high-dimensional solutions for the parametric advecting viscous shock problem using an AA autoencoder and LSTM models. The LSTM models are individually trained for each parameter value and are presented as (**a**) $Re = 50$, (**b**) $Re = 300$, (**c**) $Re = 500$, and (**d**) $Re = 400$.

In the second approach, a parametric LSTM (pLSTM) network is trained on the encoded snapshots obtained by the AA3 model. An encoded representation is obtained for every snapshot corresponding to the training parameter values, $Re_{train} = \{50, 150, 300, 500\}$. Moreover, the encoded snapshots are augmented by explicitly attaching the corresponding scaled parameter values as labels. The pLSTM network is defined using three stacked LSTM cells with 128 hidden units each and *swish* activation. The network is trained to read an input consisting of 8 time steps and predict the next element in the time series. Training is performed for 50,000 epochs using the Adam optimizer with minibatches of size 150, and with an initial learning rate of 1×10^{-3}, that decays by 20% every 2083 epochs. For optimization of the network hyperparameters, losses in the latent space predictions are computed using the NMSE loss. Similar to the previous approach, scaling the input features for the pLSTM network were not found to be beneficial.

Figure 18a,b show the comparison between the encoded true snapshots and the latent space predictions of the pLSTM model for snapshots corresponding to training parameter values $Re = 150$ and $Re = 300$, respectively. The pLSTM model can be clearly seen to approximate the time trajectory of the latent space coefficients accurately. In Figure 18c,d, the true decoder of the AA3 model is applied to the latent space predictions at four intermediate time steps and the results are compared to the high-dimensional solution snapshots. The solutions predicted by the combined AA3+pLSTM model perfectly match the high-dimensional simulation snapshots.

In the next set of numerical experiments, the trained pLSTM model is deployed to emulate the evolution of the latent space coefficients in a recursive fashion, i.e., the pLSTM model outputs for one time step are recursively rolled into the time step input window and used for pLSTM predictions at the next time step. Figure 19 shows two such examples of recursive pLSTM predictions for training parameter values $Re = 50$ and $Re = 500$, starting from randomly chosen initial time points. In panel (**a**), pLSTM predictions of the latent space evolution for $Re = 50$ are computed by randomly choosing the encoded high-dimensional solution at $t = 0.26$ as the initial data, and marching forward until $t = 2$ in a recursive fashion. Similarly, in (**b**), the initial point is chosen to be $t = 0.50$ and the latent space evolution for $Re = 500$ is computed recursively. In both cases, the predicted trajectories show remarkable agreement with the encoded high-dimensional solution trajectories. Finally, in (**c**) and (**d**), the pLSTM latent space predictions are decoded using model AA3 to compare with the high-dimensional solution snapshots at four intermediate time steps. Again, the predicted solutions are found to closely align with the true high-dimensional snapshot data.

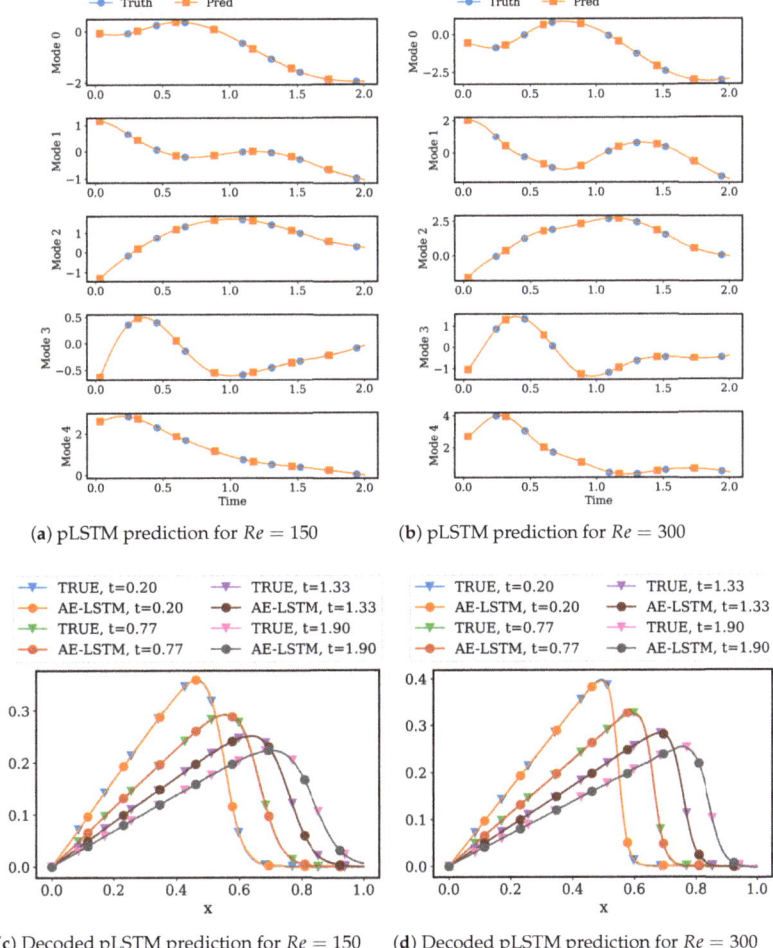

(**a**) pLSTM prediction for $Re = 150$ (**b**) pLSTM prediction for $Re = 300$

(**c**) Decoded pLSTM prediction for $Re = 150$ (**d**) Decoded pLSTM prediction for $Re = 300$

Figure 18. Comparing predictions obtained using a parametric AA autoencoder and a parametric LSTM (pLSTM) model. (**a**,**b**) show the latent space predictions using the pLSTM model and the encoded high-dimensional snapshots for training parameter values $Re = 150$ and $Re = 300$, respectively. (**c**,**d**) compare the corresponding decoded pLSTM predictions with the high-dimensional snapshots at four intermediate time steps.

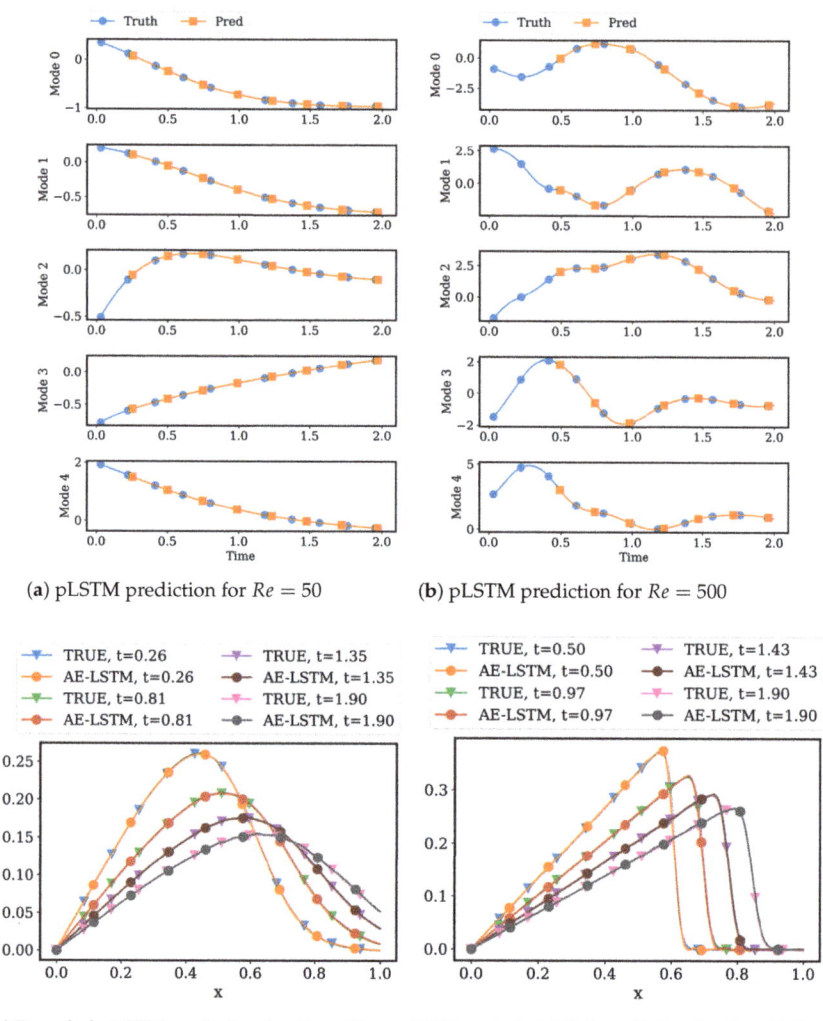

Figure 19. Comparing recursive predictions obtained using a parametric AA autoencoder and a parametric LSTM (pLSTM) model. (**a**,**b**) show the recursive latent space predictions using the pLSTM model and the encoded high-dimensional snapshots for training parameter values $Re = 50$ and $Re = 500$, respectively. (**c**,**d**) compare the corresponding decoded pLSTM predictions with the high-dimensional snapshots at four intermediate time steps.

5. Conclusions and Future Work

In this study, we propose a novel advection-aware autoencoder network that can find a low-dimensional nonlinear embedding of the salient physical features of advection-dominated transport problems. Such systems are known to exhibit instabilities and inefficient compression ratios when expressed in terms of a linear subspace defined by POD-Galerkin projection-based reduced order models. The novelty of the proposed design lies in the definition of a latent space vector that can simultaneously be efficiently mapped to the corresponding high-fidelity simulation snapshot as well as an arbitrary snapshot that effectively emulates the advective features of the high-fidelity snapshot. We demonstrate that for a linear advection problem that requires about 60 linear POD basis modes to accurately capture solution features, the AA autoencoder model can achieve a stable nonlinear embedding using a latent representation of dimension 15, and for a viscous advecting shock problem that requires about 15 POD basis modes, the AA autoencoder model can produce accurate representations with a nonlinear embedding of dimension 5. We also develop a fully, non-intrusive ROM framework by combining the AA autoencoder architecture with a separately trained LSTM network that captures the temporal dynamics in the latent space defined by the AA autoencoder model. This non-intrusive ROM formulation is extended to parametric problems by concatenating the parameter information to both the high-dimensional input snapshots for the AA autoencoder as well as the low-dimensional latent states used for the LSTM training, and then training each model independently. The proposed parametric, non-intrusive ROM is numerically evaluated on the parametric test problem involving a viscous advecting shock. The results indicate that the framework is capable of learning essential underlying features by exhaustively exploring different types of parametric design spaces. Moreover, evaluations with unseen parameter values reveal that the model is also able to produce accurate extrapolatory predictions. The numerical examples presented here demonstrate that the proposed approach is capable of handling not just uniform advection, but also nonlinear problems involving viscous shocks. The key to extending this approach to a wider class of advection-dominated problems such as solitary waves, chaotic systems, and systems governed by multiple traveling waves, lies in the appropriate selection of a shifted snapshot that optimally endows the latent space with information about the underlying advective transport. The registration-type approach of selecting an intermediate high-fidelity simulation snapshot as the shifted snapshot, as adopted here, is a versatile strategy that can be extended even to problems with multiple traveling features. For the examples studied here, the efficacy of the approach was found to be independent of the choice of the particular intermediate time step. However, a systematic sensitivity analysis is required to understand the effect of this choice for systems characterized by multiple traveling waves and chaotic dynamics.

Currently, we are engaged in developing a robust, end-to-end algorithm for concurrent training of the AA autoencoder and the dynamics model, and formulating design guidelines to help the end user choose between concurrent and separately trained ROMs. One important step towards this goal is to explore the addition of physics-based regularizing constraints to the loss function, in order to resolve some of the instabilities in the training trajectory. We are also interested in investigating the use of other ML-based and classical time series learning strategies to model the dynamical features of the latent space coefficients. Another planned direction of future work is aimed at the development of a method that can efficiently map the high-fidelity solution on to a regular, logically rectangular grid. This will allow us to construct AA autoencoder models using convolutional and deconvolutional layers that can more efficiently extract localized spatial patterns in the input features.

Author Contributions: Conceptualization, S.D. and M.W.F.; methodology, S.D. and M.W.F.; software, P.R.-C. and S.D.; validation, S.D., P.R.-C., B.S., and M.W.F.; formal analysis, S.D. and M.W.F.; data curation, S.D. and B.S.; writing—original draft preparation, S.D.; writing—review and editing, M.W.F. and P.R.-C.; visualization, S.D. and B.S.; supervision, M.W.F.; project administration, M.W.F.; funding acquisition, M.W.F. All authors have read and agreed to the published version of the manuscript.

Funding: This research received no external funding.

Data Availability Statement: All the specific details of the AA autoencoder and the LSTM frameworks used in this study may be found in the supporting source code that will be made available at https://github.com/erdc/aa_autoencoder_mca, accessed on 1 February 2022.

Acknowledgments: The authors acknowledge all the helpful comments from Orie Cecil and Adam Collins. This research was supported in part by an appointment of the first author to the Postgraduate Research Participation Program at the U.S. Army Engineer Research and Development Center, Coastal and Hydraulics Laboratory (ERDC-CHL) administered by the Oak Ridge Institute for Science and

Education through an interagency agreement between the U.S. Department of Energy and ERDC. Permission was granted by the Chief of Engineers to publish this information.

Conflicts of Interest: The authors declare no conflict of interest.

References

1. Kutz, J.N. *Data-Driven Modeling & Scientific Computation: Methods for Complex Systems & Big Data*; Oxford University Press, Inc.: Oxford, UK, 2013.
2. Holmes, P.; Lumley, J.L.; Berkooz, G. *Turbulence, Coherent Structures, Dynamical Systems and Symmetry*; Cambridge Monographs on Mechanics, Cambridge University Press: Cambridge, UK, 1996. [CrossRef]
3. Benner, P.; Gugercin, S.; Willcox, K. A Survey of Projection-Based Model Reduction Methods for Parametric Dynamical Systems. *SIAM Rev.* **2015**, *57*, 483–531. [CrossRef]
4. Brunton, S.L.; Noack, B.R.; Koumoutsakos, P. Machine Learning for Fluid Mechanics. *Annu. Rev. Fluid Mech.* **2020**, *52*, 477–508. [CrossRef]
5. Hesthaven, J.S.; Rozza, G.; Stamm, B. *Certified Reduced Basis Methods for Parametrized Partial Differential Equations*; Springer: Cham, Switzerland, 2016; pp. 1–131. [CrossRef]
6. Rowley, C.W.; Dawson, S.T. Model Reduction for Flow Analysis and Control. *Annu. Rev. Fluid Mech.* **2017**, *49*, 387–417. [CrossRef]
7. Taira, K.; Brunton, S.L.; Dawson, S.T.; Rowley, C.W.; Colonius, T.; McKeon, B.J.; Schmidt, O.T.; Gordeyev, S.; Theofilis, V.; Ukeiley, L.S. Modal analysis of fluid flows: An overview. *AIAA J.* **2017**, *55*, 4013–4041. [CrossRef]
8. Berkooz, G.; Holmes, P.; Lumley, J.L. The proper orthogonal decomposition in the analysis of turbulent flows. *Annu. Rev. Fluid Mech.* **1993**, *25*, 539–575. [CrossRef]
9. Lozovskiy, A.; Farthing, M.; Kees, C.; Gildin, E. POD-based model reduction for stabilized finite element approximations of shallow water flows. *J. Comput. Appl. Math.* **2016**, *302*, 50–70. [CrossRef]
10. Lozovskiy, A.; Farthing, M.; Kees, C. Evaluation of Galerkin and Petrov–Galerkin model reduction for finite element approximations of the shallow water equations. *Comput. Methods Appl. Mech. Eng.* **2017**, *318*, 537–571. [CrossRef]
11. Carlberg, K.; Barone, M.; Antil, H. Galerkin v. least-squares Petrov–Galerkin projection in nonlinear model reduction. *J. Comput. Phys.* **2017**, *330*, 693–734. [CrossRef]
12. Dutta, S.; Rivera-Casillas, P.; Cecil, O.; Farthing, M. pyNIROM—A suite of python modules for non-intrusive reduced order modeling of time-dependent problems. *Softw. Impacts* **2021**, *10*, 100129. [CrossRef]
13. Alla, A.; Kutz, J.N. Nonlinear model order reduction via dynamic mode decomposition. *SIAM J. Sci. Comput.* **2017**, *39*, B778–B796. [CrossRef]
14. Wu, Z.; Brunton, S.L.; Revzen, S. Challenges in Dynamic Mode Decomposition. *J. R. Soc. Interface* **2021**, *18*, 20210686. [CrossRef] [PubMed]
15. Xiao, D.; Fang, F.; Pain, C.C.; Navon, I.M. A parameterized non-intrusive reduced order model and error analysis for general time-dependent nonlinear partial differential equations and its applications. *Comput. Methods Appl. Mech. Eng.* **2017**, *317*, 868–889. [CrossRef]
16. Dutta, S.; Farthing, M.W.; Perracchione, E.; Savant, G.; Putti, M. A greedy non-intrusive reduced order model for shallow water equations. *J. Comput. Phys.* **2021**, *439*, 110378. [CrossRef]
17. Guo, M.; Hesthaven, J.S. Data-driven reduced order modeling for time-dependent problems. *Comput. Methods Appl. Mech. Eng.* **2019**, *345*, 75–99. [CrossRef]
18. Xiao, D. Error estimation of the parametric non-intrusive reduced order model using machine learning. *Comput. Methods Appl. Mech. Eng.* **2019**, *355*, 513–534. [CrossRef]
19. Hesthaven, J.S.; Ubbiali, S. Non-intrusive reduced order modeling of nonlinear problems using neural networks. *J. Comput. Phys.* **2018**, *363*, 55–78. [CrossRef]
20. Wan, Z.Y.; Vlachas, P.; Koumoutsakos, P.; Sapsis, T. Data-assisted reduced-order modeling of extreme events in complex dynamical systems. *PLoS ONE* **2018**, *13*, e0197704. [CrossRef]
21. Maulik, R.; Mohan, A.; Lusch, B.; Madireddy, S.; Balaprakash, P.; Livescu, D. Time-series learning of latent-space dynamics for reduced-order model closure. *Phys. D Nonlinear Phenom.* **2020**, *405*, 132368. [CrossRef]
22. Chen, R.T.Q.; Rubanova, Y.; Bettencourt, J.; Duvenaud, D. Neural Ordinary Differential Equations. In Proceedings of the 32nd International Conference on Neural Information Processing Systems (NIPS'18), Montréal, QC, Canada, 2–8 December 2018; pp. 6572–6583. [CrossRef]
23. Dutta, S.; Rivera-Casillas, P.; Farthing, M.W. Neural Ordinary Differential Equations for Data-Driven Reduced Order Modeling of Environmental Hydrodynamics. In Proceedings of the AAAI 2021 Spring Symposium on Combining Artificial Intelligence and Machine Learning with Physical Sciences, Virtual Meeting, 22–24 March 2021; CEUR-WS: Stanford, CA, USA, 2021.
24. Wu, P.; Sun, J.; Chang, X.; Zhang, W.; Arcucci, R.; Guo, Y.; Pain, C.C. Data-driven reduced order model with temporal convolutional neural network. *Comput. Methods Appl. Mech. Eng.* **2020**, *360*, 112766. [CrossRef]
25. Taddei, T.; Perotto, S.; Quarteroni, A. Reduced basis techniques for nonlinear conservation laws. *ESAIM Math. Model. Numer. Anal.* **2015**, *49*, 787–814. [CrossRef]
26. Greif, C.; Urban, K. Decay of the Kolmogorov N-width for wave problems. *Appl. Math. Letters* **2019**, *96*, 216–222. [CrossRef]

27. Carlberg, K.; Farhat, C.; Cortial, J.; Amsallem, D. The GNAT method for nonlinear model reduction: Effective implementation and application to computational fluid dynamics and turbulent flows. *J. Comput. Phys.* **2013**, *242*, 623–647. [CrossRef]
28. Nair, N.J.; Balajewicz, M. Transported snapshot model order reduction approach for parametric, steady-state fluid flows containing parameter-dependent shocks. *Int. J. Numer. Methods Eng.* **2019**, *117*, 1234–1262. [CrossRef]
29. Rim, D.; Moe, S.; LeVeque, R.J. Transport reversal for model reduction of hyperbolic partial differential equations. *SIAM/ASA J. Uncertain. Quantif.* **2018**, *6*, 118–150. [CrossRef]
30. Reiss, J.; Schulze, P.; Sesterhenn, J.; Mehrmann, V. The shifted proper orthogonal decomposition: A mode decomposition for multiple transport phenomena. *SIAM J. Sci. Comput.* **2018**, *40*, A1322–A1344. [CrossRef]
31. Rim, D.; Peherstorfer, B.; Mandli, K.T. Manifold approximations via transported subspaces: Model reduction for transport-dominated problems. *arXiv* **2019**, arXiv:1912.13024.
32. Cagniart, N.; Maday, Y.; Stamm, B. Model order reduction for problems with large convection effects. In *Contributions to Partial Differential Equations and Applications*; Springer International Publishing: Cham, Switzerland, 2019; pp. 131–150. [CrossRef]
33. Taddei, T. A registration method for model order reduction: Data compression and geometry reduction. *SIAM J. Sci. Comput.* **2020**, *42*, A997–A1027. [CrossRef]
34. Peherstorfer, B. Model reduction for transport-dominated problems via online adaptive bases and adaptive sampling. *SIAM J. Sci. Comput.* **2020**, *42*, A2803–A2836. [CrossRef]
35. Kashima, K. Nonlinear model reduction by deep autoencoder of noise response data. In Proceedings of the 2016 IEEE 55th Conference on Decision and Control (CDC), Las Vegas, NV, USA, 12–14 December 2016; pp. 5750–5755. [CrossRef]
36. Lee, K.; Carlberg, K.T. Model reduction of dynamical systems on nonlinear manifolds using deep convolutional autoencoders. *J. Comput. Phys.* **2020**, *404*, 108973. [CrossRef]
37. Kim, Y.; Choi, Y.; Widemann, D.; Zohdi, T. A fast and accurate physics-informed neural network reduced order model with shallow masked autoencoder. *J. Comput. Phys.* **2022**, *451*, 110841. [CrossRef]
38. Willcox, K. Unsteady flow sensing and estimation via the gappy proper orthogonal decomposition. *Comput. Fluids* **2006**, *35*, 208–226. [CrossRef]
39. Chaturantabut, S.; Sorensen, D.C. Nonlinear model reduction via Discrete Empirical Interpolation. *SIAM J. Sci. Comput.* **2010**, *32*, 2737–2764. [CrossRef]
40. Mendible, A.; Brunton, S.L.; Aravkin, A.Y.; Lowrie, W.; Kutz, J.N. Dimensionality reduction and reduced-order modeling for traveling wave physics. *Theor. Comput. Fluid Dyn.* **2020**, *34*, 385–400. [CrossRef]
41. Haasdonk, B.; Ohlberger, M. Adaptive basis enrichment for the reduced basis method applied to finite volume schemes. In Proceedings of the Fifth International Symposium on Finite Volumes for Complex Applications, Aussois, France, 8–13 June 2008; pp. 471–479.
42. Chen, P.; Quarteroni, A.; Rozza, G. A weighted empirical interpolation method: A priori convergence analysis and applications. *ESAIM Math. Model. Numer. Anal.* **2014**, *48*, 943–953. [CrossRef]
43. Amsallem, D.; Farhat, C. An online method for interpolating linear parametric reduced-order models. *SIAM J. Sci. Comput.* **2011**, *33*, 2169–2198. [CrossRef]
44. Maday, Y.; Stamm, B. Locally adaptive greedy approximations for anisotropic parameter reduced basis spaces. *SIAM J. Sci. Comput.* **2013**, *35*, A2417–A2441. [CrossRef]
45. Peherstorfer, B.; Butnaru, D.; Willcox, K.; Bungartz, H.J. Localized discrete empirical interpolation method. *SIAM J. Sci. Comput.* **2014**, *36*, A168–A192. [CrossRef]
46. Carlberg, K. Adaptive h-refinement for reduced-order models. *Int. J. Numer. Methods Eng.* **2015**, *102*, 1192–1210. [CrossRef]
47. Peherstorfer, B.; Willcox, K. Online adaptive model reduction for nonlinear systems via low-rank updates. *SIAM J. Sci. Comput.* **2015**, *37*, A2123–A2150. [CrossRef]
48. Tenenbaum, J. Mapping a manifold of perceptual observations. In *Advances in Neural Information Processing Systems*; Jordan, M., Kearns, M., Solla, S., Eds.; MIT Press: Cambridge, MA, USA, 1998; Volume 10.
49. Roweis, S.T.; Saul, L.K. Nonlinear dimensionality reduction by locally linear embedding. *Science* **2000**, *290*, 2323–2326. [CrossRef]
50. Belkin, M.; Niyogi, P. Laplacian eigenmaps for dimensionality reduction and data representation. *Neural Comput.* **2003**, *15*, 1373–1396. [CrossRef]
51. van der Maaten, L.; Hinton, G. Visualizing data using t-SNE. *J. Mach. Learn. Res.* **2008**, *9*, 2579–2605.
52. Kohonen, T. Self-organized formation of topologically correct feature maps. *Biol. Cybern.* **1982**, *43*, 59–69. [CrossRef]
53. Mika, S.; Schölkopf, B.; Smola, A.; Müller, K.R.; Scholz, M.; Rätsch, G. Kernel PCA and de-noising in feature spaces. In *Advances in Neural Information Processing Systems*; Kearns, M., Solla, S., Cohn, D., Eds.; MIT Press: Cambridge, MA, USA, 1999; Volume 11.
54. Walder, C.; Schölkopf, B. Diffeomorphic dimensionality reduction. In *Advances in Neural Information Processing Systems 21*; Koller, D., Schuurmans, D., Bengio, Y., Bottou, L., Eds.; Curran Associates Inc.: Red Hook, NY, USA, 2009; pp. 1713–1720.
55. Hinton, G.E.; Salakhutdinov, R.R. Reducing the dimensionality of data with neural networks. *Science* **2006**, *313*, 504–507. [CrossRef] [PubMed]
56. Masci, J.; Meier, U.; Cireşan, D.; Schmidhuber, J. Stacked Convolutional Auto-Encoders for Hierarchical Feature Extraction. In *Artificial Neural Networks and Machine Learning—ICANN 2011*; Honkela, T., Duch, W., Girolami, M., Kaski, S., Eds.; Springer: Berlin/Heidelberg, Germany, 2011; pp. 52–59.

57. Radford, A.; Metz, L.; Chintala, S. Unsupervised representation learning with deep convolutional generative adversarial networks. In Proceedings of the 4th International Conference on Learning Representations (ICLR 2016), San Juan, Puerto Rico, 2–4 May 2016; pp. 1–16.
58. Champion, K.; Lusch, B.; Nathan Kutz, J.; Brunton, S.L. Data-driven discovery of coordinates and governing equations. *Proc. Natl. Acad. Sci. USA* **2019**, *116*, 22445–22451. [CrossRef]
59. Dutta, S.; Rivera-Casillas, P.; Cecil, O.M.; Farthing, M.W.; Perracchione, E.; Putti, M. Data-driven reduced order modeling of environmental hydrodynamics using deep autoencoders and neural ODEs. *arXiv* **2021**, arXiv:2107.02784.
60. Maulik, R.; Lusch, B.; Balaprakash, P. Reduced-order modeling of advection-dominated systems with recurrent neural networks and convolutional autoencoders. *Phys. Fluids* **2021**, *33*, 037106. [CrossRef]
61. Wehmeyer, C.; Noé, F. Time-lagged autoencoders: Deep learning of slow collective variables for molecular kinetics. *J. Chem. Phys.* **2018**, *148*, 241703. [CrossRef]
62. Nishizaki, H. Data augmentation and feature extraction using variational autoencoder for acoustic modeling. In Proceedings of the 2017 Asia-Pacific Signal and Information Processing Association Annual Summit and Conference (APSIPA ASC), Kuala Lumpur, Malaysia, 12–15 December 2017; pp. 1222–1227. [CrossRef]
63. Bakarji, J.; Champion, K.; Kutz, J.N.; Brunton, S.L. Discovering Governing Equations from Partial Measurements with Deep Delay Autoencoders. *arXiv* **2022**, arXiv:2201.05136.
64. Erichson, N.B.; Muehlebach, M.; Mahoney, M.W. Physics-informed Autoencoders for Lyapunov-stable Fluid Flow Prediction. *arXiv* **2019**, arXiv:1905.10866.
65. Gonzalez, F.J.; Balajewicz, M. Deep convolutional recurrent autoencoders for learning low-dimensional feature dynamics of fluid systems. *arXiv* **2018**, arXiv:1808.01346.
66. Mojgani, R.; Balajewicz, M. Low-Rank Registration Based Manifolds for Convection-Dominated PDEs. In Proceedings of the Thirty-Fifth AAAI Conference on Artificial Intelligence (AAAI-21), Vancouver, BC, Canada, 2–9 February 2021; pp. 399–407.
67. Plaut, E. From Principal Subspaces to Principal Components with Linear Autoencoders. *arXiv* **2018**, arXiv:1804.10253.
68. Greff, K.; Srivastava, R.K.; Koutník, J.; Steunebrink, B.R.; Schmidhuber, J. LSTM: A search space odyssey. *IEEE Trans. Neural Netw. Learn. Syst.* **2016**, *28*, 2222–2232. [CrossRef]
69. Eivazi, H.; Veisi, H.; Naderi, M.H.; Esfahanian, V. Deep neural networks for nonlinear model order reduction of unsteady flows. *Phys. Fluids* **2020**, *32*, 105104. [CrossRef]
70. Maulik, R.; Lusch, B.; Balaprakash, P. Non-autoregressive time-series methods for stable parametric reduced-order models. *Phys. Fluids* **2020**, *32*, 087115. [CrossRef]
71. del Águila Ferrandis, J.; Triantafyllou, M.S.; Chryssostomidis, C.; Karniadakis, G.E. Learning functionals via LSTM neural networks for predicting vessel dynamics in extreme sea states. *Proc. R. Soc. A* **2021**, *477*, 20190897. [CrossRef]
72. Chattopadhyay, A.; Mustafa, M.; Hassanzadeh, P.; Kashinath, K. Deep Spatial Transformers for Autoregressive Data-Driven Forecasting of Geophysical Turbulence. In Proceedings of the 10th International Conference on Climate Informatics, Oxford, UK, 22–25 September 2020; Association for Computing Machinery: New York, NY, USA, 2020; pp. 106–112. [CrossRef]
73. Pathak, J.; Hunt, B.; Girvan, M.; Lu, Z.; Ott, E. Model-Free Prediction of Large Spatiotemporally Chaotic Systems from Data: A Reservoir Computing Approach. *Phys. Rev. Lett.* **2018**, *120*, 24102. [CrossRef]
74. Usman, A.; Rafiq, M.; Saeed, M.; Nauman, A.; Almqvist, A.; Liwicki, M. Machine learning-accelerated computational fluid dynamics. *Proc. Natl. Acad. Sci. USA* **2021**, *118*, e2101784118. [CrossRef]
75. Stabile, G.; Zancanaro, M.; Rozza, G. Efficient geometrical parametrization for finite-volume-based reduced order methods. *Int. J. Numer. Methods Eng.* **2020**, *121*, 2655–2682. [CrossRef]

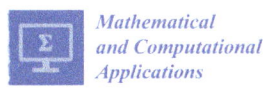

Article

Challenges in Kinetic-Kinematic Driven Musculoskeletal Subject-Specific Infant Modeling

Yeram Lim [1], Tamara Chambers [1], Christine Walck [1], Safeer Siddicky [2,3], Erin Mannen [2] and Victor Huayamave [1,*]

[1] Department of Mechanical Engineering, Embry-Riddle Aeronautical University, Daytona Beach, FL 32114, USA; ydlim.3@gmail.com (Y.L.); chambet2@my.erau.edu (T.C.); daileyc1@erau.edu (C.W.)

[2] Mechanical and Biomedical Engineering Department, Boise State University, Boise, ID 83725, USA; safeersiddicky@boisestate.edu (S.S.); erinmannen@boisestate.edu (E.M.)

[3] Department of Kinesiology and Health Education, The University of Texas at Austin, Austin, TX 78712, USA

* Correspondence: huayamav@erau.edu

Abstract: Musculoskeletal computational models provide a non-invasive approach to investigate human movement biomechanics. These models could be particularly useful for pediatric applications where in vivo and in vitro biomechanical parameters are difficult or impossible to examine using physical experiments alone. The objective was to develop a novel musculoskeletal subject-specific infant model to investigate hip joint biomechanics during cyclic leg movements. Experimental motion-capture marker data of a supine-lying 2-month-old infant were placed on a generic GAIT 2392 OpenSim model. After scaling the model using body segment anthropometric measurements and joint center locations, inverse kinematics and dynamics were used to estimate hip ranges of motion and moments. For the left hip, a maximum moment of 0.975 Nm and a minimum joint moment of 0.031 Nm were estimated at 34.6° and 65.5° of flexion, respectively. For the right hip, a maximum moment of 0.906 Nm and a minimum joint moment of 0.265 Nm were estimated at 23.4° and 66.5° of flexion, respectively. Results showed agreement with reported values from the literature. Further model refinements and validations are needed to develop and establish a normative infant dataset, which will be particularly important when investigating the movement of infants with pathologies such as developmental dysplasia of the hip. This research represents the first step in the longitudinal development of a model that will critically contribute to our understanding of infant growth and development during the first year of life.

Keywords: musculoskeletal model; infant movement; biomechanics; motion capture; OpenSim

Citation: Lim, Y.; Chambers, T.; Walck, C.; Siddicky, S.; Mannen, E.; Huayamave, V. Challenges in Kinetic-Kinematic Driven Musculoskeletal Subject-Specific Infant Modeling. *Math. Comput. Appl.* **2022**, *27*, 36. https://doi.org/10.3390/mca27030036

Academic Editors: Simona Perotto, Gianluigi Rozza and Antonia Larese

Received: 21 January 2022
Accepted: 19 April 2022
Published: 22 April 2022

Publisher's Note: MDPI stays neutral with regard to jurisdictional claims in published maps and institutional affiliations.

Copyright: © 2022 by the authors. Licensee MDPI, Basel, Switzerland. This article is an open access article distributed under the terms and conditions of the Creative Commons Attribution (CC BY) license (https:// creativecommons.org/licenses/by/ 4.0/).

1. Introduction

Human movements are complex event sequences that involve high coordination levels between musculoskeletal and neurological systems. Establishing the normative characteristics of specific human movements is particularly important when investigating individuals who have pathologies preventing natural movements. In the biomechanics field, aspects of human movements, such as segmental kinematics, kinetics, and muscle activity, are experimentally quantified and characterized by using established methodologies, such as marker-based motion capture (MOCAP), inertial measurement units, force plates, and electromyography. While these technologies are non-invasive, they require the presence of human subjects in the laboratory following specific instructions to obtain a useful dataset. For common human movements such as walking gait, once normative experimental data ranges are established for specific populations, biomechanists often turn to musculoskeletal computational models (MCM). MCM are convenient to non-invasively study the simulated dynamics of human movement, bypassing the investigation of in vivo and in vitro biomechanical parameters that may be difficult or impossible to examine using physical experiments alone.

Over the past two decades, the complexity and quality of MCM advanced at a rapid rate. In the biomechanics literature, MCMs have been used for a wide range of applications, including sports performance [1,2], clinical outcomes [3–8], occupational ergonomics [9,10], and accident reconstruction [11]. While there has been a steady advance in adult human MCM [12–14], neonatal and infant populations have been widely under-investigated in modeling.

In early infancy, when babies experience rapid development, it is vital to understand the nature of infant movements in daily body positioning environments and the effect on healthy musculoskeletal development. Novel MCMs have studied these effects during fetal movement [15–17] and pathological conditions in infancy, such as developmental dysplasia of the hip (DDH) [18–20]. DDH is an abnormal condition in infants characterized by dislocation, misalignment, or instability of the hip [21,22]. Infants are at a greater risk of DDH if they were in a breech position during delivery, are female, are the first-born, or have a family history of DDH [23]. Challenges in establishing a normative dataset are due to difficulties in recruiting infant subjects and the dearth of biomechanical studies examining movement and coordination in early infancy. Developing realistic infant MCM is crucial, particularly due to the paucity of experimental infant data in the literature. Additionally, since infants experience rapid growth in their first year of life, it is equally important to develop subject-specific infant MCM to obtain a more robust understanding of their movements and the subsequent joint and musculoskeletal development. MCM achieve the closest approximation to physiologically accurate movements when they are developed as subject-specific models. However, developing subject-specific MCM is a complicated, multi-layered process, often involving segmental anthropometric measurements, 3D imaging such as MRI/CT, 3D kinematics using MOCAP, and electromyography.

Typically, subject-specific MCMs are developed by scaling the generic model using the subject's segment anthropometries [24]. The segment lengths can be calculated using surface anatomical landmarks [24] alone or in combination with joint center locations. Kainz et al. [25] found that incorporating joint centers in the scaling process significantly increased the accuracy of the thigh and shank segment estimates when compared to scaling with surface markers alone. The hip joint center (HJC) locations are difficult to estimate because HJC locations cannot be directly identified from surface marker locations. HJC locations can be estimated using functional estimation methods or regression equations. Knee and ankle joint centers can also be estimated using functional methods. Functional approaches are implemented during MOCAP for subjects who have a sufficient hip range of motion and can easily perform the instructed functional movements [24]. However, regression equations are implemented after MOCAP for subjects who have a limited hip range of motion [24] or cannot perform the required movements. Both approaches are accepted methods of calculating HJC locations when medical imaging is not available [26], which is the case for infant populations under the age of one year where MRI/CT is unavailable.

Previous studies have attempted to quantify an infant's joint kinematics and kinetics during spontaneous movement, but an infant MCM has yet to be created. The purpose of this work was to develop a novel lower extremity computational musculoskeletal model representative of an infant, using body segment anthropometric measurements and experimental MOCAP data of a single infant taken from previously collected infant biomechanics data [27,28]. This study will provide more insight into biomechanical loadings at the hip joint during a spontaneous kick and will provide a noninvasive technique for evaluating the mechanisms contributing to infant hip development.

2. Materials and Methods

2.1. Experimental Methods

De-identified experimental data for one healthy, full-term male infant (2.4 months) was obtained from a study approved by the Institutional Review Board of the University of Arkansas for Medical Sciences [28]. The infant was weighed on an infant scale at 5.35 kg and was measured head to heel (lying supine) at 56 cm. Leg length measurements were made with a standard measuring tape, with the right and left leg measuring 23 cm and 22 cm, respectively. Marker-based MOCAP (100 Hz; Vicon, Oxford, UK) recorded movement through reflective markers placed bilaterally on the anterior and posterior of the head, anterior superior iliac spine (ASIS), posterior superior iliac spine (PSIS), greater trochanter, medial and lateral epicondyles of the knee, and the medial and lateral malleolus of the ankle. Additionally, three-marker rigid bodies were placed on the anterior and posterior of the pelvis and bilaterally on the lateral aspect of each thigh. Data were recorded over a 30 s period with the infant lying in the supine position and was allowed to move freely and naturally without any external stimulations as shown in Figure 1a.

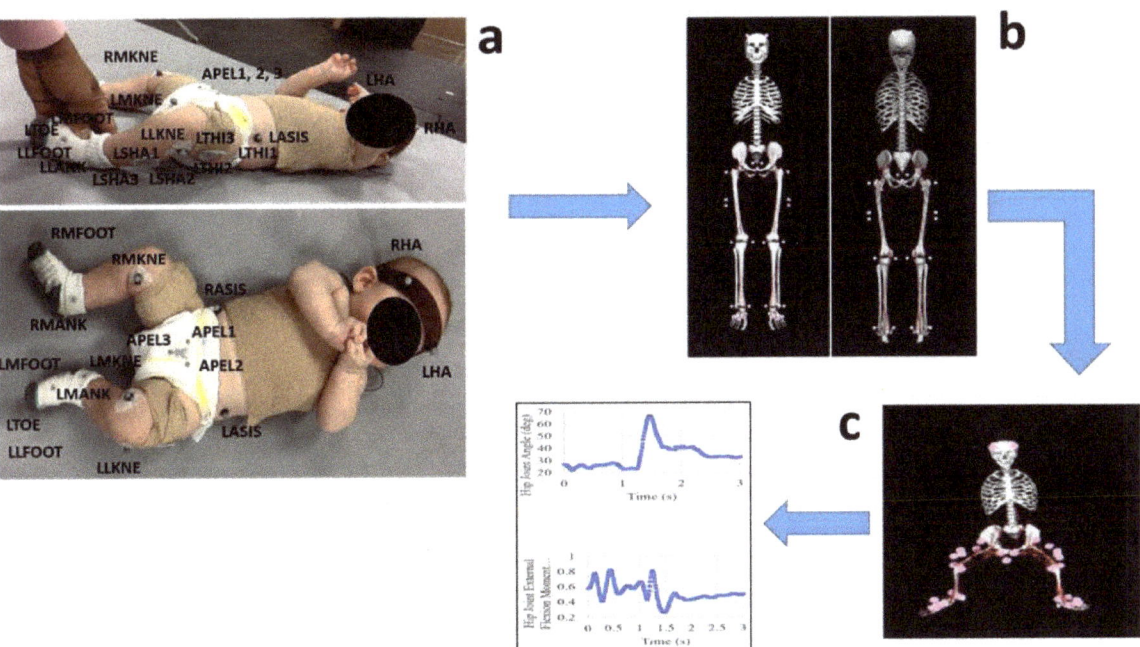

Figure 1. Methodology pipeline: (**a**) infant MOCAP data collection, (**b**) musculoskeletal scaling, and (**c**) musculoskeletal model to predict hip joint ROMs and external hip joint moments.

2.2. Musculoskeletal Model Development

Markers from the experiment were placed on a generic GAIT 2392 OpenSim model [29,30], which is used for simulating lower extremity dominant motions. This model has 23 degrees of freedom and 96 musculotendon actuators that represent 76 muscles in the lower extremity, including the pelvis, femur, tibia, fibula, talus, foot, and toes.

To create a subject-specific model representative of an infant's size, marker-based scaling was used. This scaling approach minimizes the distance between experimental and model (virtual) markers through optimization and predicts scale factors. The optimization used during scaling was a weighted least squares optimization problem given by Equation (1) [21], where q is the vector of generalized coordinates being solved for, x_i^{exp} is the experimental position of marker i, $x_i(q)$ is the position of the corresponding model marker (which depends on the coordinate values), and q_j^{exp} is the experimental value for coordinate j. The prescribed coordinates are set to their experimental values. The goal of the optimization problem was to place the model markers in the position that closely matched the subject's position while minimizing marker errors.

$$\min_q \left[\sum_{i \in \text{markers}} w_i \| x_i^{\exp} - x_i(q) \|^2 + \sum_{i \in \text{unprescribed coords}} w_j \left(q_j^{\exp} - q_j \right)^2 \right] \quad (1)$$

$$q_j = q_j^{\exp} \text{ for all prescribed coordinates j}$$

Functional joint centers for the hips, knees, and ankles were estimated and used during the scaling process. The ankle and knee joint centers were calculated using the midpoint between the medial and lateral marker positions. The hip joint centers (HJC) were estimated using regression equations based on the subject's leg length (LL) [31], as shown in Equations (2)–(4).

$$HJC_x = 11 - 0.063 \times LL \quad (2)$$

$$HJC_y = 8 + 0.086 \times LL \quad (3)$$

$$HJC_z = -9 - 0.078 \times LL \quad (4)$$

The mean absolute errors for the posterior-anterior (HJC_x), medial-lateral (HJC_y), and inferior-superior (HJC_z) position equations are 5.2 mm, 4.4 mm, and 3.8 mm, respectively. The femur and tibia were both scaled non-uniformly by using a scale factor for the medial-lateral direction and the superior-inferior direction. Since the infant was placed in the supine position, the posterior pelvic markers were not captured during MOCAP. Thus, the pelvis was scaled uniformly with respect to the medial-lateral direction.

The average upper-segment lower-segment (USLS) infant ratio [32], which compares the upper segment (torso) and lower segment (legs), was also used to validate scaling. A 5% difference was found between the reported USLS infant ratio (1.70) and the scaled musculoskeletal model USLS infant ratio (1.61). In addition, a user-defined constant ground reaction force (GRF) was defined to represent the infant's weight normal to the ground and was applied at the infant's coccyx. The GRFs in the shear directions were neglected since the infant was lying supine and the motion was only observed in the lower extremity. Typically, generic models use a synchronous GRF during scaling obtained from a calibration sequence completed by the participant (either a static pose or a series of predefined functional movements). For infant MOCAP studies, infants are generally placed in a supine position, and there is not much variation in GRFs. Therefore, we assumed a constant GRF concentrated at the coccyx. Inverse kinematics and inverse dynamics were then used to estimate hip joint range of motion (ROM) and external moments. Figure 1 shows the methodology pipeline used in this study. Figure 2 shows the subject-specific musculoskeletal infant model.

2.3. Outcome Measures

A movement of the hip joint beginning from an extended position moving through a single flexion phase and then returning to the extended position is defined as a single kick. A single, discrete kick, defined by a kick where no other kicking motion is observed within 1 s before and after the kick, was identified and isolated over a 3 s period for each hip to visualize the extended–flexed–extended pattern [33]. To compare the kick on the left and right leg, the kick-start time was adjusted to plot the kicks generated by both the right and left hips.

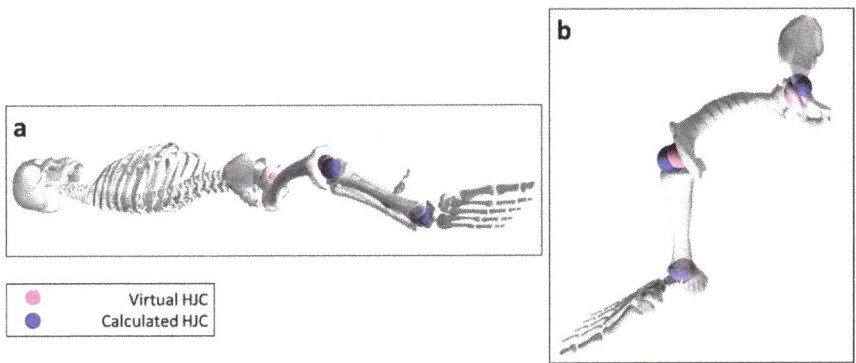

Figure 2. Subject-specific OpenSim model displaying the virtual joint center marker positions in pink and the calculated joint center marker positions in blue. (**a**) Right-side view of the musculoskeletal model in the supine position and (**b**) isolated right lower extremity. Head and torso bodies are not to scale.

3. Results

3.1. Hip Joint Range of Motion

Over a 3 s period, minimum joint angle values of 22.2° (right hip) and 23.2° (left hip) and maximum joint angle values of 66.6° (right hip) and 66.3° (left hip) were predicted as shown in Figure 3. For both kicks, the maximum hip flexion was calculated to be approximately 66.0°. Additionally, Figure 4 shows the beginning and end of the kick cycle for both the right and left hips and identifies all different stages of the motion defined by each single kick, which was classified as discrete.

Once both kicks were classified, the cyclic data for the right hip and left hip were isolated again to show the beginning and end of the kick cycle. Figure 4 shows details of flexion and extension for both hips where the kick starts with the hip moving towards the trunk as it reaches maximum flexion and then moves away from the body as the hip extends.

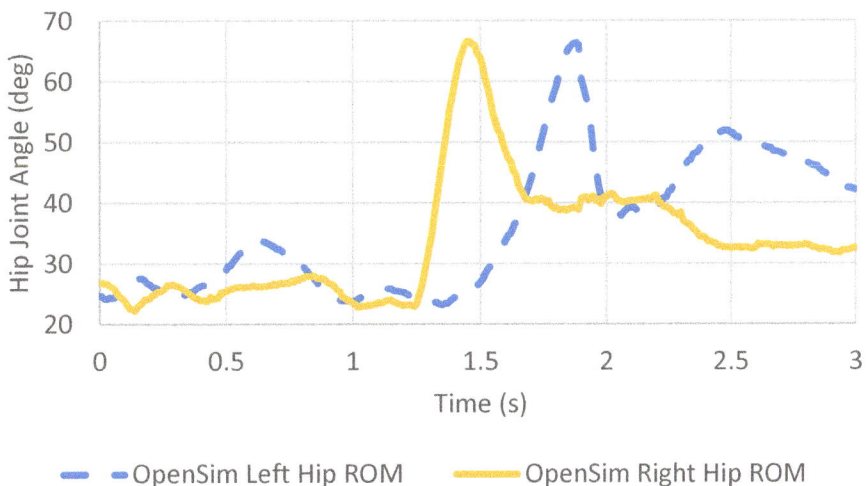

Figure 3. Hip joint ROM for left and right hips over a 3 s period.

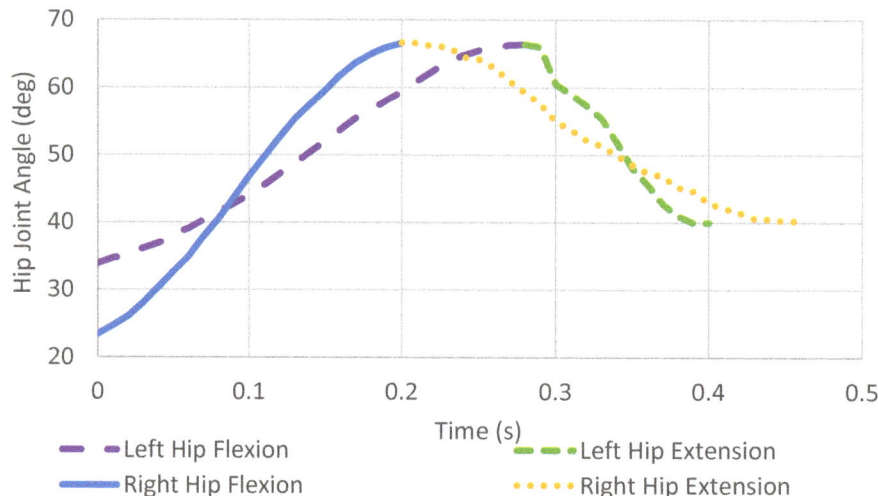

Figure 4. Hip joint angle data during discrete isolated kicks for right and left hips.

3.2. Hip Joint Moment

The correlating hip joint moment results of the discrete kicks for the right and left hip joints are shown in Figure 5 and are consistent with the observed hip ranges of motion. Figure 5 shows that the maximum torque is required at the beginning of the kick cycle when the hip joint torque must overcome the gravitational force to reach maximum flexion. Once the gravitational force is overcome, the moment decreases to slow down the motion as it reaches maximum flexion. Similarly, the moment increases as the motion reverses as it goes into extension where the maximum torque is needed at the end of the kick cycle to slow down the hip joint.

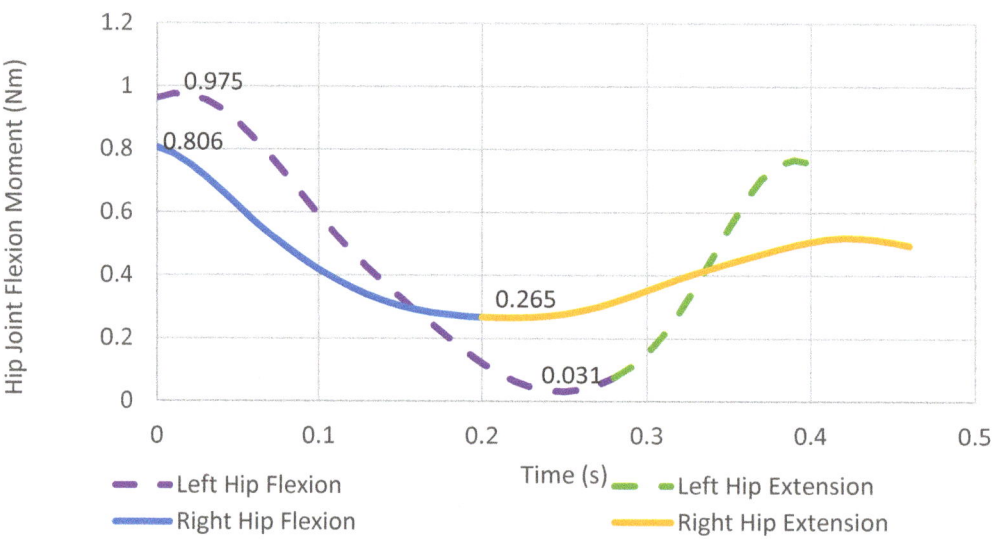

Figure 5. Hip joint moment during discrete isolated kicks for right and left hips.

4. Discussion

The purpose of this study was to create a preliminary musculoskeletal computational model representative of an infant to study the biomechanics of the lower extremity. By using experimental motion-capture data along with external forces, a novel musculoskeletal model was created using inverse kinematics and dynamics. Our preliminary results suggest that the musculoskeletal model is able to portray the biomechanics of infants when estimating hip joint ROM and moments to investigate healthy infant movement.

Our results for the hip flexion angle and hip moment were normalized by the time scale to better compare the kicks with varying kick duration to the kicks reported in the literature [33]. Figure 6 shows a trend agreement between the reported values and our results. However, reported flexion values were higher when compared to our model. There was a maximum hip joint angle difference of 30.5° (right hip) and 30.8° (left hip) when compared to the literature. This maximum difference was observed at the corresponding time of maximum hip flexion angle. This difference is attributed to dissimilarities in data collection methodologies and subject heterogeneity. In the methods used by Schneider et al. [33], the infant subject's chest and abdomen were immobilized while the lower extremity could move freely and naturally. On the other hand, our study permitted no movement restrictions, allowing the upper extremity as well as the lower extremity to move freely and naturally in coordination with the upper body. With these different approaches, we see that the kick cycle ends at a greater hip joint angle for both hips compared to the starting hip joint angle at the beginning of the kick cycle. This may be caused by the additional muscle control needed when motion is reversed from flexion to extension to stabilize the hip joint.

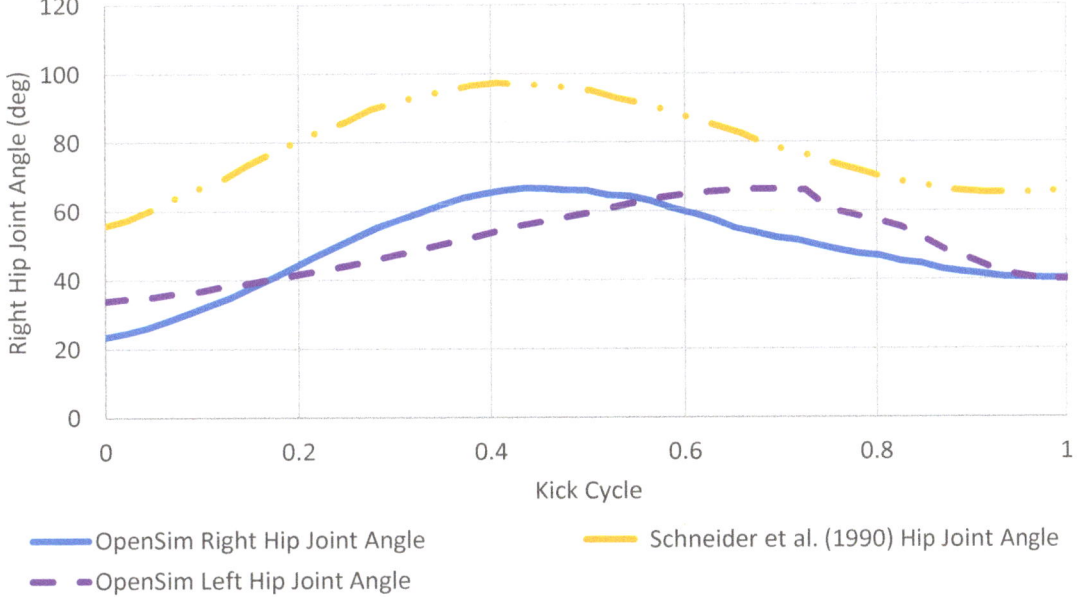

Figure 6. Time normalized hip joint ROM comparison between infant model and Schneider et al.'s [33].

Hip joint moment data obtained through the musculoskeletal model were also normalized and compared to values reported in the literature [33]. In both results, a decrease in moment was observed during hip flexion, and an increase in moment was observed during the extension of the hip, as shown in Figure 7.

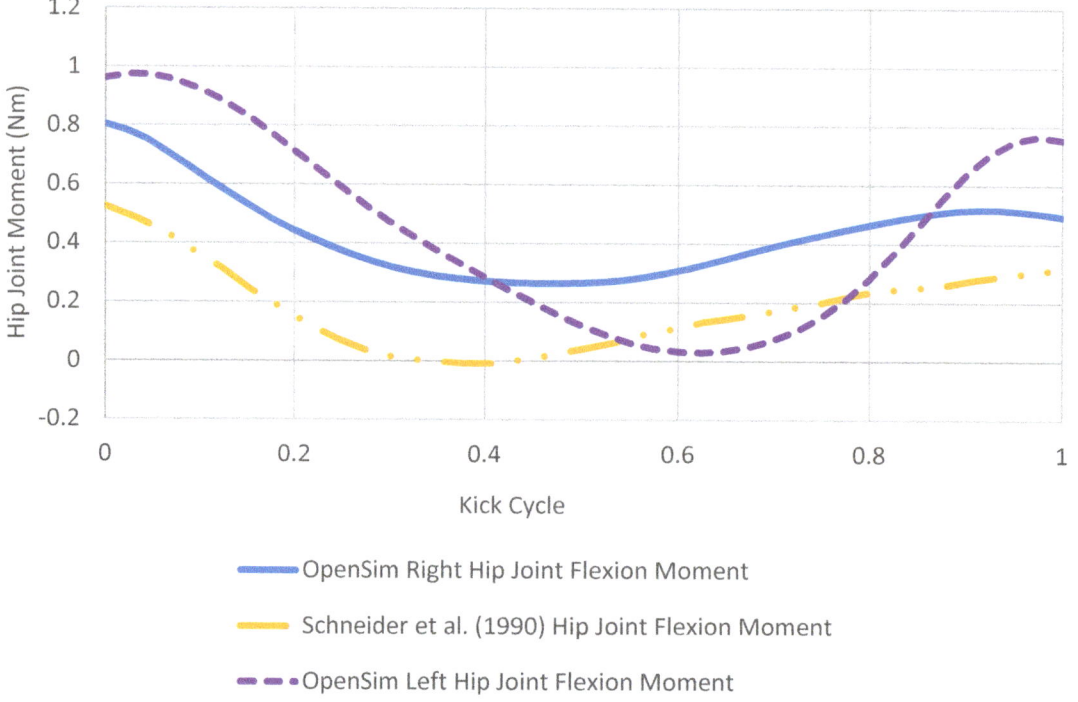

Figure 7. Normalized hip joint moment comparison between infant model and Schneider et al.'s [33].

Figure 7 shows hip joint moment for both the right and left hip decreases as the hip joint undergoes flexion and then increases as the hip returns to the extended position. The flexion phase of the kick corresponds to concentric hip flexor contraction, while the extension phase corresponds to eccentric hip flexor contraction. This trend coincides with the results found in Schneider et al.'s study in their joint moment calculations [33]. Similar to the comparison made on the hip joint angle, a trend agreement was observed when comparing moment values from the infant model to reported values from the literature [33]. There was a hip joint moment difference of 0.275 Nm (right hip) and 0.083 Nm (left hip) when compared to the literature. This maximum difference (right: 0.283 Nm; left: 0.039 Nm) was observed at the corresponding time of maximum hip flexion angle. The difference in moments can be attributed to several factors, such as not restraining the upper extremity, the difference in kick intensity, and that the present model was developed from subject-specific data.

Due to the limitation of accessible data, and the fact that infant subjects are unable to follow verbal instructions, recreating specific kicking motions is impossible. The similarity in how the joint moment behaves with respect to hip joint flexion and extension is observed in both the results of this study and results found in the literature.

Another limitation of this study is that the regression equations were based on child and adult populations ranging between the ages of 5-years-old and 40-years-old [31]. There are currently no regression equations present in the literature for estimating the location of hip joint centers in infants. Regression equations are developed using either cadaveric data or medical imaging studies [34]. To the knowledge of the authors, such data for infant populations under the age of one year old are not available. The risks of radiation exposure from taking CT scans of young, healthy infants to obtain more representative regression equations are not justifiable. Other approaches that do not include ionizing radiation, including MRI scans or ultrasound, were outside of the scope of this study. Nonetheless, amongst the regression methods, Hara et al. [31] found an improved accuracy of their regression equations when compared to previously published equations [35–37] when using the leg length as the predictor. Furthermore, the Hara method had the largest sample size of all regression methods for estimating hip joint centers. Exploring the best methods of defining the HJC for an infant population should be a focus of future research.

The paucity of infant biomechanical data in the literature is a major limitation in infant musculoskeletal modeling. Additionally, infants' inability to follow instructions in an experimental laboratory makes infant biomechanics research a unique challenge. Our model, while not yet generalizable to the infant population, represents a crucial advance in developing subject-specific infant lower-extremity MCM from experimental data.

This study represents a novel musculoskeletal model that can enable innovative research on the understudied infant population and eventually extend to pathologies such as DDH. Eventually, researchers with limited or no access to infant participants will be able to use the infant MCM model to study biomechanical loads that occur at the hip joint during dynamic movements and use the results to identify and evaluate the mechanisms that contribute to infant hip development. Future work should include how joint moments might change with age during the first year of life before infants start walking. During this period, the anatomy and neuromuscular system of infants undergo rapid changes, making the development of valid subject-specific MCM of infants a crucial step to gain a more granular and holistic understanding of infant growth and development.

5. Conclusions

A preliminary musculoskeletal computational model representative of an infant to study the biomechanics of the lower extremity was created. This novel musculoskeletal model was created using experimental MOCAP and GRF data, as well as OpenSim's inverse kinematics and inverse dynamics post-processing tools. The infant MCM can enable innovative research on the understudied infant population by providing more insight into biomechanical loadings at the hip joint during a spontaneous kick and can eventually be extended to evaluating the mechanisms contributing to pathologies such as DDH.

Author Contributions: Conceptualization, V.H. and E.M.; methodology, E.M., S.S., V.H., Y.L. and C.W.; software, Y.L., T.C. and C.W.; validation, Y.L. and T.C.; formal analysis, Y.L., T.C., V.H., E.M., S.S. and C.W.; investigation, Y.L. and T.C.; resources, E.M., S.S. and V.H.; data curation, Y.L. and T.C.; writing—original draft preparation, Y.L., T.C., V.H., E.M., S.S. and C.W.; writing—review and editing, T.C., V.H., E.M. and S.S.; visualization, Y.L. and T.C.; supervision, V.H. and E.M.; project administration, V.H.; funding acquisition, E.M. and V.H. All authors have equally contributed to this paper. All authors have read and agreed to the published version of the manuscript.

Funding: This study was supported in part by Boba, Inc., the International Hip Dysplasia Institute, NIH P20GM125503, and internal funds.

Conflicts of Interest: The authors declare no conflict of interest.

References

1. Matsuo, T.; Fleisig, G.S.; Zheng, N.; Andrews, J.R. Influence of shoulder abduction and lateral trunk tilt on peak elbow varus torque for college baseball pitchers during simulated pitching. *J. Appl. Biomech.* **2006**, *22*, 93–102. [PubMed]
2. Walck, C.; Huayamave, V.; Osbahr, D.; Furman, T.; Farnese, T. A Patient-Specific Lower Extremity Biomechanical Analysis of a Knee Orthotic during a Deep Squat Movement. *Med. Eng. Phys.* **2020**, *80*, 1–7.
3. Dao, T.; Marin, F.; Pouletaut, P.; Charleux, F.; Aufaure, P.; Ho Ba Tho, M. Estimation of accuracy of patient-specific musculoskeletal modelling: Case study on a post polio residual paralysis subject. *Comput. Methods Biomech. Biomed. Eng.* **2012**, *15*, 745–751.
4. Gaffney, B.M.; Hillen, T.J.; Nepple, J.J.; Clohisy, J.C.; Harris, M.D. Statistical shape modeling of femur shape variability in female patients with hip dysplasia. *J. Orthop. Res.* **2019**, *37*, 665–673.
5. Marra, M.A.; Vanheule, V.; Fluit, R.; Koopman, B.H.; Rasmussen, J.; Verdonschot, N.; Andersen, M.S. A subject-specific musculoskeletal modeling framework to predict in vivo mechanics of total knee arthroplasty. *J. Biomech. Eng.* **2015**, *137*, 020904.
6. Rahman, M.; Cil, A.; Bogener, J.W.; Stylianou, A.P. Lateral collateral ligament deficiency of the elbow joint: A modeling approach. *J. Orthop. Res.* **2016**, *34*, 1645–1655.

7. Song, K.; Anderson, A.E.; Weiss, J.A.; Harris, M.D. Musculoskeletal models with generic and subject-specific geometry estimate different joint biomechanics in dysplastic hips. *Comput. Methods Biomech. Biomed. Eng.* **2019**, *22*, 259–270.
8. Thomas-Aitken, H.D.; Goetz, J.E.; Dibbern, K.N.; Westermann, R.W.; Willey, M.C.; Brown, T.S. Patient age and hip morphology alter joint mechanics in computational models of patients with hip dysplasia. *Clin. Orthop. Relat. Res.* **2019**, *477*, 1235–1245.
9. Alizadeh, M.; Knapik, G.G.; Mageswaran, P.; Mendel, E.; Bourekas, E.; Marras, W.S. Biomechanical musculoskeletal models of the cervical spine: A systematic literature review. *Clin. Biomech.* **2020**, *71*, 115–124.
10. Ghezelbash, F.; Shirazi-Adl, A.; Arjmand, N.; El-Ouaaid, Z.; Plamondon, A. Subject-specific biomechanics of trunk: Musculoskeletal scaling, internal loads and intradiscal pressure estimation. *Biomech. Model. Mechanobiol.* **2016**, *15*, 1699–1712.
11. Rasmussen, J.; Tørholm, S.; de Zee, M. Computational analysis of the influence of seat pan inclination and friction on muscle activity and spinal joint forces. *Int. J. Ind. Ergon.* **2009**, *39*, 52–57.
12. Mundt, M.; Koeppe, A.; David, S.; Bamer, F.; Potthast, W.; Markert, B. Prediction of ground reaction force and joint moments based on optical motion capture data during gait. *Med. Eng. Phys.* **2020**, *86*, 29–34.
13. Kong, P.W.; Pan, J.W.; Chu, D.P.K.; Cheung, P.M.; Lau, P.W.C. Acquiring expertise in precision sport: What can we learn from an elite snooker player? *Phys. Act. Health* **2021**, *5*, 98–106.
14. Sylvester, A.D.; Lautzenheiser, S.G.; Kramer, P.A. A review of musculoskeletal modelling of human locomotion. *Interface Focus* **2021**, *11*, 20200060.
15. Verbruggen, S.W.; Loo, J.H.; Hayat, T.T.; Hajnal, J.V.; Rutherford, M.A.; Phillips, A.T.; Nowlan, N.C. Modeling the biomechanics of fetal movements. *Biomech. Model. Mechanobiol.* **2016**, *15*, 995–1004.
16. Nowlan, N.C. Biomechanics of foetal movement. *Eur. Cell Mater.* **2015**, *29*, 21.
17. Chen, H.; Song, Y.; Xuan, R.; Hu, Q.; Baker, J.S.; Gu, Y. Kinematic Comparison on Lower Limb Kicking Action of Fetuses in Different Gestational Weeks: A Pilot Study. *Healthcare* **2021**, *9*, 1057.
18. Zwawi, M.A.; Moslehy, F.A.; Rose, C.; Huayamave, V.; Kassab, A.J.; Divo, E.; Jones, B.J.; Price, C.T. Developmental dysplasia of the hip: A computational biomechanical model of the path of least energy for closed reduction. *J. Orthop. Res.* **2016**, *35*, 1799–1805. [CrossRef]
19. Huayamave, V.; Rose, C.; Serra, S.; Jones, B.; Divo, E.; Moslehy, F.; Kassab, A.J.; Price, C.T. A patient-specific model of the biomechanics of hip reduction for neonatal Developmental Dysplasia of the Hip: Investigation of strategies for low to severe grades of Developmental Dysplasia of the Hip. *J. Biomech.* **2015**, *48*, 2026–2033.
20. Huayamave, V.; Lozinski, B.; Rose, C.; Ali, H.; Kassab, A.; Divo, E.; Moslehy, F.; Price, C. Biomechanical evaluation of femoral anteversion in developmental dysplasia of the hip and potential implications for closed reduction. *Clin. Biomech.* **2020**, *72*, 179–185.
21. Bialik, V.; Bialik, G.M.; Blazer, S.; Sujov, P.; Wiener, F.; Berant, M. Developmental Dysplasia of the Hip: A New Approach to Incidence. *Pediatrics* **1999**, *103*, 93–99.
22. Dezateux, C.; Rosendahl, K. Developmental dysplasia of the hip. *Lancet* **2007**, *369*, 1541–1552.
23. Ortiz-Neira, C.L.; Paolucci, E.O.; Donnon, T. A meta-analysis of common risk factors associated with the diagnosis of developmental dysplasia of the hip in newborns. *Eur. J. Radiol.* **2012**, *81*, e344–e351.
24. Bahl, J.S.; Zhang, J.; Killen, B.A.; Taylor, M.; Solomon, L.B.; Arnold, J.B.; Lloyd, D.G.; Besier, T.F.; Thewlis, D. Statistical shape modelling versus linear scaling: Effects on predictions of hip joint centre location and muscle moment arms in people with hip osteoarthritis. *J. Biomech.* **2019**, *85*, 164–172. [PubMed]
25. Kainz, H.; Carty, C.P.; Maine, S.; Walsh, H.P.J.; Lloyd, D.G.; Modenese, L. Effects of hip joint centre mislocation on gait kinematics of children with cerebral palsy calculated using patient-specific direct and inverse kinematic models. *Gait Posture* **2017**, *57*, 154–160. [PubMed]
26. Kainz, H.; Hoang, H.X.; Stockton, C.; Boyd, R.R.; Lloyd, D.G.; Carty, C.P. Accuracy and reliability of marker-based approaches to scale the pelvis, thigh, and shank segments in musculoskeletal models. *J. Appl. Biomech.* **2017**, *33*, 354–360. [PubMed]
27. Siddicky, S.F.; Bumpass, D.B.; Krishnan, A.; Tackett, S.A.; McCarthy, R.E.; Mannen, E.M. Positioning and baby devices impact infant spinal muscle activity. *J. Biomech.* **2020**, 109741.
28. Siddicky, S.F.; Wang, J.; Rabenhorst, B.; Buchele, L.; Mannen, E.M. Exploring infant hip position and muscle activity in common baby gear and orthopedic devices. *J. Orthop. Res.* **2020**, *39*, 941–949.
29. Delp, S.L.; Anderson, F.C.; Arnold, A.S.; Loan, P.; Habib, A.; John, C.T.; Guendelman, E.; Thelen, D.G. OpenSim: Open-source software to create and analyze dynamic simulations of movement. *Biomed. Eng. IEEE Trans.* **2007**, *54*, 1940–1950.
30. Seth, A.; Hicks, J.L.; Uchida, T.K.; Habib, A.; Dembia, C.L.; Dunne, J.J.; Ong, C.F.; DeMers, M.S.; Rajagopal, A.; Millard, M. OpenSim: Simulating musculoskeletal dynamics and neuromuscular control to study human and animal movement. *PLoS Comput. Biol.* **2018**, *14*, e1006223.
31. Hara, R.; McGinley, J.; Briggs, C.; Baker, R.; Sangeux, M. Predicting the location of the hip joint centres, impact of age group and sex. *Sci. Rep.* **2016**, *6*, 37707.
32. Kliegman, R.M.; Behrman, R.E.; Jenson, H.B.; Stanton, B.M. *Nelson Textbook of Pediatrics e-Book*; Elsevier Health Sciences: Amsterdam, The Netherlands, 2007.
33. Schneider, K.; Zernicke, R.F.; Ulrich, B.D.; Jensen, J.L.; Thelen, E. Understanding movement control in infants through the analysis of limb intersegmental dynamics. *J. Mot. Behav.* **1990**, *22*, 493–520.

34. Kainz, H.; Hajek, M.; Modenese, L.; Saxby, D.J.; Lloyd, D.G.; Carty, C.P. Reliability of functional and predictive methods to estimate the hip joint centre in human motion analysis in healthy adults. *Gait Posture* **2017**, *53*, 179–184.
35. Bell, A.L.; Brand, R.A.; Pedersen, D.R. Prediction of hip joint centre location from external landmarks. *Hum. Mov. Sci.* **1989**, *8*, 3–16.
36. Davis Iii, R.B.; Ounpuu, S.; Tyburski, D.; Gage, J.R. A gait analysis data collection and reduction technique. *Hum. Mov. Sci.* **1991**, *10*, 575–587.
37. Harrington, M.E.; Zavatsky, A.B.; Lawson, S.E.M.; Yuan, Z.; Theologis, T.N. Prediction of the hip joint centre in adults, children, and patients with cerebral palsy based on magnetic resonance imaging. *J. Biomech.* **2007**, *40*, 595–602.

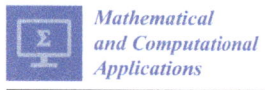

Article

A Trust Region Reduced Basis Pascoletti-Serafini Algorithm for Multi-Objective PDE-Constrained Parameter Optimization

Stefan Banholzer, Luca Mechelli and Stefan Volkwein *

Department of Mathematics and Statistics, University of Konstanz, Universitätsstraße 10, 78464 Konstanz, Germany; stefan.banholzer@uni-konstanz.de (S.B.); luca.mechelli@uni-konstanz.de (L.M.)
* Correspondence: stefan.volkwein@uni-konstanz.de

Abstract: In the present paper non-convex multi-objective parameter optimization problems are considered which are governed by elliptic parametrized partial differential equations (PDEs). To solve these problems numerically the Pascoletti-Serafini scalarization is applied and the obtained scalar optimization problems are solved by an augmented Lagrangian method. However, due to the PDE constraints, the numerical solution is very expensive so that a model reduction is utilized by using the reduced basis (RB) method. The quality of the RB approximation is ensured by a trust-region strategy which does not require any offline procedure, in which the RB functions are computed in a greedy algorithm. Moreover, convergence of the proposed method is guaranteed and different techniques to prevent the excessive growth of the number of basis functions are explored. Numerical examples illustrate the efficiency of the proposed solution technique.

Keywords: non-convex multi-objective optimization; partial differential equations; Pascoletti-Serafini method; augmented Lagrangian; reduced basis method; trust-region strategy

Citation: Banholzer, S.; Mechelli, L.; Volkwein, S. A Trust Region Reduced Basis Pascoletti-Serafini Algorithm for Multi-Objective PDE-Constrained Parameter Optimization. *Math. Comput. Appl.* **2022**, 27, 39. https://doi.org/10.3390/mca27030039

Academic Editors: Simona Perotto, Gianluigi Rozza and Antonia Larese

Received: 18 January 2022
Accepted: 29 April 2022
Published: 3 May 2022

Publisher's Note: MDPI stays neutral with regard to jurisdictional claims in published maps and institutional affiliations.

Copyright: © 2022 by the authors. Licensee MDPI, Basel, Switzerland. This article is an open access article distributed under the terms and conditions of the Creative Commons Attribution (CC BY) license (https://creativecommons.org/licenses/by/4.0/).

1. Introduction

Multi-objective optimization plays an important role in many applications, e.g., in industry, medicine or engineering. One of the mentioned examples is the minimization of costs with simultaneous quality optimization in production or the minimization of CO_2 emission in energy generation and simultaneous cost minimization. These problems lead to multi-objective optimization problems (MOPs), where we want to achieve an optimal compromise with respect to all given objectives at the same time. Normally, the different objectives are contradictory such that there exists an infinite number of optimal compromises. The set of these compromises is called the *Pareto set*. The goal is to approximate the Pareto set in an efficient way, which turns out to be more expensive than solving a single objective optimization problem.

Since MOPs are of great importance, there exist several algorithms to solve them. Among the most popular methods are scalarization methods, which transform MOPs into scalar problems. For example, in the weighted sum method [1–3], convex combinations of the original objectives are optimized. However, in our case the multi-objective optimization problem

$$\min \hat{J}(u) = \big(\hat{J}_1(u), \ldots, \hat{J}_k(u)\big)^T \quad \text{subject to (s.t.)} \quad u \in \mathcal{U}_{\text{ad}} \qquad \textbf{(MOP)}$$

is non-convex with a bounded, non-empty, convex and closed set $\mathcal{U}_{\text{ad}}$. To solve (**MOP**) a suitable scalarization method in that case is the Pascoletti-Serafini (PS) scalarization [4,5]: For a chosen *reference point* $z \in \mathbb{R}^k$ and a given *target direction* $r \in \mathbb{R}^k$ with $r_i > 0$ for all $i \in \{1, \ldots, k\}$ the Pascoletti-Serafini problem is given by

$$\min t \quad \text{s.t.} \quad (t, u) \in \mathbb{R} \times \mathcal{U}_{\text{ad}} \text{ and } \hat{J}(u) - z \leq t\, r. \qquad (\mathbf{P}^{\text{PS}}_{z,r})$$

In the present paper ($\mathbf{P}_{z,r}^{PS}$) is solved by an augmented Lagrangian approach. However, in our case the evaluation of the objective $\hat{f}$ requires the solution of an elliptic partial differential equation (PDE) for the given parameter u. This implies further that for the computation of the gradients $\nabla \hat{f}_i$, $i = 1, \ldots, k$, adjoint PDEs have to be solved; cf. [6]. Here, surrogate models offer a promising tool to reduce the computational effort significantly [7]. Examples are dimensional reduction techniques such as the Reduced Basis (RB) method [8,9]. In an offline phase, a low-dimensional surrogate model of the PDE is constructed by using, e.g., the greedy algorithm, cf. [8,10,11]. In the online phase, only the RB model is used to solve the PDE, which saves a lot of computing time.

Since the early 2000s the combination of model order reduction with trust-region algorithms in the setting of PDE-constrained optimization is present in the literature, cf. [12,13]. The idea in these methods is to replace the usual quadratic model function in each trust-region step with the reduced-order approximation of the cost function. More recent publications followed and enhanced this approach by using a-posteriori error estimates of the cost function and its gradient, cf. [14,15]. These works were the starting point for the trust-region reduced basis methods developed in [16–18]. Let us mention that [19,20] have proposed similar methods for the combination of reduced-order and trust-region methods based on previous works on trust-region algorithms for PDE-constrained optimization under uncertainty, cf. [21,22]. In contrast to the approach followed by [14–18], these methods do not use rigorous a-posteriori error estimates but rather asymptotic error indicators which still allow for a global convergence result. Here we propose an extension of the method in [16] for solving multi-objective PDE-constrained parameter optimization problems, which is based on a combination of the trust-region reduced basis method presented in [17,18] and the PS method. In particular, we discuss different strategies to handle the increasing number of reduced basis functions, which is crucial in order to guarantee good performances of the algorithm. Notice that our approach is designed for applications, where we have to solve the multi-objective PDE-constrained parameter optimization problem once. For that reason, our trust-region reduced basis method does not rely on any offline computations. These proposed strategies are not only interesting in the field of multi-objective optimization by the PS method, but can also be used in other applications where many PDE-constrained optimization problems must be solved and it is hence crucial to keep the number of reduced basis functions small enough, as, e.g., in model predictive control; cf. [23].

The paper is organized as follows: In Section 2 we introduce a general MOP and explain the PS method, in particular, a hierarchical version of the PS algorithm which turns out to be very efficient in the numerical realization. The concrete PDE-constrained MOP is investigated in Section 3. The trust-region RB method and its combination with the PS method is described in Section 4. Convergence is ensured and the algorithmic realization of the approach is explained. Numerical examples are discussed in detail in Section 5. Finally, we draw some conclusions.

2. Multi-Objective Optimization

Let $(\mathcal{U}, \langle \cdot, \cdot \rangle_\mathcal{U})$ be a real Hilbert space, $\mathcal{U}_{ad} \subset \mathcal{U}$ non-empty, convex and closed, $k \geq 2$ arbitrary and $\hat{f}_1, \ldots, \hat{f}_k : \mathcal{U}_{ad} \subset \mathcal{U} \to \mathbb{R}$ be given real-valued functions. In this manuscript, we assume also that $\mathcal{U}_{ad}$ is bounded. This is an assumption we will require later for the convergence of our method. Note that one can derive similar results of this section if $\mathcal{U}_{ad}$ is unbounded by introducing additional assumptions; cf. [16]. To shorten the notation, we write $\hat{f} := (\hat{f}_1, \ldots, \hat{f}_k)^T : \mathcal{U}_{ad} \to \mathbb{R}^k$. In the following, we deal with the multi-objective optimization problem

$$\min \hat{f}(u) \quad \text{s.t.} \quad u \in \mathcal{U}_{ad}. \tag{MOP}$$

Definition 1. (a) The functions $\hat{J}_1, \ldots, \hat{J}_k$ are called cost or objective functions. Analogously, the vector-valued function $\hat{J} \colon \mathcal{U}_{ad} \to \mathbb{R}^k$ is named the (multi-objective) cost or (multi-objective) objective function.
(b) The Hilbert space $\mathcal{U}$ is named the admissible space, the set $\mathcal{U}_{ad}$ is called the admissible set and a vector $u \in \mathcal{U}_{ad}$ is called admissible.
(c) The space $\mathbb{R}^k$ is named the objective space and the image set $\hat{J}(\mathcal{U}_{ad})$ is called the objective set. A vector $y = \hat{J}(u) \in \hat{J}(\mathcal{U}_{ad})$ is called objective point.

Definition 2 (Partial ordering on $\mathbb{R}^k$). On $\mathbb{R}^k$ we define the partial ordering $\leq$ as

$$x \leq y :\iff (\forall i \in \{1, \ldots, k\} \colon x_i \leq y_i)$$

for all $x, y \in \mathbb{R}^k$. Moreover, we define

$$x < y :\iff (\forall i \in \{1, \ldots, k\} \colon x_i < y_i).$$

For convenience, we write

$$x \lneq y :\iff (x \leq y \ \& \ x \neq y)$$

for all $x, y \in \mathbb{R}^k$ and define the two sets $\mathbb{R}^k_\leq := \{y \in \mathbb{R}^k \mid y \leq 0\}$, $\mathbb{R}^k_{\lneq} := \{y \in \mathbb{R}^k \mid y \lneq 0\}$. Analogously, the relations $\geq, >$ and $\gneq$ as well as the sets $\mathbb{R}^k_\geq$ and $\mathbb{R}^k_{\gneq}$ are defined.

Definition 3 (Pareto optimality).
(a) An admissible vector $\bar{u} \in \mathcal{U}_{ad}$ and its corresponding objective point $\bar{y} := \hat{J}(\bar{u}) \in \hat{J}(\mathcal{U}_{ad})$ are called (locally) weakly Pareto optimal if there is no $\tilde{u} \in \mathcal{U}_{ad}$ (in a neighborhood of $\bar{u}$) with $\hat{J}(\tilde{u}) < \hat{J}(\bar{u})$. The sets

$$\mathcal{U}_{opt,w} := \{u \in \mathcal{U}_{ad} \mid u \text{ is weakly Pareto optimal}\} \subset \mathcal{U}_{ad},$$
$$\mathcal{U}_{opt,w,loc} := \{u \in \mathcal{U}_{ad} \mid u \text{ is locally weakly Pareto optimal}\} \subset \mathcal{U}_{ad}$$

are said to be the weak Pareto set and the locally weak Pareto set, respectively. The sets

$$\mathcal{J}_{opt,w} := \hat{J}(\mathcal{U}_{opt,w}) \subset \mathbb{R}^k, \quad \mathcal{J}_{opt,w,loc} := \hat{J}(\mathcal{U}_{opt,w,loc}) \subset \mathbb{R}^k,$$

are the weak Pareto front and the locally weak Pareto front, respectively.
(b) An admissible vector $\bar{u} \in \mathcal{U}_{ad}$ and its corresponding objective point $\bar{y} := \hat{J}(\bar{u}) \in \hat{J}(\mathcal{U}_{ad})$ are called (locally) Pareto optimal if there is no $\tilde{u} \in \mathcal{U}_{ad}$ (in a neighborhood of $\bar{u}$) with $\hat{J}(\tilde{u}) \lneq \hat{J}(\bar{u})$. The sets

$$\mathcal{U}_{opt} := \{u \in \mathcal{U}_{ad} \mid u \text{ is Pareto optimal}\} \subset \mathcal{U}_{ad},$$
$$\mathcal{U}_{opt,loc} := \{u \in \mathcal{U}_{ad} \mid u \text{ is locally Pareto optimal}\} \subset \mathcal{U}_{ad}$$

are called the Pareto set and the local Pareto set, respectively. The sets

$$\mathcal{J}_{opt} := \hat{J}(\mathcal{U}_{opt}) \subset \mathbb{R}^k, \quad \mathcal{J}_{opt,loc} := \hat{J}(\mathcal{U}_{opt,loc}) \subset \mathbb{R}^k$$

are called the Pareto front and the local Pareto front, respectively.

If we talk about the different notions of (local) (weak) Pareto optimality in one sentence, we use the notation $\mathcal{U}_{opt,(w),(loc)}$ to keep the sentence compact. Analogously, $\mathcal{U}_{opt,(w),loc}$, $\mathcal{U}_{opt,(loc)}$, $\mathcal{J}_{opt,(w),(loc)}$ etc. are to be understood. An example with the different concepts of Pareto optimality can be found in [16] (Example 1.2.6).

The next theorem about a sufficient condition for the existence of Pareto optimal points goes back to [24]. It also appears in a similar form in [25,26].

Theorem 1. *Suppose that there is $y \in \hat{J}(\mathcal{U}_{ad}) + \mathbb{R}^k_{\geq}$ such that the set $(y - \mathbb{R}^k_{\geq}) \cap (\hat{J}(\mathcal{U}_{ad}) + \mathbb{R}^k_{\geq})$ is compact. Then it holds $\mathcal{J}_{opt} \neq \emptyset$.*

Proof. This is a slight generalization of [1] (Theorem 2.10) using the argument that adding $\mathbb{R}^k_{\geq}$ to the set $\hat{J}(\mathcal{U}_{ad})$ does not change the Pareto front $\mathcal{J}_{opt}$. □

Given any $y = \hat{J}(u) \in \hat{J}(\mathcal{U}_{ad})$ with $y \notin \mathcal{J}_{opt}$, it follows directly from the definition of Pareto optimality that there is $\bar{y} = \hat{J}(\bar{u}) \in \hat{J}(\mathcal{U}_{ad})$ with $\bar{y} \lneq y$. However, even if the Pareto front $\mathcal{J}_{opt}$ is not empty (e.g., since the assumptions of Theorem 1 are satisfied), it is not clear that there is $\bar{y} \in \mathcal{J}_{opt}$ with $\bar{y} \lneq y$. If this property holds for all $y \in \hat{J}(\mathcal{U}_{ad}) \setminus \mathcal{J}_{opt}$, the set $\mathcal{J}_{opt}$ is said to be *externally stable*; cf. [1,26].

Definition 4. *The set $\mathcal{J}_{opt}$ is said to be* externally stable *if for every $y \in \hat{J}(\mathcal{U}_{ad})$ there is $\bar{y} \in \mathcal{J}_{opt}$ with $\bar{y} \leq y$. This is equivalent to $\hat{J}(\mathcal{U}_{ad}) \subset \mathcal{J}_{opt} + \mathbb{R}^k_{\geq}$.*

Especially for the investigation of suitable solution methods for solving (**MOP**), we are interested in guaranteeing that the Pareto front is externally stable. The next result provides a sufficient condition for this property.

Theorem 2. *If for every $y \in \hat{J}(\mathcal{U}_{ad}) + \mathbb{R}^k_{\geq}$ the set $(y - \mathbb{R}^k_{\geq}) \cap (\hat{J}(\mathcal{U}_{ad}) + \mathbb{R}^k_{\geq})$ is compact, then $\mathcal{J}_{opt}$ is externally stable.*

Proof. For a proof of a similar version of this theorem, we refer to [1] (Theorem 2.21). □

Among the methods to solve multi-objective optimization problems, the ones based on scalarization techniques are frequently appearing in the literature. Let us mention here the weighted-sum method [1,3], the Euclidian reference point method [27] and the PS method [4,5]. Since in our case the set $\hat{J}(\mathcal{U}_{ad}) + \mathbb{R}^k_{\geq}$ is non-convex, we apply the PS method which is proven to be able to solve a non-convex (**MOP**).

2.1. The PS Method

For a chosen *reference point* $z \in \mathbb{R}^k$ and a given *target direction* $r \in \mathbb{R}^k_{>}$ the PS problem is given by

$$\min t \quad \text{s.t.} \quad (t, u) \in \mathbb{R} \times \mathcal{U}_{ad} \text{ and } \hat{J}(u) - z \leq t\, r. \tag{$\mathbf{P}^{PS}_{z,r}$}$$

Analogously, we can define the PS problem as a scalarization problem. For $z \in \mathbb{R}^k$ and $r \in \mathbb{R}^k_{>}$ we define the scalarization function

$$g_{z,r} \colon \mathbb{R}^k \to \mathbb{R}, \quad x \mapsto g_{z,r}(x) := \max_{1 \leq i \leq k} \frac{1}{r_i}(x_i - z_i),$$

and the *PS scalarized function*

$$\hat{J}^{g_{z,r}}(u) := g_{z,r}(\hat{J}(u)) = \max_{1 \leq i \leq k} \frac{1}{r_i}(\hat{J}_i(u) - z_i) \quad \text{for } u \in \mathcal{U}_{ad}.$$

Then the *reformulated PS problem* is given by

$$\min \hat{J}^{g_{z,r}}(u) \quad \text{s.t.} \quad u \in \mathcal{U}_{ad}. \tag{$\mathbf{RP}^{PS}_{z,r}$}$$

The following theorem proved in [16] (Theorem 1.7.3) ensures the equivalence between ($\mathbf{P}^{PS}_{z,r}$) and ($\mathbf{RP}^{PS}_{z,r}$).

Theorem 3. *Let $z \in \mathbb{R}^k$ and $r \in \mathbb{R}^k_{>}$ be arbitrary. On the one hand, if $(\bar{u}, \bar{t})$ is a global (local) solution of ($\mathbf{P}^{PS}_{z,r}$), then $\bar{u}$ is a global (local) solution of ($\mathbf{RP}^{PS}_{z,r}$) with minimal function value $\bar{t}$. On the other hand, if $\bar{u}$ is a global (local) solution of ($\mathbf{RP}^{PS}_{z,r}$), then $(\bar{u}, \bar{t})$ with $\bar{t} := \max_{1 \leq i \leq k}(\hat{J}_i(\bar{u}) - z_i)/r_i$ is a global (local) solution of ($\mathbf{P}^{PS}_{z,r}$).*

Assumption 1. *The cost functions $\hat{f}_1, \ldots, \hat{f}_k$ are weakly lower semi-continuous and bounded from below.*

Theorem 4. *Let Assumption 1 be satisfied and $z \in \mathbb{R}^k$ as well as $r \in \mathbb{R}^k_>$ be arbitrary. Then ($\mathbf{RP}^{PS}_{z,r}$) has a global solution $\bar{u} \in \mathcal{U}_{\text{opt}}$.*

Proof. A proof of this statement can be found in [16] (Corollary 1.7.12). □

The previous result also shows that the existing global solution of ($\mathbf{RP}^{PS}_{z,r}$) belongs to the Pareto set. To guarantee a good reconstruction of the Pareto set by the PS method, one needs that, given a (weakly) Pareto optimal point, it is possible to choose the parameters z and r such that this point solves ($\mathbf{RP}^{PS}_{z,r}$). This is stated in [16] (Theorem 1.7.13), which we report here for clearness.

Theorem 5. *Let $\bar{u} \in \mathcal{U}_{\text{opt,w}}$ be arbitrary. Then for every $r \in \mathbb{R}^k_>$ and every $\bar{t} \in \mathbb{R}$ we have that $\bar{u}$ is a global solution of ($\mathbf{RP}^{PS}_{z,r}$) for the reference point $z := \hat{f}(\bar{u}) - \bar{t}r$. If even $\bar{u} \in \mathcal{U}_{\text{opt}}$, any other global solution $\tilde{u}$ of ($\mathbf{RP}^{PS}_{z,r}$) satisfies $\hat{f}(\tilde{u}) = \hat{f}(\bar{u})$.*

Remark 1. *We refer the reader to [16] (Lemma 1.7.15) for the derivation of first-order necessary optimality condition for a global solution of ($\mathbf{P}^{PS}_{z,r}$).*

Thus, the PS method can compute in principle every (locally) (weak) Pareto optimal point so that many algorithms based on PS method have been proposed. Here we only mention the ones which are related to (but differ from) our proposed technique. Our main idea is to keep the parameter r fixed, while varying the reference point z. This was also proposed in [4], but the method turns out to be, on the one hand, not numerically efficient for $k > 2$ and, on the other hand, not numerically applicable in some cases for $k > 2$. In [28], the authors provide assumptions on the Pareto front to ensure that the so-called trade-off limits (i.e., points on the Pareto front which cannot be improved in at least one component), are given by the solution to subproblems. Their idea was then to find these trade-off points first and then compute the rest of the Pareto front. A similar idea but with the use of Centroidal Voronoi Tessellation was presented by [29]. Finally, [30] shows and fixes some problematic behavior associated to the algorithm in [28]. We follow the idea of the mentioned contributions of hierarchically solving subproblems of (**MOP**), but with the focus of finding a set of reference points, by looking at subproblems, for which we can obtain Pareto optimal points. We are then not interested in finding 'boundary' points (i.e., the trade-off limits) of the Pareto front and then filling its 'interior' as in [28–30], but rather to partly generalize this approach. In what follows, we characterize which reference points are necessary and/or sufficient for computing the entire (local) (weak) Pareto front. First, we recall the following well-defined solution mappings of ($\mathbf{RP}^{PS}_{z,r}$); cf. [16] (Definition 1.7.16).

Definition 5. *We define the set-valued mappings*

$$\mathcal{Q}_{\text{opt,w}} : \mathbb{R}^k \rightrightarrows \mathcal{U}_{\text{opt,w}}, \quad z \mapsto \{u \in \mathcal{U}_{\text{ad}} \mid u \text{ is a global solution of } (\mathbf{RP}^{PS}_{z,r})\},$$

$$\mathcal{Q}_{\text{opt,w,loc}} : \mathbb{R}^k \rightrightarrows \mathcal{U}_{\text{opt,w,loc}}, \quad z \mapsto \{u \in \mathcal{U}_{\text{ad}} \mid u \text{ is a local solution of } (\mathbf{RP}^{PS}_{z,r})\},$$

$$\mathcal{Q}_{\text{opt,(loc)}} : \mathbb{R}^k \rightrightarrows \mathcal{U}_{\text{opt,(loc)}}, \quad z \mapsto \mathcal{Q}_{\text{opt,w,(loc)}}(z) \cap \mathcal{U}_{\text{opt,(loc)}}.$$

From Theorem 3, it follows that $\mathcal{Q}_{\text{opt,(w),(loc)}}(\mathbb{R}^k) = \mathcal{U}_{\text{opt,(w),(loc)}}$, i.e., by solving ($\mathbf{RP}^{PS}_{z,r}$) for all $z \in \mathbb{R}^k$, we obtain all (locally), (weakly) Pareto optimal points. Furthermore, if Assumption 1 is satisfied, we infer from Theorem 4 that $\mathcal{Q}_{\text{opt,(w),(loc)}}(z) \neq \emptyset$ for all $z \in \mathbb{R}^k$. We also introduce the notion of a (locally) (weakly) Pareto sufficient set for the PS method.

Definition 6. *A set $Z \subset \mathbb{R}^k$ is called* (locally) (weakly) Pareto sufficient *if we have $\mathcal{Q}_{\text{opt},(w),(\text{loc})}(Z) = \mathcal{U}_{\text{opt},(w),(\text{loc})}$.*

Hence, a (locally) (weakly) Pareto sufficient set contains the reference points which allow us to compute the entire (local) (weak) Pareto front. Clearly, the set $\mathbb{R}^k$ is (locally) (weakly) Pareto sufficient, but this fact is not computationally useful. The next lemma gives a first condition towards this computational efficiency.

Lemma 1. *Let $Z \subset \mathbb{R}^k$ be arbitrary. Z is (locally) (weakly) Pareto sufficient, if*

$$\forall \bar{u} \in \mathcal{U}_{\text{opt},(w),(\text{loc})} \colon \exists t \in \mathbb{R} \colon \hat{f}(\bar{u}) - tr \in Z. \tag{1}$$

Proof. Let $Z \subset \mathbb{R}^k$ be such that (1) holds. Let $\bar{u} \in \mathcal{U}_{\text{opt},(w),(\text{loc})}$ be arbitrary. We need to show that there is a $z \in Z$ with $\bar{u} \in \mathcal{Q}_{\text{opt},(w),(\text{loc})}(z)$. Indeed, by (1) there is $t \in \mathbb{R}$ with $z := \hat{f}(\bar{u}) - tr \in Z$ and by Theorem 5 we already have $\bar{u} \in \mathcal{Q}_{\text{opt},(w),(\text{loc})}(z)$. □

To proceed we introduce the concepts of ideal point and shifted ideal point, which will first be used to define a set of shifted coordinate planes D. On this set we can then define a set of reference points $\mathcal{Z}^D_{\text{opt},(w),(\text{loc})}$ which turns out to be an optimal Pareto sufficient set (The word 'optimal' here means that removing any point from the set will cause the loss of the Pareto sufficient property).

Definition 7.

(a) We define the ideal objective point $y^{\text{id}} \in \mathbb{R}^k \cup \{-\infty\}$ by $y^{\text{id}}_i := \inf_{u \in \mathcal{U}_{\text{ad}}} \hat{f}_i(u)$ for all $i \in \{1, \ldots, k\}$.

(b) For an arbitrary vector $\tilde{d} \in \mathbb{R}^k_{>}$ define the shifted ideal point $\tilde{y}^{\text{id}} := y^{\text{id}} - \tilde{d}$. Let $D_i \subset \mathbb{R}^k$ be given by $D_i := \{y \in \mathbb{R}^k \mid y \geq \tilde{y}^{\text{id}}, y_i = \tilde{y}^{\text{id}}_i\}$ for all $i \in \{1, \ldots, k\}$. Then the set $D \subset \mathbb{R}^k$ is defined by $D := \bigcup_{i=1}^k D_i$.

(c) We define $\mathcal{Z}^D_{\text{opt},(w),(\text{loc})} := \{z \in D \mid \exists \bar{u} \in \mathcal{U}_{\text{opt},(w),(\text{loc})} \colon \exists t \in \mathbb{R} \colon z = \hat{f}(\bar{u}) - tr\}$.

(d) For any $y \in \mathbb{R}^k$ we set $t^D(y) := \min_{i \in \{1,\ldots,k\}} (y_i - \tilde{y}^{\text{id}}_i)/r_i \in \mathbb{R}$.

Remark 2. *It is proved in [16] (Lemma 1.7.24) that*

$$\mathcal{Z}^D_{\text{opt},(w),(\text{loc})} = \{\hat{f}(\bar{u}) - t^D(\hat{f}(\bar{u}))\, r \mid \bar{u} \in \mathcal{U}_{\text{opt},(w),(\text{loc})}\}.$$

Furthermore, the set $\mathcal{Z}^D_{\text{opt},(w),(\text{loc})}$ is (locally) (weakly) Pareto sufficient and there is a Lipschitz continuous bijection between $\mathcal{Z}^D_{\text{opt}}$ and the Pareto front $\mathcal{J}_{\text{opt}}$. Unfortunately, there is no bijection between $\mathcal{Z}^D_{\text{opt},(w),(\text{loc})}$ and $\mathcal{J}_{\text{opt},(w),(\text{loc})}$, but the set $\mathcal{Z}^D_{\text{opt},(w),(\text{loc})}$ is still (locally) (weakly) Pareto sufficient. Therefore, it is anyway possible to use it for the computation of the Pareto front.

2.2. Hierarchical PS Method

Due to Definition 7 and Remark 2 the set $\mathcal{Z}^D_{\text{opt},(w),(\text{loc})}$ can only by computed once the set $\mathcal{U}_{\text{opt},(w),(\text{loc})}$ is available. Clearly, this characterization of $\mathcal{Z}^D_{\text{opt},(w),(\text{loc})}$ is not useful for a numerical algorithm since the availability of $\mathcal{U}_{\text{opt},(w),(\text{loc})}$ means that we have already solved (**MOP**). Fortunately, in [16,31] it is shown that the Pareto set has a hierarchical structure. This means that the (weak) Pareto front and the (weak) Pareto sets of (**MOP**) are contained in the set of all (weak) Pareto fronts and (weak) Pareto sets of all of its subproblems. This particular structure of the Pareto set can be exploited to set up a hierarchical algorithm for obtaining a superset of $\mathcal{Z}^D_{\text{opt},(w),(\text{loc})}$ without having to compute the entire (local) (weak) Pareto set $\mathcal{U}_{\text{opt},(w),(\text{loc})}$ first. We start the explanation of the hierarchical algorithm by introducing the notion of a subproblem and related notations.

Definition 8. *For the index set $I \subset \{1,\ldots,k\}$ we denote by $\hat{J}^I$ the multi-objective cost function $(\hat{J}_i)_{i \in I} \colon \mathcal{U}_{\mathsf{ad}} \to \mathbb{R}^I$, and call the problem*

$$\min \hat{J}^I(u) \quad \text{s.t.} \quad u \in \mathcal{U}_{\mathsf{ad}} \qquad (\mathbf{MOP}_I)$$

a subproblem of (**MOP**). *For $I, K \subset \{1,\ldots,k\}$ with $K \subset I$,*

(a) *and for every $y \in \mathbb{R}^I$ we denote by $y^K := (y_i)_{i \in K} \in \mathbb{R}^K$ the canonical projection to $\mathbb{R}^K$.*
(b) *the set $\mathcal{U}^I_{\mathsf{opt},(\mathsf{w}),(\mathsf{loc})} := \{u \in \mathcal{U}_{\mathsf{ad}} \mid u \text{ is (loc.) (weak.) Pareto optimal for } (\mathbf{MOP}_I)\}$ denotes the (local) (weak) Pareto set and the set $\mathcal{J}^I_{\mathsf{opt},(\mathsf{w}),(\mathsf{loc})} := \hat{J}^I(\mathcal{U}^I_{\mathsf{opt},(\mathsf{w}),(\mathsf{loc})}) \subset \mathbb{R}^I$ denotes the (local) (weak) Pareto front of the subproblem $(\mathbf{MOP}_I)$.*
(c) *the (local) (weak) nadir objective point for the subproblem $(\mathbf{MOP}_I)$ is defined by*

$$y_i^{\mathsf{nad},I,(\mathsf{w}),(\mathsf{loc})} := \sup\{y_i \mid y \in \mathcal{J}^I_{\mathsf{opt},(w),(loc)}\} \quad \text{for all } i \in I.$$

Given a subproblem $(\mathbf{MOP}_I)$ it is straight-forward to define the PS problem for this setting.

Definition 9. *Let $I \subset \{1,\ldots,k\}$ be arbitrary. For a given reference point $z \in \mathbb{R}^I$ and target direction $r \in \mathbb{R}^I_>$, we define the PS problem for $(\mathbf{MOP}_I)$ by*

$$\min t \quad \text{s.t.} \quad (t,u) \in \mathbb{R} \times \mathcal{U}_{\mathsf{ad}} \text{ and } \hat{J}^I(u) - z \leq tr^I. \qquad (\mathbf{P}^{\mathsf{PS}}_{I,z,r})$$

Again, it is possible to show that $(\mathbf{P}^{\mathsf{PS}}_{I,z,r})$ is equivalent (in the sense of Theorem 3) to the problem

$$\min\left(\max_{i \in I} \frac{1}{r_i}\bigl(\hat{J}_i(u) - z_i\bigr)\right) \quad \text{s.t.} \quad u \in \mathcal{U}_{\mathsf{ad}}. \qquad (\mathbf{RP}^{\mathsf{PS}}_{I,z,r})$$

Let us mention that the statements proved in Section 2.1 can be adapted for the PS method for the subproblems. Similarly, we can also generalize the definition of the shifted coordinate plane D and the (locally) (weakly) Pareto sufficient set of reference points $\mathcal{Z}^D_{\mathsf{opt},(\mathsf{w}),(\mathsf{loc})}$ to this setting.

Definition 10. *Let $I \subset \{1,\ldots,k\}$ be arbitrary. Given the vector $\tilde{d} \in \mathbb{R}^k_>$ and the shifted ideal point $\tilde{y}^{\mathsf{id}} \in \mathbb{R}^k$, which were both introduced in Definition 7, let $D^I_i \subset \mathbb{R}^I$ be given by*

$$D^I_i := \{y \in \mathbb{R}^I \mid y \geq (\tilde{y}^{\mathsf{id}})^I, \ y_i = \tilde{y}^{\mathsf{id}}_i\} \quad \text{for } i \in I.$$

Then the set $D^I \subset \mathbb{R}^I$ is defined by $D^I := \bigcup_{i \in I} D_i$. Moreover, for all $K \subset \{1,\ldots,k\}$ we define the sets

$$\mathcal{Z}^{D^I,K}_{\mathsf{opt},(\mathsf{w}),(\mathsf{loc})} := \{z \in D^I \mid \exists \bar{u} \in \mathcal{U}^K_{\mathsf{opt},(\mathsf{w}),(\mathsf{loc})} \colon \exists t \in \mathbb{R} \colon z = \hat{J}^I(\bar{u}) - tr^I\}.$$

To ease the notation, we write $\mathcal{Z}^{D^I}_{\mathsf{opt},(\mathsf{w}),(\mathsf{loc})} := \mathcal{Z}^{D^I,I}_{\mathsf{opt},(\mathsf{w}),(\mathsf{loc})}$. If $I = \{1,\ldots,k\}$ we set $\mathcal{Z}^{D,K}_{\mathsf{opt},(\mathsf{w}),(\mathsf{loc})} := \mathcal{Z}^{D^I,K}_{\mathsf{opt},(\mathsf{w}),(\mathsf{loc})}$ and $\mathcal{Z}^D_{\mathsf{opt},(\mathsf{w}),(\mathsf{loc})} := \mathcal{Z}^{D^I,I}_{\mathsf{opt},(\mathsf{w}),(\mathsf{loc})}$. Finally, for any $y \in \mathbb{R}^I$ we set $t^{D^I}(y) := \min_{i \in I} \frac{y_i - \tilde{y}^{\mathsf{id}}_i}{r_i} \in \mathbb{R}$.

Note that also Remark 2 can be rewritten for the subproblems.

The main ingredient of the hierarchical PS method is the result that a superset of $\mathcal{Z}^{D^I}_{\mathsf{opt},(\mathsf{w}),(\mathsf{loc})}$ can be computed by using the sets $\mathcal{U}^K_{\mathsf{opt},(\mathsf{w}),(\mathsf{loc})}$ for all $K \subsetneq I$. In other words, in contrast to Definition 10 only the Pareto optimal solutions to all subproblems–but not the problem itself–are needed to compute the (locally) (weakly) Pareto sufficient set of reference points $\mathcal{Z}^{D^I}_{\mathsf{opt},(\mathsf{w}),(\mathsf{loc})}$ for the subproblem $(\mathbf{MOP}_I)$. The very technical details of the analytical derivation and verification of this result are omitted here to ease and shorten the presentation. For a reader interested in the details we refer to [16] (Sections 1.7.4.2–1.7.4.4). Building on this result, the idea of the hierarchical PS method is to iteratively solve subproblems with increasing number of cost functions. During this procedure the required reference points for the current subproblem can be computed by using the Pareto optimal solutions of all of its subproblems as described above.

Before we formulate the hierarchical algorithm, we give the necessary numerical condition in order to compute a numerical approximation of the set $\mathcal{Z}^{D^I}_{\text{opt},(w),(\text{loc})}$ by using the numerical solution to all subproblems.

To do so, we introduce a grid on D^I as follows.

Definition 11. *Let $I \subset \{1, \ldots, k\}$ be arbitrary. For a given grid size $h > 0$ and any $i \in I$, we define*

$$\mathcal{Z}_i^{h,I} := \left\{ z \in D_i^I \,\middle|\, \forall j \in I \setminus \{i\} : \left(\exists k \geq 0 : z_j = \bar{y}_j^{\text{id}} + \frac{h}{2} + kh \right) \& \left(z_j \leq y_j^{\text{nad},I,w} - \bar{r}^i r_j \right) \right\}.$$

Furthermore, we set $\mathcal{Z}^{h,I} := \bigcup_{i \in I} \mathcal{Z}_i^{h,I}$. If $I = \{1, \ldots, k\}$, we write $\mathcal{Z}^h := \mathcal{Z}^{h,I}$.

The idea is to only choose reference points that lie on the grid $\mathcal{Z}^{h,I}$ and do not satisfy the condition

$$\exists K \subsetneq I : \exists (\bar{u}, \bar{t}, \bar{z}) \in \mathcal{UTZ}^{\text{num}}(K) : z^K = \bar{z}^K \,\&\, z^{I\setminus K} \geq \hat{f}^{I\setminus K}(\bar{u}) - \bar{t} r^{I\setminus K}, \quad (2)$$

where $\mathcal{UTZ}^{\text{num}}(K)$ is a numerical approximation of $\mathcal{UTZ}(K) = \{(u, \tilde{d}_j, \tilde{y}_j^{\text{id}}) \mid u \in \tilde{\mathcal{U}}_{\text{opt},w}(I)\}$. An explanation for excluding points based on (2) can be found in [16] (Section 1.7.4.5). Finally, we describe the proposed numerical hierarchical PS method in Algorithm 1.

Remark 3. *In [32], the author introduce three different quality criteria for the numerical implementation of a scalarization method, which we discuss here for the presented hierarchical PS method.*

(a) **Coverage:** *Every part of the Pareto set and front has to be represented in the sets $\mathcal{U}^{\text{num}}_{\text{opt},w}$ and $\mathcal{J}^{\text{num}}_{\text{opt},w}$, respectively. This can be measured by*

$$\text{cov}(\mathcal{J}_{\text{opt},(w),(\text{loc})}) := \max_{\bar{y} \in \mathcal{J}_{\text{opt},(w),(\text{loc})}} \min_{y \in \mathcal{J}^{\text{num}}_{\text{opt},(w),(\text{loc})}} \|\bar{y} - y\|.$$

In the case of Algorithm 1, we have that $\text{cov}(\mathcal{J}_{\text{opt},(w),(\text{loc})}) = \mathcal{O}(h)$ (cf. [16] (Remark 1.7.69-(a))).

(b) **Uniformity:** *The points on the Pareto set and front should be distributed (almost) equidistantly; cf. [16] (Remark 1.7.69-(b)).*

(c) **Cardinality:** *The number of points contained in the numerical approximation should be reasonable. In the case of Algorithm 1 is not possible to estimate a-priori the number of elements computed by the method. It is possible to show a bound which can be computed when the nadir objective point $y^{\text{nad},(w)}$ is known (cf. [16], Remark 1.7.69-(c)).*

Algorithm 1 Solving (MOP) numerically by the hierarchical PS method
1: **for** $j = 1 : k$ **do**
2:    Set $I := \{j\}$;
3:    Compute $\mathcal{U}^{\text{num}}_{\text{opt,w}}(I) = \{u \mid u \text{ minimizes } \hat{J}_j\}$;
4:    Choose $\tilde{d}_j$, compute y^{id}_j and set $\tilde{y}^{\text{id}}_j = y^{\text{id}}_j - \tilde{d}_j$;
5:    Set $\mathcal{UTZ}^{\text{num}}(I) = \{(u, \tilde{d}_j, \tilde{y}^{\text{id}}_j) \mid u \in \mathcal{U}^{\text{num}}_{\text{opt,w}}(I)\}$;
6: **end for**
7: **for** $i = 2 : k$ **do**
8:    **for all** $I \subset \{1, \ldots, k\}$ with $|I| = i$ **do**
9:       Initialize $\mathcal{U}^{\text{num}}_{\text{opt,w}}(I) = \bigcup_{K \subsetneq I} \mathcal{U}^{\text{num}}_{\text{opt,w}}(K)$ and $\mathcal{UTZ}^{\text{num}}(I) = \emptyset$;
10:       Compute the reference points $Z^{\text{num}}(I) = \{z \in \mathcal{Z}^{h,I} \mid \neg(2)\}$;
11:       **while** $Z^{\text{num}}(I) \neq \emptyset$ **do**
12:          Choose $z \in Z^{\text{num}}(I)$ and remove z from $Z^{\text{num}}(I)$;
13:          Solve $(\mathbf{P}^{\text{PS}}_{I,z,r})/(\mathbf{RP}^{\text{PS}}_{I,z,r})$;
14:          Set $\mathcal{U}^{\text{num}}_{\text{opt,w}}(I) \leftarrow \mathcal{U}^{\text{num}}_{\text{opt,w}}(I) \cup \mathcal{Q}^I_{\text{opt,w}}(z)$;
15:          Set
            $\mathcal{UTZ}^{\text{num}}(I) \leftarrow \mathcal{UTZ}^{\text{num}}(I) \cup \{(\bar{u}, \bar{t}, z) \mid (\bar{u}, \bar{t}) \text{ gl. sol. of } (\mathbf{P}^{\text{PS}}_{I,z,r})\}$;
16:          Add solutions of PSPs with respect to redundant reference points: Set
            $\mathcal{UTZ}^{\text{num}}(I) \leftarrow \mathcal{UTZ}^{\text{num}}(I) \cup \{(\bar{u}, \bar{t}, \tilde{z}) \mid (\bar{u}, \bar{t}) \text{ gl. sol. of } (\mathbf{P}^{\text{PS}}_{I,z,r}),$
            $\tilde{z} \in Z^{\text{num}}(I) \cap [z - (\bar{t}r^I - (\hat{J}^I(\bar{u}) - z)), z]\}$;
17:          Remove redundant reference points: Set
            $Z^{\text{num}}(I) \leftarrow Z^{\text{num}}(I) \setminus [z - (\bar{t}r^I - (\hat{J}^I(\bar{u}) - z)), z]$ for all $\bar{u} \in \mathcal{Q}^I_{\text{opt,(w)}}(z)$;
18:       **end while**
19:    **end for**
20: **end for**
21: **if** *computeParetoFront == true* **then**
22:    Remove all $u \in \mathcal{U}^{\text{num}}_{\text{opt,w}}(\{1, \ldots, k\})$ with $u \notin \mathcal{U}_{\text{opt}}$ by a non-dominance test;
23: **end if**

3. The Non-Convex Parametric PDE-Constrained MOP

Before defining our exemplary MOP, we introduce the PDE model which will later serve as an equality constraint. Let $\Omega \subset \mathbb{R}^d$, $d \in \{2, 3\}$, be a bounded domain with Lipschitz-continuous boundary $\Gamma = \partial \Omega$. Furthermore, let $\Omega_1, \ldots, \Omega_m$ be a pairwise disjoint decomposition of the domain Ω and set $\Gamma_i := \partial \Omega_i \cap \partial \Omega$ for all $i = 1, \ldots, m$. Then we are interested in the following elliptic diffusion-reaction equation with Robin boundary condition:

$$-\nabla \cdot \left(\sum_{i=1}^{m} u^\kappa_i \chi_{\Omega_i}(x) \nabla y(x) \right) + u^r r(x) y(x) = f(x) \quad \text{a.e. in } \Omega, \tag{3a}$$

$$u^\kappa_i \frac{\partial y}{\partial n}(s) + \alpha y(s) = \alpha y_a(s) \quad \text{a.e. on } \Gamma_i. \tag{3b}$$

For every $i \in \{1, \ldots, m\}$, the parameter $u^\kappa_i > 0$ represents the diffusion coefficient on the subdomain Ω_i. By $r \in L^\infty(\Omega)$, we denote a reaction function, which is supposed to satisfy $r > 0$ a.e. in Ω and is controlled by the scalar parameter $u^r > 0$. On the right-hand side of (3a), we have the source term $f \in L^2(\Omega)$. The constant $\alpha > 0$ in (3b) models the heat exchange with the outside of the domain Ω, where a temperature of $y_a \in L^2(\Gamma)$ is assumed. In total, the parameter space is given by $\mathcal{U} = \mathbb{R}^m \times \mathbb{R}$ and any parameter $u \in \mathcal{U}$ can be written as the vector $u = (u^\kappa, u^r)^T$ with $u^\kappa = (u^\kappa_1, \ldots, u^\kappa_m)^T \in \mathbb{R}^m$. Setting $H = L^2(\Omega)$ and $V = H^1(\Omega)$ the weak formulation of (3) is

$$a(u; y, \varphi) = \mathcal{F}(\varphi) \quad \text{for all } \varphi \in V \tag{4}$$

for any $u \in \mathcal{U}$. In (4) the parameter-dependent symmetric bilinear form $a(u; \cdot, \cdot)\colon V \times V \to \mathbb{R}$ is given by

$$a(u; \varphi, \psi) := \sum_{i=1}^{m} u_i^{\kappa} \int_{\Omega_i} \nabla \varphi(x) \cdot \nabla \psi(x) \, dx + u^r \int_{\Omega} r(x) \varphi(x) \psi(x) \, dx + \alpha \int_{\Gamma} \varphi(s) \psi(s) \, ds$$

for all $\varphi, \psi \in V$ and $u \in \mathcal{U}$. The linear functional $\mathcal{F} \in V'$ is defined by

$$\mathcal{F}(\varphi) := \int_{\Omega} f(x) \varphi(x) \, dx + \alpha \int_{\Gamma} y_a(s) \varphi(s) \, ds \quad \text{for all } \varphi \in V.$$

Lemma 2. (a) For all $u \in \mathcal{U}$ it holds

$$\|a(u; \cdot, \cdot)\|_{L(V,V')} \leq C \|u\|_{\mathcal{U}}$$

with a constant $C > 0$, which does not depend on u.

(b) For all $u \in \mathcal{U}$ with $u^{\kappa} > 0$ in $\mathbb{R}$ and $u^r > 0$, it holds

$$a(u; \varphi, \varphi) \geq \min(u_1^{\kappa}, \ldots, u_m^{\kappa}, u^r) \|\varphi\|_V^2 \quad \text{for all } \varphi \in V.$$

(c) The mapping $\mathcal{F} \in V'$ is well-defined.

Proof. All statements follow from similar arguments of [33] (Lemma 1.4), where related operators were considered in the parabolic case. □

Theorem 6. Let $u \in \mathcal{U}$ with $u > 0$ be arbitrary. Then there is a unique solution $y = y(u) \in V$ of (3). Moreover, the estimate

$$\|y\|_V \leq C \left(\|f\|_{L^2(\Omega)} + \|y_a\|_{L^2(\Gamma)} \right) \quad (5)$$

holds with a constant $C > 0$, which depends continuously on u, but is independent of f and y_a.

Proof. The claims follow from the Lax-Milgram theorem (cf. [34]) and Lemma 2. □

Definition 12. Let $u_{\min}^{\kappa} \in (0, \infty)^m$ and $u_{\min}^r > 0$ be arbitrary. Then we define the closed set

$$\mathcal{U}_{\text{eq}} := \{u \in \mathcal{U} \mid u^{\kappa} \geq u_{\min}^{\kappa}, \ u^r \geq u_{\min}^r\}.$$

In view of Theorem 6, it is possible to define the solution operator $\mathcal{S}\colon \mathcal{U}_{\text{eq}} \to V$, which maps any parameter $u \in \mathcal{U}_{\text{eq}}$ to the unique solution $y = \mathcal{S}(u) \in V$ of (4).

Remark 4. Due to Lemma 2, we can conclude that $a(u; \varphi, \varphi) \geq \alpha_{\min} \|\varphi\|_V^2$ for all $\varphi \in V$ and $u \in \mathcal{U}_{\text{eq}}$, where $\alpha_{\min} := \min((u_{\min}^{\kappa})_1, \ldots, (u_{\min}^{\kappa})_m, u^r) > 0$. In particular, the constant C in (5) can be chosen independently of u if we restrict ourselves to parameters $u \in \mathcal{U}_{\text{eq}}$.

Theorem 7. The solution operator $\mathcal{S}\colon \mathcal{U}_{\text{eq}} \to V$ is twice continuously Fréchet differentiable. For the first derivative $\mathcal{S}'\colon \mathcal{U}_{\text{eq}} \to L(\mathcal{U}, V)$, we have that for any $u \in \mathcal{U}_{\text{eq}}$ and $h \in \mathcal{U}$ the function $y^h := \mathcal{S}'(u)h \in V$ solves the equation

$$a(u; y^h, \varphi) = -\partial_u a(u; \mathcal{S}(u), \varphi)h \quad \text{for all } \varphi \in V.$$

The second derivative $\mathcal{S}'' : \mathcal{U}_{eq} \to L(\mathcal{U}, L(\mathcal{U}, V))$ is given as follows: For any $u \in \mathcal{U}_{eq}$ and $h_1, h_2 \in \mathcal{U}$, the function $y^{h_1, h_2} := \mathcal{S}''(u)(h_1, h_2)$ solves the equation

$$a(u; y^{h_1, h_2}, \varphi) = -\partial_u a(u; \mathcal{S}'(u)h_1, \varphi)h_2 - \partial_u a(u; \mathcal{S}'(u)h_2, \varphi)h_1 \quad \text{for all } \varphi \in V.$$

Remark 5. By $\partial_u a$ we denote the partial derivative of the mapping a w.r.t. the parameter u. Since a is linear in u, it holds

$$\partial_u a(u; \varphi, \psi) h = a(h; \varphi, \psi), \quad \partial_u^2 a(u; \varphi, \psi) = 0 \in L(\mathcal{U}, \mathcal{U}')$$

for all $u, h \in \mathcal{U}$ and all $\varphi, \psi \in V$. In particular, we can identify $\partial_u a(u; \varphi, \psi) \in \mathcal{U}'$ by

$$\partial_u a(u; \varphi, \psi) = \begin{pmatrix} \int_{\Omega_1} \nabla \varphi(x) \cdot \nabla \psi(x) \, dx \\ \vdots \\ \int_{\Omega_m} \nabla \varphi(x) \cdot \nabla \psi(x) \, dx \\ \int_\Omega r(x) \varphi(x) \psi(x) \, dx \end{pmatrix} \in \mathcal{U}$$

by using the Riesz representation theorem.

We are now ready to state the multiobjective parametric PDE-constrained optimization problem (MPPOP). Let $k \in \mathbb{N}$ be fixed and

$$\sigma_\Omega^{(1)}, \ldots, \sigma_\Omega^{(k)} \geq 0 \quad \text{as well as} \quad \sigma_\mathcal{U}^{(1)}, \ldots, \sigma_\mathcal{U}^{(k)} \geq 0$$

be non-negative weights. Furthermore, denote by $y_\Omega^{(1)}, \ldots, y_\Omega^{(k)} \in H$ the desired states and by $u_d^{(1)}, \ldots, u_d^{(k)} \in \mathcal{U}$ the desired parameters. Then we define the multiobjective essential cost functions $\hat{J}_1, \ldots, \hat{J}_k : \mathcal{U}_{eq} \to \mathbb{R}$ by

$$\hat{J}_i(u) := \frac{\sigma_\Omega^{(i)}}{2} \|\mathcal{S}(u) - y_\Omega^{(i)}\|_H^2 + \frac{\sigma_\mathcal{U}^{(i)}}{2} \|u - u_d^{(i)}\|_\mathcal{U}^2 \quad \text{for all } u \in \mathcal{U}_{eq} \text{ and } i \in \{1, \ldots, k\}.$$

Moreover, u_a, u_b with $u_a \leq u_b$ are lower and upper bounds on the parameter u which we assume to be finite. We define $\mathcal{U}_{ad} := \{u \in \mathcal{U} \mid u_a \leq u \leq u_b\}$ and we assume that $\mathcal{U}_{ad} \subset \mathcal{U}_{eq}$ holds. Note that $\mathcal{U}_{ad}$ is a closed, convex and bounded set because of the finiteness assumption on u_a and u_b. We are interested in solving

$$\min_{u \in \mathcal{U}_{ad}} \hat{J}(u) = \min_{u \in \mathcal{U}_{ad}} \left(\hat{J}_1(u), \ldots, \hat{J}_k(u) \right)^T. \quad \textbf{(MPPOP)}$$

Note that, thanks to the assumptions on $\mathcal{U}_{ad}$ and $\sigma_\mathcal{U}^{(i)}$, the costs $\hat{J}_1, \ldots, \hat{J}_k$ satisfy Assumption 1. This problem fits into the framework of non-convex multiobjective optimization and Algorithm 1 can be applied. The non-convexity comes from the way the bilinear form depends on the parameter u. This makes, in fact, the solution mapping non-linear and thus the MPPOP non-convex. To close this section, we derive the expression of the gradient and Hessian of the cost functionals $\hat{J}_1, \ldots, \hat{J}_k$. We define the i-th adjoint equation and its solution operator as

Definition 13. For $i = 1, \ldots, k$, the solution operator of the i-th adjoint equation is $\mathcal{A}_i : \mathcal{U}_{eq} \to V$, where for any given $u \in \mathcal{U}_{eq}$, $p^{(i)} := \mathcal{A}_i(u)$ solves the equation

$$a(u; \varphi, p^{(i)}) = \langle \sigma_\Omega^{(i)} (\mathcal{S}(u) - y_\Omega^{(i)}), \varphi \rangle_H \quad \text{for all } \varphi \in V. \quad (6)$$

As shown in [16], this operators satisfy the two following results:

Lemma 3. *The solution operator $\mathcal{A}_i : \mathcal{U}_{eq} \to V$ is continuously Fréchet differentiable for all $i = 1, \ldots, k$. For all $i = 1, \ldots k$, for the first derivative $\mathcal{A}'_i : \mathcal{U}_{eq} \to L(\mathcal{U}, V)$, we have that for any $u \in \mathcal{U}_{eq}$ and $h \in \mathcal{U}$ the function $p^{(i),h} := \mathcal{A}'_i(u)h \in V$ solves the equation*

$$a(u; \varphi, p_i^{(i),h}) = -\partial_u a(u; \varphi, \mathcal{A}_i(u))h + \sigma_\Omega \langle \mathcal{S}'(u)h, \varphi \rangle_{V',V} \quad \text{for all } \varphi \in V. \tag{7}$$

Corollary 1. *Let $\mathcal{U}_{ad} \subset \mathcal{U}_{eq}$, $u \in \mathcal{U}_{ad}$ and $h \in \mathcal{U}$ be arbitrary. Then for $i = 1, \ldots, k$ the cost functions $\hat{J}_i$ are twice continuously Fréchet differentiable and it holds*

$$\nabla \hat{J}_i(u) = -\partial_u a(u; \mathcal{S}(u), \mathcal{A}_i(u)) + \sigma_\mathcal{U}(u - u_d^{(i)}) \in \mathcal{U},$$
$$\nabla^2 \hat{J}_i(u)h = -\partial_u a(u; \mathcal{S}'(u)h, \mathcal{A}_i(u)) - \partial_u a(u; \mathcal{S}(u), \mathcal{A}'_i(u)h) + \sigma_\mathcal{U}^{(i)} h \in \mathcal{U}.$$

where we use the representation of $\partial_u a(u; \mathcal{S}(u), \mathcal{A}_i(u)) \in \mathcal{U}'$ in $\mathcal{U}$, cf. Remark 5.

The RB Method for MPPOP

One of the limitations of solving the MPPOP directly with the PS method is the high computational cost. Algorithm 1, in fact, requires to solve the state and adjoint equation a large number of times in order to efficiently approximate the Pareto set. Unfortunately, the numerical evaluation of the state and adjoint solution operators is costly due to the high number of degrees of freedom required to apply, for example, the FE method. For this reason, we use the RB method. In the following we explain how the RB method can be applied to our model. From Theorem 6, we know that the weak form of the state equation admits a unique solution for any control $u \in \mathcal{U}_{eq}$. This allows us to define the solution operator $\mathcal{S} : \mathcal{U}_{eq} \to V$. Now, let us consider the so-called solution manifold $\mathcal{M} := \{\mathcal{S}(u) \mid u \in \mathcal{U}_{eq}\} \subset V$. The goal of the RB method is to provide a low-dimensional subspace $V^\ell \subset V$, which is a good approximation of $\mathcal{M}$. The subspace V^ℓ is defined as the span of linearly independent snapshots $\mathcal{S}(u_1), \ldots, \mathcal{S}(u_\ell)$ for selected parameters $u_1, \ldots, u_\ell \in \mathcal{U}_{eq}$. Clearly, V^ℓ has dimension ℓ and the snapshots constitute its basis. Let us postpone the discussion on how to select good parameters for generating V^ℓ. Given an RB space V^ℓ, we obtain the reduced-order state equation by a Galerkin projection:

$$a(u; y^\ell, \psi) = \mathcal{F}(\psi) \quad \text{for all } \psi \in V^\ell. \tag{8}$$

Also for the reduced-order equation, we have unique solvability for all parameters $u \in \mathcal{U}_{eq}$. The solution map $\mathcal{S}^\ell : \mathcal{U}_{eq} \to V^\ell$, which maps any parameter $u \in \mathcal{U}_{eq}$ to the unique solution $y^\ell = \mathcal{S}^\ell(u) \in V^\ell$ of (8), is then well-defined. We can similarly define a reduced-order adjoint equation and essential cost functional. For $i = 1, \ldots, k$, we define the essential reduced-order cost functions $\hat{J}_i^\ell : \mathcal{U}_{eq} \to \mathbb{R}$ by

$$\hat{J}_i^\ell(u) := \frac{\sigma_\Omega^{(i)}}{2} \|\mathcal{S}^\ell(u) - y_\Omega^{(i)}\|_H^2 + \frac{\sigma_\mathcal{U}^{(i)}}{2} \|u - u_d^{(i)}\|_\mathcal{U}^2,$$

the reduced-order adjoint equation by

$$a(u; \psi, p^{(i),\ell}) = \langle \sigma_\Omega^{(i)} (\mathcal{S}^\ell(u) - y_\Omega^{(i)}), \psi \rangle_H \quad \text{for all } \psi \in V^\ell \tag{9}$$

and the reduced-order adjoint solution operator $\mathcal{A}_i^\ell : \mathcal{U}_{eq} \to V$. Following Corollary 1, it is possible to represent the gradient and the Hessian of the essential reduced-order cost functions $\hat{J}_i^\ell$ for $i = 1, \ldots, k$ by simply replacing the operators $\mathcal{S}$ and $\mathcal{A}_i$ by their respective reduced-order versions $\mathcal{S}^\ell$ and $\mathcal{A}_i^\ell$. There are still two aspects which remain to be clarified: first, how to generate an RB space which guarantees a good approximation of the state and adjoint solution manifolds and, second, how to estimate a-posteriori (i.e., without explicitly evaluating the full-order solution operators $\mathcal{S}$ and $\mathcal{A}$) the error of such an approximation.

For the first aspect, one can think of building an RB space either prior to solving the reduced-order optimization problem or while solving it. The first approach is the so-called offline/online decomposition; cf. [35]. This technique exploits a greedy algorithm in the offline phase, which iteratively searches for the parameter for which the approximation error between the full- and reduced-order state and adjoint variables is the largest. Then, the RB space is enriched (by solving the full-order state and adjoint equations at the respective parameter and orthonormalizing the newly computed snapshots with respect to the current RB basis) until a pre-defined tolerance for the approximation error is reached. Once the RB space is computed, the online phase can start: the optimization problem is solved fast on the reduced-order level. Although this technique is still widely used in literature, it shows many disadvantages in the context of optimization. At first, it suffers from the curse of dimensionality: for a high-dimensional parameter space it is too costly to explore the entire parameter space with a greedy procedure. At second, it is counter-intuitive to prepare an RB space which is accurate enough for any parameter, when usually the optimization method follows a (short) pattern in the parameter space to find the solution or when the Pareto set is contained in some local regions of the parameter space, as often in the case of non-convex multiobjective problems. While it is true that the computational costs of an offline phase could be amortized in the context of multiobjective optimization for a reasonably small dimension of the parameter space due to the vast amount a scalarized PS problems that need to be solved in the online phase, the disadvantage of the offline-online splitting in this setting is the lack of control of the accuracy of the Pareto optimal solutions. Indeed, to the best of our knowledge there are no suitable error indicators for the greedy algorithm to guarantee a certified accuracy of the reduced-order Pareto optimal points w.r.t. full-order ones. Luckily, the focus has shifted recently towards adapting the RB space while proceeding with the optimization method. This procedure is followed, e.g., by the methods presented in [14,15,17,18]. The advantage of these methods with respect to methods based on an offline-online splitting is that they compute first-order critical points of the full-order optimization problems. Let us specify that in [14,17,18] the authors proposed and progressively improved an RB method combined with a TR algorithm, based on more general results presented in [15]. Such a method constructs the RB space adaptively while the optimizer is computing the optimal solution. Our focus here is on further improving the method in [17], which can be considered the most general among the TR-RB methods.

For any of the above-mentioned methods, a-posteriori error estimates are crucial to compute upper bounds of the approximation error made by the RB space in reconstructing the solution for a given parameter without any full-order solution at hand. In case of optimization, one is also interested in estimating the error in reconstructing the cost functional and its gradient. For our model, we can use the following estimates:

Theorem 8. *Let $u \in \mathcal{U}_{ad}$ be arbitrary and denote by $\alpha(u)$ the coercivity constant of the bilinear form $a(u;\cdot,\cdot)$. By Remark 4, it holds $\alpha(u) \geq \alpha_{\min} > 0$. Let the residual $r_{st}(u;\cdot) \in V'$ be given by $r_{st}(u;\varphi) := \mathcal{F}(\varphi) - a(u;\mathcal{S}^\ell(u),\varphi)$ for all $\varphi \in V$. Then it holds*

$$\left\|\mathcal{S}(u) - \mathcal{S}^\ell(u)\right\|_V \leq \Delta_{st}(u) := \frac{\|r_{st}(u;\cdot)\|_{V'}}{\alpha(u)}. \tag{10}$$

For $i = 1,\ldots,k$ the residual $r_{adj}^{(i)}(u;\cdot) \in V'$ of the adjoint equations is given by $r_{adj}^{(i)}(u;\varphi) := \langle \sigma_\Omega^{(i)}(\mathcal{S}^\ell(u) - y_\Omega^{(i)}),\varphi\rangle_H - a(u;\varphi,\mathcal{A}_i^\ell(u))$ for all $\varphi \in V$. Then it holds

$$\left\|\mathcal{A}_i(u) - \mathcal{A}_i^\ell(u)\right\|_V \leq \Delta_{adj}^{(i)}(u) := \frac{\|r_{adj}^{(i)}(u;\cdot)\|_{V'} + \sigma_\Omega^{(i)}\Delta_{st}(u)}{\alpha(u)}.$$

Furthermore, for $i = 1, \ldots, k$ we have

$$|\hat{J}_i(u) - \hat{J}_i^\ell(u)| \leq \Delta_{\mathsf{st}}(u)\|r_{\mathsf{adj}}^{(i)}(u;\cdot)\|_{V'} + \sigma_\Omega^{(i)} \Delta_{\mathsf{st}}(u)^2 =: \Delta_{\hat{J}_i^\ell}(u),$$

$$\|\nabla \hat{J}_i(u) - \nabla \hat{J}_i^\ell(u)\|_{\mathcal{U}} \leq \|\partial_u a(u;\cdot,\cdot)\| \left(\|\mathcal{S}^\ell(u)\|_V \Delta_{\mathsf{adj}}^{(i)}(u) + \Delta_{\mathsf{st}}(u) \Delta_{\mathsf{adj}}^{(i)}(u) \right.$$
$$\left. + \Delta_{\mathsf{st}}(u) \|\mathcal{A}_i^\ell(u)\|_V \right) =: \Delta_{\nabla \hat{J}_i^\ell}(u).$$

Proof. A proof of the a-posteriori error estimates for the state and adjoint can be found in [35]. For the cost function and the gradient, we refer to [18] (Proposition 2.5). □

Note that we only need the reduced-order state and adjoint state to evaluate the a-posteriori error estimates. For our example, the computation of the coercivity constant $\alpha(u)$ is cheap, see Lemma 2. In more general examples, this might not be the case. Thus, one often uses a quickly computable lower bound $\alpha_{\mathsf{LB}}(u)$ instead. Possible methods for computing such a lower bound are, e.g., the min-theta approach (cf. [35]) or the Successive Constraint Method (SCM) (cf. [36]). In situations in which the computation or the estimation of the coercivity constant is complicated, the TR-RB algorithms presented in [19,20] have the advantage that they do not require the computation or estimation of the coercivity constant but only rely on asymptotic error estimates consisting of residual based error indicators. Note finally that the computation of the terms $\|r_{\mathsf{st}}(u;\cdot)\|_{V'}$ and $\|r_{\mathsf{adj}}^{(i)}(u;\cdot)\|_{V'}$ is not possible in an infinite-dimensional setting. Even after discretization with the FE method, the cost of computing such a term depends on the dimension of the full-order model, which contradicts the request of having a computationally cheap estimate. However, in our case, the parameter-separability of the bilinear form $a(u;\cdot,\cdot)$ can be exploited to preassemble certain quantities in such a way that the computational cost for evaluating $\|r_{\mathsf{st}}(u;\cdot)\|_{V'}$ and $\|r_{\mathsf{adj}}^{(i)}(u;\cdot)\|_{V'}$ only depends on the dimension of the RB space; see, e.g., [36]. Finally, we apply the RB method to (**MPPOP**): for a given RB space V^ℓ the reduced-order MPPOP reads

$$\min \hat{J}^\ell(u) = \left(\hat{J}_1^\ell(u), \ldots, \hat{J}_k^\ell(u) \right)^T \quad \text{s.t.} \quad u \in \mathcal{U}_{\mathsf{ad}}. \qquad (\textbf{MPPOP}^\ell)$$

For an arbitrary reference point $z \in \mathbb{R}^k$ and target direction $r \in \mathbb{R}^k$, the reduced-order PS problem reads

$$\min_{(u,t)} t \quad \text{s.t.} \quad (t,u) \in \mathbb{R} \times \mathcal{U}_{\mathsf{ad}} \text{ and } \hat{J}_i^\ell(u) - z_i \leq t, \quad i = 1,\ldots,k. \qquad (\textbf{P}_{z,r}^{\mathsf{PS},\ell})$$

One could then outline an algorithm similar to Algorithm 1 by using an offline/online splitting. Because of the above-mentioned disadvantages, we focus on combining the PSPs with the TR-RB method from [17] and extend it with respect to the method in [16]. The TR method introduces new aspects to the RB implementation, such as the adaptive construction of the RB space; see the next section for further details.

4. The TR-RB Method

We briefly introduce the method from [17] and clarify how to apply this in combination with the PS method. In Section 4.2 we highlight our extension to this method and how this can reduce the computational time. The basic idea of a TR method is to compute a first-order critical point of a costly optimization problem by iteratively solving some cheap-to-solve approximations in local regions of the admissible space, where these model approximations can be trusted (i.e., are accurate enough). In such a way, one can derive a global method, which converges in a finite number of steps. For each outer iteration $j \geq 0$ of the TR method, the cheap approximation of the objective is generally indicated by $m^{(j)}$ and the trust regions are described by a radius $\delta^{(j)}$. To simplify the exposition, let us stick with the case $\mathcal{U} = \mathbb{R}^m \times \mathbb{R}$, as in Section 3. The TR method solves then, for each $j \geq 0$, the following constrained optimization sub-problems

$$\min_{v \in \mathcal{U}} m^{(j)}(v) \quad \text{s.t.} \quad \|v\|_2 \leq \delta^{(j)}, \tilde{u} := u^{(j)} + v \in \mathcal{U}_{\text{ad}}. \tag{11}$$

Under suitable assumptions, problem (11) admits a unique solution $\bar{v}^{(j)}$, which is used to compute the next outer iteration $u^{(j+1)} = u^{(j)} + \bar{v}^{(j)}$. To further simplify the presentation of the algorithm in [17], let us present it for a general cost functional $\mathcal{J}$. Later in this section we will give more details about its application to the MPPOP and the PS method. The TR-RB version of problem (11) is

$$\min_{\tilde{u} \in \mathcal{U}_{\text{ad}}} \mathcal{J}^{\ell,(j)}(\tilde{u}) \quad \text{s.t.} \quad q^{(j)}(\tilde{u}) := \frac{\Delta_{\mathcal{J}^{\ell,(j)}}(\tilde{u})}{\mathcal{J}^{\ell,(j)}(\tilde{u})} \leq \delta^{(j)}, \tag{12}$$

where $\mathcal{J}^{\ell,(j)}(\tilde{u})$ is the reduced-order cost functional w.r.t. the reduced-order model at the j-th iteration and $\Delta_{\mathcal{J}^{\ell,(j)}}(\tilde{u})$ is an estimate for the error $|\mathcal{J}(\tilde{u}) - \mathcal{J}^{\ell,(j)}(\tilde{u})|$. Looking at (12), one clearly sees that the role of the model function $m^{(j)}$ is played by the reduced-order model cost functional. This is perfectly in line with the TR spirit of having a cheap-to-solve approximation of the original optimization problem. The trust regions are defined instead through the RB error estimator, which is in fact the way we use to check the quality of the approximation. Let us mention at this point that there are also different approaches to this. In [19,20] the authors incorporated the usual trust-region constraints as seen in (11) into a TR-RB algorithm. In [18] also the importance of introducing a correction term on the RB level is discussed to improve the performance of the method. We point out that this only has to be done if one chooses two separate RB spaces for state and adjoint equations (see also [17]). This will not be the case for our application. In Algorithm 2, we report the method from [17]. In what follows, we guide the reader through the features of the algorithm. At first, we need to initialize the reduced-order model at the initial guess $u^{(0)}$. This means computing $\mathcal{S}(u^{(0)})$ and $\mathcal{A}_i(u^{(0)})$ for $i = 1, \ldots, k$ and generating the RB space $V^{\ell,(0)}$ as their span. Similarly, updating the RB space $V^{\ell,(j)}$ at the point $u^{(j+1)}$ means computing the full-order quantities $\mathcal{S}(u^{(j+1)})$ and $\mathcal{A}_i(u^{(j+1)})$ for $i = 1, \ldots, k$ and adding them to the RB space by a Gram-Schmidt orthonormalization.

In Line 3 of Algorithm 2, it is required to compute the so-called approximated generalized Cauchy (AGC) point. We report here its definition according to [15,18].

Definition 14. *Let $\kappa \in (0,1)$ and $\kappa_{\text{arm}} \in (0,1)$ be backtracking parameters. For the current iterate $u^{(j)}$ define $d^{(j)} := \nabla \mathcal{J}^{\ell,(j)}(u^{(j)})$. Let $\alpha \in \mathbb{N}$ be the smallest number for which the two conditions*

$$\mathcal{J}^{\ell,(j)}\left(P_{\mathcal{U}_{\text{ad}}}(u^{(j)} - \kappa^\alpha d^{(j)})\right) - \mathcal{J}^{\ell,(j)}(u^{(j)}) \leq -\frac{\kappa_{\text{arm}}}{\kappa^\alpha} \|P_{\mathcal{U}_{\text{ad}}}(u^{(j)} - \kappa^\alpha d^{(j)}) - u^{(j)}\|_{\mathcal{U}}^2, \tag{13}$$

$$q^{(j)}\left(P_{\mathcal{U}_{\text{ad}}}(u^{(j)} - \kappa^\alpha d^{(j)})\right) \leq \delta^{(j)} \tag{14}$$

are satisfied, where $P_{\mathcal{U}_{\text{ad}}} : \mathcal{U} \to \mathcal{U}_{\text{ad}}$ is the canonical projection onto the closed and convex set $\mathcal{U}_{\text{ad}}$. Then we define the AGC point as $u_{\text{AGC}}^{(j)} := P_{\mathcal{U}_{\text{ad}}}(u^{(j)} - \kappa^\alpha d^{(j)})$.

The TR-RB subproblem (12) is then solved in Line 4 using a projected Newton-CG algorithm with the AGC point as a warm start and the following termination criteria

$$\|u - P_{\mathcal{U}_{\text{ad}}}(u - \nabla \mathcal{J}^{\ell,(j)}(u))\|_{\mathcal{U}} \leq \tau_{\text{sub}}, \quad \beta_{\text{bound}} \delta^{(j)} \leq q^{(j)}(u) \leq \delta^{(j)}. \tag{15}$$

The first condition in (15) is the standard first-order criticality condition with tolerance $\tau_{\text{sub}} \in (0,1)$ and the second one was already introduced in [14] to avoid too many iterations close to the TR boundary, which is generally an area where we are already starting to trust the model function less. The parameter β_{bound} is usually chosen to be close to one exactly for this purpose.

Algorithm 2 TR-RB algorithm

1: Initialize the reduced-order model at $u^{(0)}$, set $j = 0$ and Loop_flag=True;
2: **while** Loop_flag **do**
3:    Compute the AGC point $u_{\text{AGC}}^{(j)}$;
4:    Compute $u^{(j+1)}$ as solution of (12) with stopping criteria (15);
5:    **if** $\mathcal{J}^{\ell,(j)}(u^{(j+1)}) + \Delta_{\mathcal{J}^{\ell,(j)}}(u^{(j+1)}) < \mathcal{J}^{\ell,(j)}(u_{\text{AGC}}^{(j)})$ **then**
6:      Accept $u^{(j+1)}$, set $\delta^{(j+1)} = \delta^{(j)}$, compute $\varrho^{(j)}$ and $g(u^{(j+1)})$;
7:      **if** $g(u^{(j+1)}) \leq \tau_{\text{FOC}}$ **then**
8:        Set Loop_flag=False;
9:      **else**
10:        **if** $\varrho^{(j)} \geq \eta_\varrho$ **then**
11:          Enlarge the TR radius $\delta^{(j+1)} = \beta_1^{-1}\delta^{(j)}$;
12:        **end if**
13:        **if not** Skip_enrichment_flag(j) **then**
14:          Update the RB model at $u^{(j+1)}$;
15:        **end if**
16:      **end if**
17:    **else if** $\mathcal{J}^{\ell,(j)}(u^{(j+1)}) - \Delta_{\mathcal{J}^{\ell,(j)}}(u^{(j+1)}) > \mathcal{J}^{\ell,(j)}(u_{\text{AGC}}^{(j)})$ **then**
18:      **if** $\beta_1\delta^{(j)} \leq \delta_{\min}$ **or** Skip_enrichment_flag($j-1$) **then**
19:        Update the RB model at $u^{(j+1)}$;
20:      **end if**
21:      Reject $u^{(j+1)}$, shrink the radius $\delta^{(j+1)} = \beta_1\delta^{(j)}$ and go to 4;
22:    **else**
23:      Compute $\mathcal{J}(u^{(j+1)})$, $g(u^{(j+1)})$, $\varrho^{(j)}$ and set $\delta^{(j+1)} = \beta_1^{-1}\delta^{(j)}$;
24:      **if** $g(u^{(j+1)}) \leq \tau_{\text{FOC}}$ **then**
25:        Set Loop_flag=False;
26:      **else**
27:        **if** Skip_enrichment_flag(j) **and** $\varrho^{(j)} \geq \eta_\varrho$ **then**
28:          Accept $u^{(j+1)}$;
29:        **else if** $\mathcal{J}(u^{(j+1)}) \leq \mathcal{J}^{\ell,(j)}(u_{\text{AGC}}^{(j)})$ **then**
30:          Accept $u^{(j+1)}$ and update the RB model;
31:          **if** $\varrho^{(j)} < \eta_\varrho$ **then**
32:            Set $\delta^{(j+1)} = \delta^{(j)}$;
33:          **end if**
34:        **else**
35:          **if** $\beta_1\delta^{(j)} \leq \delta_{\min}$ **or** Skip_enrichment_flag($j-1$) **then**
36:            Update the RB model at $u^{(j+1)}$;
37:          **end if**
38:          Reject $u^{(j+1)}$, set $\delta^{(j+1)} = \beta_1\delta^{(j)}$ and go to 4;
39:        **end if**
40:      **end if**
41:    **end if**
42:    Set $j = j + 1$;
43: **end while**

An important aspect of TR methods is the decision to accept or reject the step $u^{(j+1)}$. Generally, one asks for the so-called sufficient decrease condition $\mathcal{J}^{\ell,(j+1)}(u^{(j+1)}) \leq \mathcal{J}^{\ell,(j)}(u_{\text{AGC}}^{(j)})$; cf. [15]. Note that this condition requires to update the RB space before being sure that the step will be accepted. If it is rejected, then we performed a costly update without the possibility of exploiting it. Because of this fact, Ref. [14] proposed a sufficient

(Line 5) and a necessary (Line 17) condition for the sufficient decrease condition. In [18] it is also noted that the full-order quantities $\mathcal{J}(u^{(j+1)})$ and $\nabla\mathcal{J}(u^{(j+1)})$ are cheaply available after updating the RB space. Additionally, Ref. [17] introduced the possibility of skipping a redundant enrichment, which is particularly useful at the late stage of the method, where we are close to the optimum. This will prevent the dimension of the RB space from growing too fast, so that the cheap-to-solve property is preserved. The three conditions to be checked in order to decide whether to skip the update of the RB space are contained in the following skipping parameter

$$\text{Skip\_enrichment\_flag}(j) := \left(q^{(j)}(u^{(j+1)}) \leq \beta_q \delta^{(j+1)}\right) \text{ and}$$
$$\left(\frac{|g(u^{(j+1)}) - g^{\ell,(j)}(u^{(j+1)})|}{g^{\ell,(j)}(u^{(j+1)})} \leq \tau_g\right) \text{ and}$$
$$\left(\frac{\|\nabla\mathcal{J}^{\ell,(j)}(u^{(j+1)}) - \nabla\mathcal{J}(u^{(j+1)})\|_{\mathcal{U}}}{\|\nabla\mathcal{J}^{\ell,(j)}(u^{(j+1)})\|_{\mathcal{U}}} \leq \min\{\tau_{\text{grad}}, \beta_{\text{grad}}\delta^{(j+1)}\}\right).$$

where $\beta_q, \beta_{\text{grad}}, \tau_g, \tau_{\text{grad}} \in (0,1)$ are given parameters and

$$g(u) := \|u - P_{\mathcal{U}_{\text{ad}}}(u - \nabla\mathcal{J}(u))\|_{\mathcal{U}}, \quad g^{\ell,(j)}(u) := \|u - P_{\mathcal{U}_{\text{ad}}}(u - \nabla\mathcal{J}^{\ell,(j)}(u))\|_{\mathcal{U}}.$$

Note also that $g(u) = 0$ is nothing else than the standard first-order condition for optimization problems with constraints on the parameter set. This is the reason why Algorithm 2 terminates when $g(u^{(j+1)}) < \tau_{\text{FOC}}$ holds with $\tau_{\text{FOC}} \in (0,1)$. For more details on the skipping condition, we refer to [17]. Typically, TR methods also have the option of shrinking (enlarging) the TR radius $\delta^{(j)}$ with some factor $\beta_1 \in (0,1)$ ($\beta_1^{-1} > 1$, respectively). In the case of Algorithm 2, we shrink the radius if a point is rejected. We also compute the ratio

$$\varrho^{(j)} := \frac{\mathcal{J}(u^{(j)}) - \mathcal{J}(u^{(j+1)})}{\mathcal{J}^{\ell,(j)}(u^{(j)}) - \mathcal{J}^{\ell,(j)}(u^{(j+1)})}.$$

If this ratio is greater than a parameter $\eta_\varrho \in [0.75, 1]$, then the radius is enlarged. Algorithm 2 is proved to be convergent given some technical assumptions on the problem. We summarize everything in the following theorem (cf. [17]).

Theorem 9. *Suppose that* $\mathcal{U}_{\text{ad}} = [u^a, u^b] \subset \mathbb{R}^P$ *for some* $u^a, u^b \in \mathbb{R}^P$ *with* $u^a \leq u^b$. *Assume that* $\mathcal{J}$ *and* $\mathcal{J}^{\ell,(j)}$ ($j \in \mathbb{N}$) *are strictly positive,* $\mathcal{J}$ *is continuously Fréchet differentiable and* $\mathcal{J}^{\ell,(j)}$ *is even twice continuously Fréchet differentiable for all* $j \in \mathbb{N}$. *Moreover,* $\nabla\mathcal{J}^{\ell,(j)}$ *is uniformly Lipschitz-continuous with respect to* j. *Suppose that there is* $\delta_{\min} > 0$ *such that for every* $j \in \mathbb{N}$ *there exists a TR radius* $\delta^{(j)} \geq \delta_{\min}$, *for which there is a solution* $u^{(j+1)}$ *of the TR-RB subproblem* (12) *which is accepted by* Algorithm 2. *Assume that the family of functions* $(q^{(j)})_{j \in \mathbb{N}}$ *is uniformly continuous w.r.t. the parameter* u *and the index* j. *Then every accumulation point* $\bar{u}$ *of the sequence of iterates* $(u^{(j)})_{j \in \mathbb{N}}$ *is a first-order critical point for the full-order optimization problem, i.e., it holds*

$$\|\bar{u} - P_{\mathcal{U}_{\text{ad}}}(\bar{u} - \nabla\mathcal{J}(\bar{u}))\|_{\mathcal{U}} = 0.$$

In particular, Algorithm 2 *terminates after finitely many steps.*

Although many of the assumptions in Theorem 9 are quite technical for the proof, one can show that they are reasonable in the case of the RB method; cf. [17].

4.1. The TR-RB Algorithm Applied to the PS Method

In this section we show how Algorithm 2 can be applied to the PS method. To this end, we recall the following lemma from [16].

Lemma 4. There are constants $C_J, C_{\nabla J}, C_{\nabla^2 J} > 0$ such that for any $j \in \{1, \ldots, k\}$, any $u \in \mathcal{U}_{ad}$ and any choice of the RB space V^ℓ it holds

$$|\hat{J}_i^\ell(u)| \leq C_J, \quad \|\nabla \hat{J}_i^\ell(u)\|_\mathcal{U} \leq C_{\nabla J}, \quad \|\nabla^2 \hat{J}_i^\ell(u)\|_{L(\mathcal{U})} \leq C_{\nabla^2 J}.$$

Lemma 4 immediately implies that the reduced-order gradient is uniformly Lipschitz-continuous with respect to ℓ. We have to solve ($\mathbf{P}_{z,r}^{PS}$). We follow the approach in [16], where the target direction $r = (1, \ldots, 1)$ is chosen and an augmented Lagrangian method is used. Provided a penalty parameter $\mu > 0$, the augmented Lagrangian for ($\mathbf{P}_{z,r}^{PS}$) is

$$\mathcal{L}_A((u,t,s), \lambda; \mu) := t + \sum_{i=1}^{k} \lambda_i c_i(u,t,s) + \frac{\mu}{2} \sum_{i=1}^{k} c_i(u,t,s)^2 \tag{16}$$

with $c_i(u,t,s) = \hat{J}_i(u) - z_i - t + s_i$. The idea is to iteratively solve the subproblems

$$\min \mathcal{L}_A((u,t,s), \lambda; \mu) \quad \text{s.t.} \quad (u,t,s) \in \mathcal{U}_{ad} \times \mathbb{R} \times \mathbb{R}^k_{\geq} \tag{17}$$

approximately and then update the Lagrange multiplier λ and the penalty parameter μ until the termination criteria

$$\|c(u,t,s)\|_{\mathbb{R}^k} < \tau_{EC}, \tag{18}$$

$$\|(u,t,s) - P_{ad}((u,t,s) - \nabla_{(u,t,s)} \mathcal{L}_A((u,t,s), \lambda; \mu))\|_{\mathcal{U} \times \mathbb{R} \times \mathbb{R}^k} < \tau_{FOC} \tag{19}$$

are satisfied for some tolerances $\tau_{EC}, \tau_{FOC} \in (0,1)$, where $P_{ad}: \mathcal{U} \times \mathbb{R} \times \mathbb{R}^k \to \mathcal{U}_{ad} \times \mathbb{R} \times \mathbb{R}^k_{\geq}$ is the canonical projection onto $\mathcal{U}_{ad} \times \mathbb{R} \times \mathbb{R}^k_{\geq}$. For further details, we refer to [16] (Appendix B). We want to combine then the augmented Lagrangian method with the TR-RB algorithm to solve problem ($\mathbf{P}_{z,r}^{PS}$). To do so, we apply Algorithm 2 to solve each subproblem (17). We first define the reduced-order augmented Lagrangian

$$\mathcal{L}_A^\ell((u,t,s), \lambda; \mu) := t + \sum_{i=1}^{k} \lambda_i c_i^\ell(u,t,s) + \frac{\mu}{2} \sum_{i=1}^{k} c_i^\ell(u,t,s)^2, \tag{20}$$

with $c_i^\ell(u,t,s) = \hat{J}_i^\ell(u) - z_i - t + s_i$, which leads to the reduced-order subproblem

$$\min \mathcal{L}_A^\ell((u,t,s), \lambda; \mu) \quad \text{s.t.} \quad (u,t,s) \in \mathcal{U}_{ad} \times \mathbb{R} \times \mathbb{R}^k_{\geq}. \tag{21}$$

Note that in this case the admissible set $\mathcal{U}_{ad} \times \mathbb{R} \times \mathbb{R}^k_{\geq}$ is unbounded, which collides with the first assumption of Theorem 9. Nevertheless, Ref. [16] showed that the ($\mathbf{P}_{z,r}^{PS}$) problem is also equivalent to

$$\min t \quad \text{s.t.} \quad (t,u) \in [t^{\min}, t^{\max}] \times \mathcal{U}_{ad} \text{ and } \hat{J}(u) - z \leq t. \tag{22}$$

There is still the problem that the admissible set for the slack variables s is given by $[0, \infty)^k$. However, computing the partial derivative of the augmented Lagrangian $\mathcal{L}_A$ with respect to s_i, we obtain

$$\partial_{s_i} \mathcal{L}_A((u,t,s), \lambda; \mu) = \lambda_i + \mu(\hat{J}_i(u) - z_i - t + s_i) \geq \lambda_i + \mu(-z_i - t^{\max} + s_i).$$

Thus, $\mathcal{L}_A$ is strictly monotonically increasing in s_i for $s_i > -\lambda_i/\mu + z_i + t_{\max} =: s_i^{\max}$. Thus, given the Lagrange multiplier λ and the penalty parameter μ, we can restrict the slack variable s_i to the interval $[0, s_i^{\max}]$. This will not cause any modification to the solvability and the solution of the augmented Lagrangian subproblem. By setting $\mathcal{X}_{ad} := \mathcal{U}_{ad} \times [t^{\min}, t^{\max}] \times [0, s^{\max}]$, the equivalent formulation for the augmented Lagrangian subproblem corresponding to (22) reads

$$\min_{(u,t,s)\in \mathcal{X}_{\text{ad}}} \mathcal{L}_A((u,t,s),\lambda;\mu). \tag{23}$$

Similarly, the reduced-order augmented Lagrangian subproblem is given by

$$\min \mathcal{L}_A^\ell((u,t,s),\lambda;\mu) \quad \text{s.t.} \quad (u,t,s) \in \mathcal{X}_{\text{ad}}. \tag{24}$$

Therefore, the goal is to apply Algorithm 2 to solve the subproblem (23). To this end, we define $x = (u,t,s) \in \mathcal{U} \times \mathbb{R} \times \mathbb{R}^k$, $\mathcal{J}(x) = \mathcal{L}_A(x,\lambda;\mu)$ and $\mathcal{J}^{\ell,(j)}(x) = \mathcal{L}_A^{\ell,(j)}(x,\lambda;\mu)$ for any reference point $z \in \mathbb{R}^k$, any Lagrange multiplier $\lambda \in \mathbb{R}_\geq^k$ and any penalty parameter $\mu > 0$. Furthermore, using the a-posteriori estimates of the individual objectives (cf. Theorem 8), we have that

$$|\mathcal{J}(x) - \mathcal{J}^{\ell,(j)}(x)| \leq \sum_{j=1}^k \left(\lambda_j + c|\hat{f}_j^{\ell,(j)}(u) - z_j - t + s_j|\right)\Delta_{\hat{f}_j^{\ell,(j)}}(u)$$
$$+ \sum_{j=1}^k \frac{c}{2}\left(\Delta_{\hat{f}_j^{\ell,(j)}}(u)\right)^2 =: \Delta_{\mathcal{J}}^{\ell,(j)}(u)$$

for all $u \in \mathcal{U}_{\text{ad}}$, which can be used as a-posteriori error estimate in the TR-RB algorithm. According to Theorem 9, we still need to show the strict positivity of the costs $\mathcal{J}$ and $\mathcal{J}^{\ell,(j)}$ and the uniform Lipschitz continuity of the gradient $\nabla \mathcal{J}^{\ell,(j)}$. For the first, we note that the objectives $\mathcal{J}$ and $\mathcal{J}^{\ell,(j)}$ are bounded from below by $C := t^{\min} - \sum_{i=1}^k \lambda_i^2/(2\mu_i)$. Since C depends only on fixed parameters of the optimization problems, we can add $C + 1$ to the cost functions to obtain strict positivity. Obviously, this will not change the minimizers. The second property is a bit more technical and we prove it in the following lemma.

Lemma 5. *Let the Lagrange multiplier λ and the penalty parameter μ be given. Then the function $\mathcal{J}(\cdot) := \mathcal{L}_A(\cdot,\lambda;\mu)$ is twice continuously Fréchet-differentiable for all $j \in \mathbb{N}$ and the gradient $\nabla \mathcal{J}^{\ell,(j)}$ is uniformly Lipschitz continuous with respect to j.*

Proof. Due to Corollary 1 the cost functions $\hat{f}_1,\ldots,\hat{f}_k$ are twice continuously Fréchet-differentiable. Thus, the function $(u,t,s) \mapsto \mathcal{L}_A((u,t,s),\lambda;\mu)$ is also twice continuously Fréchet-differentiable as a composition of twice continuously Fréchet-differentiable functions. Similarly, the reduced-order augmented Lagrangians $\mathcal{L}_A^{\ell,(j)}((\cdot,\cdot,\cdot),\lambda;\mu)$ are also twice continuously Fréchet-differentiable for all $j \in \mathbb{N}$. We have that

$$\nabla^2 \mathcal{L}_A^{\ell,(j)}((u,t,s),\lambda;\mu)(h^u,h^t,h^s) =$$
$$\begin{pmatrix} \sum_{j=1}^k \left((\lambda_j + \mu c_j^{\ell,(j)})\nabla^2 \hat{f}_j^{\ell,(j)}(u)h^u + \mu(d_j^{\ell,(j)} - h^t + h_j^s)\nabla \hat{f}_j^{\ell,(j)}(u)\right) \\ k\mu h^t - \mu \sum_{j=1}^k (d_j^{\ell,(j)} + h_j^s) \\ \mu(d_1^{\ell,(j)} + h_1^s - h^t) \\ \vdots \\ \mu(d_k^{\ell,(j)} + h_k^s - h^t) \end{pmatrix}$$

for any $h = (h^u, h^t, h^s) \in \mathcal{U} \times \mathbb{R} \times \mathbb{R}^k$, where $c_j^{\ell,(j)} := \hat{f}_j^{\ell,(j)}(u) - z_j - t + s_j$ and $d_j^{\ell,(j)} := \langle \nabla \hat{f}^{\ell,(j)}(u), h^u\rangle_\mathcal{U}$ for $j \in \{1,\ldots,k\}$. Using Lemma 4, we obtain that the Hessian matrix $\nabla^2 \mathcal{L}_A^{\ell,(j)}((u,t,s),\lambda;\mu)$ can be bounded independently of (u,t,s) and j. Using the mean value theorem, we can conclude that the gradients $\nabla \mathcal{L}_A^{\ell,(j)}((\cdot,\cdot,\cdot),\lambda;\mu)$ are Lipschitz-continuous with constant C_L uniformly in j. □

As a consequence of Theorem 9, we have that Algorithm 2 applied to solve the augmented Lagrangian subproblem (23) converges after finitely many steps to a first-order critical point of (23).

Remark 6. *Algorithm 2 constructs and updates the RB space during the optimization procedure. In the case of the PS method, we are free to choose what to do for the space constructed during the TR-RB procedure. For example, we can use it for the next augmented Lagrangian subproblem (and also for the next reference point). We explored different ideas (see also [16]), but we report here only the two most interesting and efficient ones:*

(1) *Use one common RB space for all the subproblems and reference points, i.e., use a single space V^ℓ (which is, of course, updated in the process) for solving the MOP. This strategy acquires efficiency in terms of reconstructing the full-order parameter space during the iterations. Therefore, thanks to the possibility of skipping an enrichment (which is the costly part in Algorithm 2), we expect more and more speed-up, together with accuracy, as the algorithm proceeds.*

(2) *Use multiple (local) RB spaces. This idea is already exploited by [16,37,38]. In this case, we do not use the previously obtained RB space for the next minimization problem. We generate instead k initial spaces $V_1^\ell, \ldots, V_k^\ell$, resulting from the minimization (Note that this procedure does not require extra computational cost, since we need to solve these problems for the hierarchical PS method anyway) of the objectives $\hat{J}_1, \ldots, \hat{J}_k$. Then at the beginning of every PS problem, we can decide to use the space V_i^ℓ for which $q^{(0)}(u^{(0)}) < \beta_q \delta^{(0)}$ and $\dim V_i^\ell \leq \ell_{max}$, with $\ell_{max} \in \mathbb{N}$ being a predefined maximal number of basis functions. If several spaces satisfy these conditions then we select the one for which the value $q^{(0)}(u^{(0)})$ is the smallest. If instead there is no space fulfilling these conditions, we initialize a new space V_{k+1}^ℓ by using the full-order quantities $\mathcal{S}(u^{(0)})$ and $\mathcal{A}_i(u^0)$ for $i = 1, \ldots, k$.*

Although these two techniques are already efficient, we noticed that there is a common problem: the number of RB basis functions might grow too fast and prevent a good speed-up for the solution. In particular, this is the case for the first strategy. To fix this issue, we propose different strategies to remove basis functions from V^ℓ in Section 4.2. This approach was not considered in [14,16–18] and to our knowledge it has not been addressed in the literature yet. In reduced-order optimization, instead, this is meaningful, since the reduced-order model might grow too fast; see, e.g., [33], in the case of proper orthogonal decomposition.

4.2. How to Reduce the Number of Basis Functions

We point out that what is described in this section can also generally be applied to Algorithm 2 from [17] without any relation to the PS method. In particular, the strategies for reducing the number of basis functions presented in this section cannot only be used for PDE-constrained multi-objective optimization problems, but also for any other problem formulation containing PDE-constrained optimization problems. Therefore, we use again the general notation $\mathcal{J}$ for the cost, as it was done in the beginning of this section. The methodology to remove a basis function comes from the observation that some basis elements might not be used during the optimization process. Suppose that we start from a point $u^{(0)}$ very far from the optimum. Clearly, after j iterations the point $u^{(j)}$ is in a completely different region of the admissible set compared to the one of the starting point. Hence, the basis functions built for $u^{(0)}$ might give a negligible contribution in spanning the reduced-order model at the point $u^{(j)}$. If this is the case, we can expect that these functions will not play any further role also for the subsequent points and therefore they can be removed to reduce the dimension of the RB space. Our methodologies for removing basis functions are then based on Remark 6 and try to check which basis functions give a negligible contribution for the current iteration of the TR-RB algorithm. Notice that every technique we propose from now on will be applied after updating the RB space in the TR-RB algorithm. The aim is to modify the updated RB space in order to provide a new RB space, where the number of basis functions is reduced.

Technique T1. The first proposed technique is based on the computation of the so-called Fourier coefficients. Given $v \in V$ and a set of orthonormal basis functions $\{\psi_n\}_{n=1}^{\ell} \subset V^{\ell}$, the n-th Fourier coefficient is defined as $c_{\mathcal{F}}^{(n)}(v) := \langle v, \psi_n \rangle_V$. Now, T1 consists in computing $c_{\mathcal{F}}^{(n)}(\mathcal{S}(u^{(j+1)}))$ and $c_{\mathcal{F}}^{(n)}(\mathcal{A}_i(u^{(j+1)}))$, $i = 1, \ldots, k$, for $n = 1, \ldots, \ell$ and remove the basis function ψ_n for which

$$\zeta^{(n)} := \max\left\{ \frac{c_{\mathcal{F}}^{(n)}(\mathcal{S}(u^{(j+1)}))^2}{\sum_{\eta=1}^{\ell} c_{\mathcal{F}}^{(\eta)}(\mathcal{S}(u^{(j+1)}))^2}, \max_{i=1,\ldots,k} \left\{ \frac{c_{\mathcal{F}}^{(n)}(\mathcal{A}_i(u^{(j+1)}))^2}{\sum_{\eta=1}^{\ell} c_{\mathcal{F}}^{(\eta)}(\mathcal{A}_i(u^{(j+1)}))^2} \right\} \right\}$$

is below a certain tolerance. Note, in fact, that the Fourier coefficients indicate the order of magnitude of the contribution of a given basis function in reconstructing the new snapshots that we want to add to update the RB. Strategy T1 is also based on the assumption that the snapshots, which we want to include in an update, are the most relevant for the new TR subproblem, because they correspond to the last accepted optimization step $u^{(j+1)}$. The advantage of T1 is that the required Fourier coefficients are already available from the Gram-Schmidt orthogonalization performed during the update of the RB space. There is, anyway, a possible drawback of T1 due to the tolerance we set: it can happen that also important basis functions are removed although one thinks that the tolerance is small enough. Because of this, we would like to have a criteria to decide in an unbiased way which basis functions should be removed.

Technique T2. This approach is based on the idea that once a point $u^{(j+1)}$ is accepted by the TR-RB algorithm and the RB space is updated, we will compute a provisional AGC point $u_{\text{AGC}}^{(j+1),\text{prov}}$ (cf. Definition 14) with respect to the previously updated RB space. One robustness criteria that we demand is that after removing basis functions, this provisional AGC point is still inside the new TR (Note that the TR depends on the reduced-order model due to the inequality constraint in (12) and, therefore, changes if we remove basis functions), although it might not coincide with the actual AGC point $u_{\text{AGC}}^{(j+1)}$ that we compute after removing basis functions according to Line 3 in Algorithm 2 (Note that the reduced-order cost function changes by removing a basis function, so that also the first term in (13) differs after this removal). If we do not demand this robustness criteria, we can expect a deterioration of the TR performances due to lack of accuracy of the RB model in the steepest descent direction. Another important aspect is to guarantee the convergence of the TR-RB method, which implies checking that the conditions for accepting the point $u^{(j+1)}$ are still fulfilled, although we removed basis functions.

In summary, the difference with respect to T1 is then to remove basis functions starting from the one with the smallest value of $\zeta^{(n)}$ and proceeding in ascending order until one of the following conditions is satisfied

$$\frac{\Delta_{\mathcal{J}^{\ell-\text{rem},(j+1)}}(u_{\text{AGC}}^{(j+1),\text{prov}})}{\mathcal{J}^{\ell-\text{rem},(j+1)}(u_{\text{AGC}}^{(j+1),\text{prov}})} > \beta_q \delta^{(j+1)}, \tag{25a}$$

$$\frac{\Delta_{\nabla\mathcal{J}^{\ell-\text{rem},(j+1)}}(u_{\text{AGC}}^{(j+1),\text{prov}})}{\|\nabla\mathcal{J}^{\ell-\text{rem},(j+1)}(u_{\text{AGC}}^{(j+1),\text{prov}})\|_{\mathcal{U}}} > \min\{\tau_{\text{grad}}, \beta_{\text{grad}} \delta^{(j+1)}\}, \tag{25b}$$

$$\frac{\|\nabla\mathcal{J}^{\ell-\text{rem},(j+1)}(u^{(j+1)}) - \nabla\mathcal{J}(u^{(j+1)})\|_{\mathcal{U}}}{\|\nabla\mathcal{J}^{\ell-\text{rem},(j+1)}(u^{(j+1)})\|_{\mathcal{U}}} > \min\{\tau_{\text{grad}}, \beta_{\text{grad}} \delta^{(j+1)}\}, \tag{25c}$$

$$\frac{|g(u^{(j+1)}) - g^{\ell-\text{rem},(j+1)}(u^{(j+1)})|}{g^{\ell-\text{rem},(j+1)}(u^{(j+1)})} > \tau_g, \tag{25d}$$

$$\mathcal{J}^{\ell-\text{rem},(j+1)}(u^{(j+1)}) > \mathcal{J}^{\ell,(j)}(u_{\text{AGC}}^{(j)}), \tag{25e}$$

$$\mathcal{J}^{\ell-\text{rem},(j+1)}(u_{\text{AGC}}^{(j+1),\text{prov}}) - \mathcal{J}(u^{(j+1)}) > -\kappa_{\text{arm}} \|u_{\text{AGC}}^{(j+1),\text{prov}} - u^{(j+1)}\|_{\mathcal{U}}^2. \tag{25f}$$

If one of the conditions (25) holds we re-add the basis function to the RB space and finish the removal continuing with the TR-RB procedure. T2 is summarized in Algorithm 3.

Algorithm 3 Summary of T2

1: Follow the steps in Algorithm 2 until the RB model is updated at $u^{(j+1)}$;
2: Compute a provisional AGC point $u_{AGC}^{(j+1),\text{prov}}$ by using the reduced-order cost function w.r.t. the updated RB model;
3: Compute $\zeta^{(n)}$ for $n \in \{1, \ldots, \ell\}$;
4: **while** None of the conditions in (25) is fullfiled **do**
5:     Out of all remaining basis functions, remove the one with the smallest value of $\zeta^{(n)}$ from the RB space;
6: **end while**
7: Add the last removed basis function to the RB space;
8: Proceed with Algorithm 2 with the RB space obtained performing Steps 2–7;

Let us explain the meaning of (25). At first, the superindex $\ell - \text{rem}$ indicates that the space used to compute the quantity is the RB space obtained after removing a basis function. Condition (25a) is to check that the provisional AGC point will remain inside an accurate-enough region of the TR. Condition (25b) is in the spirit of (25a) but for the gradient of the objective. Conditions (25c) and (25d) are based on the skipping enrichment criteria and are checked to ensure convergence and robustness of the method after the removal. For a similar issue we need to check that the sufficient decrease condition is fulfilled as well (cf. (25e)). Finally, (25f) is to enforce that the provisional AGC point is still a Cauchy point. In such a way, we are sure that Algorithm 2 converges even after performing the basis removal (cf. [17,18]). In this sense, T2 introduces an unbiased way to deal with the technique introduced in T1. There are still a few aspects one should comment on before implementing T2. At first, note that all the above-mentioned conditions are cheaply computable, since they are based either on reduced-order quantities or the appearing full-order quantities are available because of the RB update. At second, conditions (25a) and (25b) request efficient and reliable error estimators. Although for the PS method the efficiency of $\Delta_{\mathcal{J}}^{\ell,(j)}$ is acceptable, it is not the same for an error estimator $\Delta_{\nabla \mathcal{J}}^{\ell,(j)}$ based on the a-posteriori estimates of the gradients of the individual objectives. These estimators generally produce a huge overestimation, which makes them useless in practice. We notice, in fact, that condition (25b) is immediately triggered in the case of the PS method and we can not remove any basis function. This is the reason why we solved this issue by two different related approaches:

Technique T2a. We replace the numerator of (25b) by

$$\|\nabla \mathcal{J}^{\ell-\text{rem},(j)}(u_{AGC}^{(j+1),\text{prov}}) - \nabla \mathcal{J}(u_{AGC}^{(j+1),\text{prov}})\|_{\mathcal{U}},$$

which is the true error we wanted to estimate, but it is unfortunately costly. It requires the computation of the full-order quantities $\mathcal{S}(u_{AGC}^{(j+1),\text{prov}})$ and $\mathcal{A}_i(u_{AGC}^{(j+1),\text{prov}})$, $i = 1, \ldots, k$.

Technique T2b. We replace the numerator of (25b) by

$$\|\nabla \mathcal{J}^{\ell-\text{rem},(j)}(u_{AGC}^{(j+1),\text{prov}}) - \nabla \mathcal{J}^{\ell,(j+1)}(u_{AGC}^{(j+1),\text{prov}})\|_{\mathcal{U}}$$

which is a cheap approximation of the true error that we suppose to be reliable only after enough steps of Algorithm 2, however.

Clearly, if one has a good estimation of the gradient at hand, T2 can be still used in its original form.

Technique T3. Another drawback of T2 is the fact that we first need to remove the basis function in order to check (25). This implies that when we stop the removal, we need to add back the last basis function which was removed, because it is containing important information; cf. Line 7 of Algorithm 3. This results in a waste of time for the modified Algorithm 2. We decide to add the option of introducing numerical tolerances for each of the conditions (25). In such a way, the modified algorithm will generally stop before an important basis function is removed at the price of possibly leaving one or a few redundant basis functions in the RB space. We think that this is a meaningful modification regarding the time that is wasted reintroducing the removed basis function into the RB space; cf. Section 5. We indicate this last strategy as T3.

5. Numerical Experiments

In this section we test Algorithm 2 and compare it with the results obtained in [16] (Section 3.2.2). We use the same numerical setting, which we briefly report here. Let the domain Ω be the two-dimensional unit square, split into four different subdomains $\Omega_1 = (0, 0.5) \times (0, 0.5)$, $\Omega_2 = (0, 0.5) \times (0.5, 1)$, $\Omega_3 = (0.5, 1) \times (0, 0.5)$ and $\Omega_4 = (0.5, 1) \times (0.5, 1)$. For each Ω_i, we consider a corresponding diffusion coefficient $u_i^\kappa \in \mathbb{R}$ in (3) for $i = 1, \ldots, 4$. The reaction term $r(x)$ is set to be constantly equal to 1 for any $x \in \Omega$. We impose homogeneous Neumann boundary conditions (i.e., $\alpha = 0$) and a source term $f(x) = \sum_{i=1}^{4} c_i \chi_{\Omega_i}(x)$ with $c_1 \approx 2.76$, $c_2 \approx -0.96$, $c_3 \approx 0.51$ and $c_4 \approx -1.66$ generated randomly in order to obtain a problem with a non-convex Pareto front. For the spatial discretization of the state equation, we apply the Finite Element (FE) method with 1340 nodes and piecewise linear basis functions. For (**MPPOP**) we choose the following three objectives

$$\hat{J}_1(u) := \frac{1}{2} \|\mathcal{S}(u) - y_\Omega^{(1)}\|_H^2 + \frac{\varepsilon}{2} \|u - u_d^{(1)}\|_\mathcal{U}^2,$$

$$\hat{J}_2(u) := \frac{1}{2} \|\mathcal{S}(u) - y_\Omega^{(2)}\|_H^2 + \frac{\varepsilon}{2} \|u - u_d^{(2)}\|_\mathcal{U}^2, \quad \hat{J}_3(u) := \frac{0.05}{2} \|u - u_d^{(3)}\|_\mathcal{U}^2$$

with $\varepsilon = 0.002$, the desired states

$$y_\Omega^{(1)}(x) := \chi_{(0,0.5) \times (0,1)}(x), \quad y_\Omega^{(2)}(x) := \chi_{(0.5,1) \times (0,1)}(x),$$

and the desired parameter values

$$u_d^{(1)} = u_d^{(2)} := (2, 0, 0, 0, 0.3)^T, \quad u_d^{(3)} := (2, 1, 1, 1, 0.3)^T.$$

The lower and upper parameter bounds are given by

$$u_a = (2, 0.1, 0.1, 0.1, 0.3)^T \quad \text{and} \quad u_b = (2, 4, 4, 4, 0.3)^T,$$

respectively. This implies that $u_1^\kappa = 2$ and $u^r = 0.3$ are seen as constants and we only optimize over the three parameters u_2^κ, u_3^κ and u_4^κ. Note furthermore, that the desired parameters $u_d^{(1)} = u_d^{(2)}$ are not admissible. In fact, as for the parameters of the source term, they were chosen such that the resulting Pareto front is non-convex.

For the choice of the initial value for PSPs corresponding to reference points for the entire problem $(\hat{J}_1, \hat{J}_2, \hat{J}_3)$ we do the following: Let $\bar{u}^i$ be the minimizer of $\hat{J}_i$ for $i = 1, 2, 3$. Recall that the sets D_i have been introduced in Definition 7-(ii). Then, if $z \in D_i$, we choose $\bar{u}^i$ as the initial value for solving ($\mathbf{P}_{z,r}^{PS}$). We additionally choose the shifting vectors $\bar{d} = 0.001 \cdot (1, 1, 1)^T$, while the grid size h for the reference point grid is set to $h_{PSM} = 0.003$.

5.1. Parameter Choices for the TR-RB Algorithm

There are many parameters used in the TR-RB algorithm, which we will specify and briefly comment on in this section.

- The initial TR radius is chosen as $\delta^{(0)} = 0.1$, the tolerance for increasing the TR radius is set to $\eta_\varrho = 0.75$ and the factor for shrinking the TR radius to $\beta_1 = 0.5$. For the minimal TR radius we use $\delta_{\min} = 1 \times 10^{-16}$.

- For the Armijo backtracking strategy, we use the constants $\kappa_{\mathrm{arm}} = 1 \times 10^{-4}$ and $\kappa = 0.5$.
- The tolerance of the first-order condition is set to $\tau_{\mathrm{FOC}} = \tau_{\mathrm{FOC,sub}}^{(i)}$, where $\tau_{\mathrm{FOC,sub}}^{(i)}$ is the tolerance for the first-order condition of the current augmented Lagrangian subproblem. Moreover, we choose $\tau_{\mathrm{sub}} = 0.5\,\tau_{\mathrm{FOC}}$ as the tolerance of the first-order condition of the TR-subproblem and $\beta_{\mathrm{bound}} = 0.9$ as the constant in (15).
- For checking the necessity of updating the RB space, we choose $\tau_g = 1$, $\tau_{\mathrm{grad}} = 0.1$, $\beta_{\mathrm{grad}} = 0.2$ and $\beta_q = 0.005$.
- The tolerance chosen in T1 (cf. Section 4.2) for the Fourier coefficient is 10^{-6}. Similarly, we choose the same tolerance for T3 in order to break the removal algorithm before deleting important basis functions, i.e., we subtract it on the right-hand side of (25a)–(25f).

We notice in our numerical experiments that the method without basis removal is quite robust in terms of computational time and required PDE solves with respect to all the parameters except for the ratio between the first-order conditions of the current augmented Lagrangian subproblem and the TR-subproblem $\tau_{\mathrm{FOC}}/\tau_{\mathrm{sub}}$. In our experiments we choose this ratio to be 2, but we observe that a too large ratio (already 5 is sufficient) slows down the method considerably. The reason is that the TR-subproblems are solved with too much accuracy in this case which needs a lot of numerical effort but does not benefit the overall optimization. Regarding the techniques introduced in Section 4.2, T1 heavily depends on the choice of the tolerance for truncating the Fourier coefficient. The smaller the tolerance the less basis functions are removed. Anyway, if we remove too many basis functions (e.g., tolerance of 10^{-4}), T1 becomes less stable and the method needs more iterations to converge which generally corresponds in more enrichment steps which slow it down. Conversely, removing few basis functions (e.g., 10^{-8}) implies no significant differences between T1 and the method without removal. Contrarily to T1, techniques T2, T2a and T2b are only based on the same parameters which influence the behavior of the algorithm without removal. Their performances are also robust with respect to all these parameters in terms of basis functions removed. For T3 the same discussion applies, but this method is also sensitive to the tolerance chosen to break the removal algorithm before deleting presumed important basis functions. On one hand, if this tolerance is too high (e.g., 10^{-2}), the method will not remove a significant number of basis functions to influence the performances of the algorithm. On the other hand, if this is too low (e.g., 10^{-8}) T3 will be essentially equivalent to T2.

5.2. Numerical Ressults

In this section, we focus mainly on the comparison of our proposed TR-RB variants, briefly commenting on full-order versus reduced-order model. For detailed comments and results on the PS method applied on the FE and RB level, we refer to [16] (Section 3.2.2). At first, to validate our approach, we show in Figure 1 the obtained Pareto fronts by using the method in [16] (left) and our method (right). As one can see, there is no visible difference. The approximation error is, in fact, of the order of 10^{-6} for a Pareto point computed by all the proposed techniques (i.e., T1, T2a, T2b and T3) on average. This can be essentially explained by the fact that the termination criteria for Algorithm 2 relies on the full-order model. Therefore, any computed point is first-order critical for the FE model, up to the chosen stopping tolerance. Let us remark that this is not typical for model order reduction, where generally there is an additional approximation error due to the reduced-order model inaccuracy.

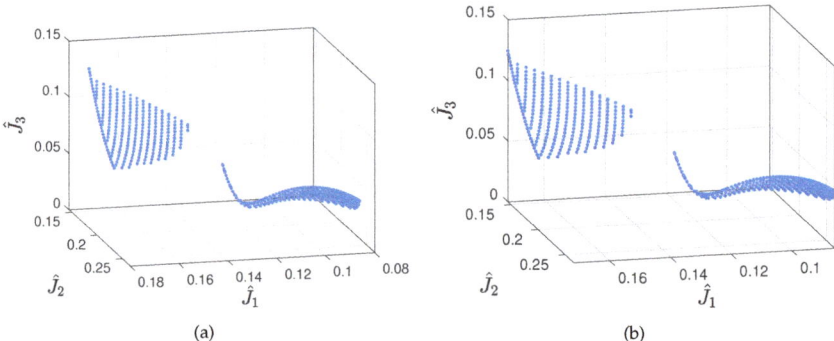

Figure 1. (**a**) Algorithm 2 no Removal local RB spaces. (**b**) Algorithm 2 T3 local RB spaces.

In Figure 2 we compare the computational time of Algorithm 2 for all the proposed techniques (cf. Section 4.2) against the full-order FE model and the algorithm in [16]. Concerning the FE method, we can save between 41% and 59% of the computational time. Considering the fact that we do not have an approximation error in reconstructing the Pareto points, we get the same result in approximately half of the time by using any of the TR-RB variants. This speed-up will also increase with an increasing number of degrees of freedom for the FE method, since the number of required FE solves of the PDE is significally smaller for the TR-RB algorithms than for the FE method; cf. Table 1.

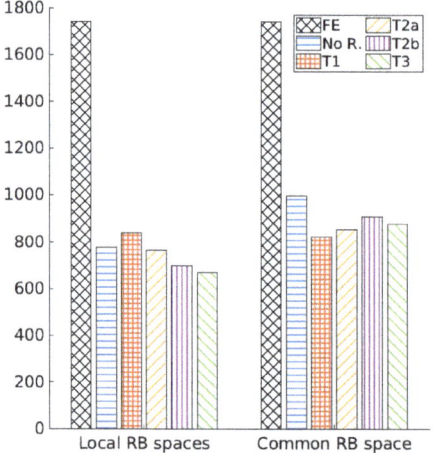

Figure 2. Computational times in seconds for Algorithm 2 with or without basis removal and using the two strategies in Remark 6 for initializing the RB space.

Table 1. Total PDE and only FE solves for the tested methods.

Method	# Total PDE Solves	# FE Solves
FE	433378	433378
Common RB Space No R.	493254	20743
Common RB Space T1	493282	20786
Common RB Space T2a	497032	20838
Common RB Space T2b	497032	20752
Common RB Space T3	493985	20792
Local RB Space No R.	497072	20773
Local RB Space T1	497589	20893
Local RB Space T2a	507064	21226
Local RB Space T2b	507064	20857
Local RB Space T3	502911	21023

Furthermore, we get a speed-up of the TR-RB algorithm by using the proposed techniques for reducing the number of basis functions in almost all cases. Depending on the strategy from Remark 6, one technique performs better than the others. Here we try to explain this phenomena in detail. Let us focus on the common RB space first. In this case, every technique helps in saving computational time. This is clearly the effect of removing redundant basis functions, which are particularly frequently included using a large common RB space. This is the reason why T1 appears to be the most effective, since it is the cheapest among the techniques (as we said it does not imply additional cost to be checked). T2a is more robust, but it comes with the price of evaluating the full-order gradient at the new AGC point and thus results to be slower than T1. Apparently, T2b should overcome this problem, but the inaccuracy of the RB space in the beginning yields a bad approximation of (25b), resulting in removing too many basis functions which leads to a worse approximation for the consecutive steps. This worsening of the approximation results in a way larger number of enrichment steps towards the end of the algorithm, which also negatively influences the computational time. T3 is comparable with T2a, meaning that for this example we are removing many basis functions in only a few instances, rather than frequently removing a few basis functions. Figure 3b confirms the above remarks for the case of a common RB space. In this figure we report the number of basis functions obtained at the end of Algorithm 2 while this is applied to compute each Pareto optimal point in the PS method.

Now, let us focus on the left group of columns in Figure 2 (and thus on Figure 3a), which corresponds to the computational times in the case of using local RB spaces (cf. Remark 6). This case is a bit more delicate, since the use of local RB spaces makes it more difficult to interpret the results. Here the problem of T1 is emerging. The fact that this technique removes a number of basis functions without any robustness criteria implies that the method slows down. In the case of local spaces, in fact, we do not have the same amount of redundant basis functions as it can occur for a common RB space. Therefore, we should only remove the basis functions which are actually redundant. As one can note in Figure 3a, T1 removes a significantly larger amount of basis functions in comparison to the other techniques. Here the criteria introduced in T2a play their role in a positive way. We can counteract the effect of T1 in such a way that the computational time is comparable to the one in [16]. The further simplification introduced in T2b helps to get an additional speed-up. In contrast to the common RB space, here we have local spaces which provide a sufficiently good accuracy for approximating (25b) also in the beginning of the optimization. This is then beneficial for the algorithm, since the cost of computing the criteria in T2b is way cheaper than T2a, where we need full-order solves of the state and adjoint equation to compute the gradient at the new AGC point. Additionally, T3 further improves T2a and T2b in terms of computational time, since in the case of local RB spaces it is more probable that we indeed remove only a few basis functions but more frequently than in the case of one common RB space. In this case, it is important to have tolerances that let us stop before removing an important basis function and save time for reintroducing it in the RB space.

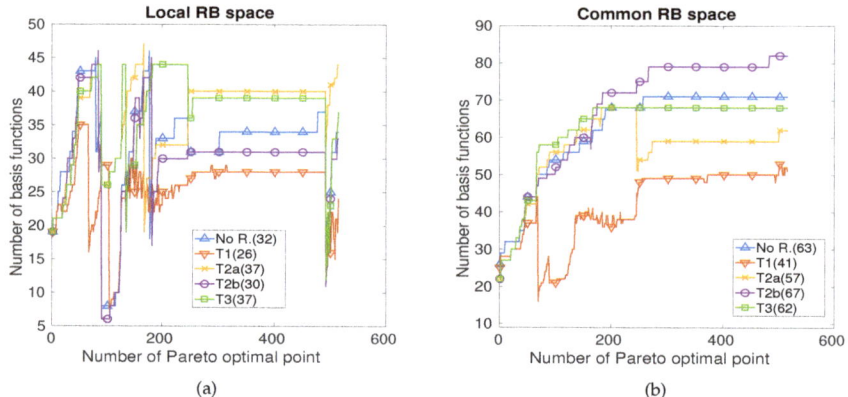

Figure 3. Number of basis functions used to compute each Pareto optimal point. (**a**) Local RB space. (**b**) Common RB space. In brackets: average number of basis functions.

In conclusion, comparing our fastest method (i.e., Algorithm 2 with local RB spaces and T3) to the slowest (i.e., using [16] with a common RB space) we get essentially the same results (the approximation error is 10^{-6}) saving approximately 30% of the computational time, which is roughly 300 s. This shows how one should invest time and resources in providing efficient techniques for reducing the number of basis functions in the RB space, while using an adaptive TR-RB algorithm. Particularly in the case of multiobjective optimization, this becomes crucial for a large number of cost functionals k. To obtain the same resolution of the Pareto front as in Figure 1 for a large k, we will need to solve the PSPs for many more points, implying higher risk of having redundant basis functions.

6. Conclusions

We showed the applicability and convergence of the TR-RB algorithm in the context multi-objective PDE-constrained parameter optimization problem. We presented and analyzed novel ways of reducing the dimension of the RB space during the optimization procedure. To our knowledge, basis reduction strategies have not been proposed yet for the RB method, although it is common for other model order reduction techniques. Such a removal significantly improved the performances of the TR-RB algorithm in the context of multiobjective optimization, leading faster to an accurate solution than the already existing techniques. The presented example contained only three parameters to be optimized. However, based on the results in [17] (Section 4.4) for an example with 28 parameters and on the various examples in [39] (Sections 3.5.4–3.5.6), we expect all of the TR-RB methods to scale well with an increasing number of parameters. As for the multi-objective optimization by the PS method, the numerical effort grows exponentially with the number of cost functions k, but is independent of the number of parameters m if $m \geq k - 1$. Moreover, the presented removal techniques of reduced basis functions can also be extended to other applications in which sequential parametric PDE-constrained optimization problems must be solved. In future work, one can try to extend the convergence theory for the presented TR-RB algorithm to a larger class of PDEs than the one presented here, as, e.g., parabolic PDEs [14] or non-affine parameter-to-state couplings. Due to the general formulation of the convergence result we are optimistic that this is possible. Moreover, one can try to achieve further improvements concerning robustness of the method and deriving tighter a-posteriori error estimators, in particular for the gradient of the cost function. This is also of great interest in the RB community. Another interesting idea could be to incorporate the usual trust-region condition based on the (Euclidean) distance from the current iterate into the presented TR-RB algorithm. In [19] the usual trust-region condition was actually performing slightly better than a residual-based error estimate as the trust-region constraint for some of the considered problems. Despite the fact that we use not only a residual-based error estimate but an error estimate of the actual cost function, a comparison between the different approaches is definitely of interest.

Author Contributions: Conceptualization, S.B., L.M. and S.V.; methodology, S.B., L.M. and S.V.; software, S.B. and L.M.; formal analysis, S.B., L.M. and S.V.; investigation, S.B., L.M. and S.V.; writing—original draft preparation, S.B., L.M. and S.V.; writing—review and editing, S.B., L.M. and S.V.; funding acquisition, S.V. All authors have read and agreed to the published version of the manuscript.

Funding: The authors acknowledge funding by the Deutsche Forschungsgemeinschaft (DFG) for the project Localized Reduced Basis Methods for PDE-constrained Parameter Optimization under contract VO 1658/6-1.

Acknowledgments: The authors thank Tim Keil, Mario Ohlberger and Felix Schindler from University of Münster (Germany) for the fruitful exchange of ideas on the topic.

Conflicts of Interest: The authors declare no conflict of interest. The funders had no role in the design of the study; in the collection, analyses, or interpretation of data; in the writing of the manuscript, or in the decision to publish the results.

Abbreviations

The following abbreviations are used in this manuscript:

AGC	Approximated generalized Cauchy
CG	Conjugate gradient
FE	Finite element
MOP	Multiobjective optization problem
MPPOP	Multiobjective parametric PDE-constrained optimization problem
PDE	Partial Differential Equation
PS	Pascoletti-Serafini
RB	Reduced basis
s.t.	subject to
TR	Trust-region

References

1. Ehrgott, M. *Multicriteria Optimization*, 2nd ed.; Springer: Berlin/Heidelberg, Germany, 2005.
2. Miettinen, K. *Nonlinear Multiobjective Optimization*; Kluwer Academic Publishers: Cambridge, MA, USA, 1999.
3. Zadeh, L. Optimality and non-scalar-valued performance criteria. *IEEE Trans. Autom. Control* **1963**, *8*, 59–60. [CrossRef]
4. Eichfelder, G. *Adaptive Scalarization Methods in Multiobjective Optimization*; Springer: Berlin/Heidelberg, Germany, 2008.
5. Pascoletti, A.; Serafini, P. Scalarizing vector optimization problems. *J. Optim. Theory Appl.* **1984**, *42*, 499–524. [CrossRef]
6. Hinze, M.; Pinnau, R.; Ulbrich, M.; Ulbrich, S. *Optimization with PDE Constraints*; Springer Science + Business Media B.V.: Berlin/Heidelberg, Germany, 2009.
7. Schilders, W.H.; Van der Vorst, H.A.; Rommes, J. *Model Order Reduction*; Springer: Berlin/Heidelberg, Germany, 2008.
8. Hesthaven, J.S.; Rozza, G.; Stamm, B. *Certified Reduced Basis Methods for Parametrized Partial Differential Equations*; SpringerBriefs in Mathematics: Heidelberg, Germany, 2016.
9. Patera, A.T.; Rozza, G. *Reduced Basis Approximation and a Posteriori Error Estimation for Parametrized Partial Differential Equations*; MIT Pappalardo Graduate Monographs in Mechanical Engineering: Cambridge, MA, USA, 2007.
10. Banholzer, S.; Gebken, B.; Reichle, L.; Volkwein, S. ROM-based inexact subdivision methods for PDE-constrained multiobjective optimization. *Math. Comput. Appl.* **2021**, *26*, 32. [CrossRef]
11. Iapichino, L.; Ulbrich, S.; Volkwein, S. Multiobjective PDE-constrained optimization using the reduced-basis method. *Adv. Comput. Math.* **2017**, *43*, 945–972. [CrossRef]
12. Schu, M. Adaptive Trust-Region POD Methods and Their Application in Finance. Ph.D. Thesis, University of Trier, Trier, Germany, 2012. Available online: https://ubt.opus.hbz-nrw.de/opus45-ubtr/frontdoor/deliver/index/docId/574/file/PhD_Thesis_Schu.pdf (accessed on 28 April 2022).
13. Arian, E.; Fahl, M.; Sachs, W.S. *Trust-Region Proper Orthogonal Decomposition for Flow Controls*; Techincal Report No. 2000–2025; Institute for Computer Applications in Science and Engineering, NASA Langley Research Center: Hampton, VA, USA, 2000.

14. Qian, E.; Grepl, M.; Veroy, K.; Willcox, K. A certified trust region reduced basis approach to PDE-constrained optimization. *SIAM J. Sci. Comput.* **2017**, *39*, S434–S460. [CrossRef]
15. Yue, Y.; Meerbergen, K. Accelerating optimization of parametric linear systems by model order reduction. *SIAM J. Optimiz.* **2013**, *23*, 1344–1370. [CrossRef]
16. Banholzer, S. ROM-Based Multiobjective Optimization with PDE Constraints. Ph.D. Thesis, University of Konstanz, Konstanz, Germany, 2021. Available online: http://nbn-resolving.de/urn:nbn:de:bsz:352-2-1g98y1ic7inp29 (accessed on 28 April 2022).
17. Banholzer, S.; Keil, T.; Mechelli, L.; Ohlberger, M.; Schindler, F.; Volkwein, S. An adaptive projected Newton non-conforming dual approach for trust-region reduced basis approximation of PDE-constrained parameter optimization. *arXiv* **2020**, arXiv:2012.11653.
18. Keil, T.; Mechelli, L.; Ohlberger, M.; Schindler, F.; Volkwein, S. A non-conforming dual approach for adaptive trust-region reduced basis approximation of PDE-constrained optimization. *ESAIM M2AN* **2021**, *55*, 1239–1269. [CrossRef]
19. Yano, M.; Huang, T.; Zahr, M.J. A globally convergent method to accelerate topology optimization using on-the-fly model reduction. *Comput. Methods Appl. Mech. Eng.* **2021** *375*, 113635. [CrossRef]
20. Zahr, M.J.; Carlberg, K.T.; Kouri, D.P. An efficient, globally convergent method for optimization under uncertainty using adaptive model reduction and sparse grids. *SIAM/ASA J. Uncertain. Quantif.* **2019**, *7*, 877–912. [CrossRef]
21. Kouri, D.P.; Heinkenschloss, M.; Ridzal, D.; van Bloemen Waanders, B.G. A trust-region algorithm with adaptive stochastic collocation for PDE optimization under uncertainty. *SIAM J. Sci. Comput.* **2013**, *35*, A1847–A1879. [CrossRef]
22. Kouri, D.P.; Heinkenschloss, M.; Ridzal, D.; van Bloemen Waanders, B.G. Inexact objective function evaluations in a trust-region algorithm for PDE-constrained optimization under uncertainty. *SIAM J. Sci. Comput.* **2014**, *36*, A3011–A3029. [CrossRef]
23. Grüne, L.; Pannek, J. *Nonlinear Model Predictive Control: Theory and Algorithms*, 2nd ed.; Springer: London, UK, 2016.
24. Borwein, J.M. On the existence of Pareto efficient points. *Math. Oper. Res.* **1983**, *8*, 64–73. [CrossRef]
25. Hartley, R. On cone-efficiency, cone-convexity and cone-compactness. *SIAM J. Appl. Math.* **1978**, *34*, 211–222. [CrossRef]
26. Sawaragi, Y.; Nakayama, H.; Tanino, T. *Theory of Multiobjective Optimization*; Elsevier: Amsterdam, The Netherlands, 1985.
27. Wierzbicki, A.P. The Use of Reference Objectives in Multiobjective Optimization. In *Multiple Criteria Decision Making Theory and Application*; Springer: Berlin/Heidelberg, Germany, 1980; pp. 468–486.
28. Mueller-Gritschneder, D.; Graeb, H.; Schlichtmann, U. A successive approach to compute the bounded Pareto front of practical multiobjective optimization problems. *SIAM J. Optim.* **2009**, *20*, 915–934. [CrossRef]
29. De Motta, R.S.; Afonso, S.M.B.; Lyra, P.R.M. A modified NBI and NC method for the solution of N-multiobjective optimization problems. *Struct. Multidiscip. Optim.* **2012**, *46*, 239–259. [CrossRef]
30. Khaledian, K.; Soleimani-damaneh, M. A new approach to approximate the bounded Pareto front. *Math. Method Oper. Res.* **2015**, *82*, 211–228. [CrossRef]
31. Lowe, T.J.; Thisse, J.-F.; Ward, J.E.; Wendell, R.E. On efficient solutions to multiple objective mathematical programs. *Manag. Sci.* **1984**, *30*, 1346–1349. [CrossRef]
32. Sayın, S. Measuring the quality of discrete representations of efficient sets in multiple objective mathematical programming. *Math. Program.* **2000**, *87*, 543–560. [CrossRef]
33. Mechelli, L. POD-Based State-Constrained Economic Model Predictive Control of Convection-Diffusion Phenomena. Ph.D. Thesis, University of Konstanz, Konstanz, Germany, 2019. Available online: http://nbn-resolving.de/urn:nbn:de:bsz:352-2-2zoi8n9sxknm1 (accessed on 28 April 2022).
34. Evans, L.C. *Partial Differential Equations*; American Mathematical Society: Providence, RI, USA, 2010.
35. Haasdonk, B. Reduced basis methods for parametrized PDEs—A tutorial introduction for stationary and instationary problems. In *Model Order Reduction and Approximation: Theory and Algorithms*; Benner, P., Ohlberger, M., Cohen, A., Willcox, K., Eds.; SIAM: Philadelphia, PA, USA, 2017; pp. 65–136.
36. Rozza, G.; Huynh, D.B.P.; Patera, A.T. Reduced basis approximation and a posteriori error estimation for affinely parametrized elliptic coercive partial differential equations. *Arch. Comput. Method E* **2008**, *15*, 229–275. [CrossRef]
37. Beermann, D.; Dellnitz, M.; Peitz, S.; Volkwein, S. Set-oriented multi-objective optimal control of PDEs using proper orthogonal decomposition. In *Reduced-Order Modeling (ROM) for Simulation and Optimization*; Keiper, W., Milde, A., Volkwein, S., Eds.; Springer International Publishing: Berlin/Heidelberg, Germany, 2018; pp. 47–72.
38. Haasdonk, B.; Dihlmann, M.; Ohlberger, M. A training set and multiple bases generation approach for parameterized model reduction based on adaptive grids in parameter space. *Math. Comput. Model. Dyn.* **2011**, *17*, 423–442. [CrossRef]
39. Keil, T. Adaptive Reduced Basis Methods for Multiscale Problems and Large-Scale PDE-Constrained Optimization. Ph.D. Thesis, WWU Münster, Münster, Germany, 2022.

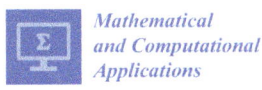

Article

Enhancing Quasi-Newton Acceleration for Fluid-Structure Interaction

Kyle Davis [1,*], Miriam Schulte [1] and Benjamin Uekermann [2]

[1] Simulation of Large Systems, Institute for Parallel and Distributed Systems (IPVS), University of Stuttgart, 70569 Stuttgart, Germany; miriam.schulte@ipvs.uni-stuttgart.de
[2] Usability and Sustainability of Simulation Software, Institute for Parallel and Distributed Systems (IPVS), University of Stuttgart, 70569 Stuttgart, Germany; benjamin.uekermann@ipvs.uni-stuttgart.de
* Correspondence: kyle.davis@ipvs.uni-stuttgart.de

Abstract: We propose two enhancements of quasi-Newton methods used to accelerate coupling iterations for partitioned fluid-structure interaction. Quasi-Newton methods have been established as flexible, yet robust, efficient and accurate coupling methods of multi-physics simulations in general. The coupling library preCICE provides several variants, the so-called IQN-ILS method being the most commonly used. It uses input and output differences of the coupled solvers collected in previous iterations and time steps to approximate Newton iterations. To make quasi-Newton methods both applicable for parallel coupling (where these differences contain data from different physical fields) and to provide a robust approach for re-using information, a combination of information filtering and scaling for the different physical fields is typically required. This leads to good convergence, but increases the cost per iteration. We propose two new approaches—pre-scaling weight monitoring and a new, so-called QR3 filter, to substantially improve runtime while not affecting convergence quality. We evaluate these for a variety of fluid-structure interaction examples. Results show that we achieve drastic speedups for the pure quasi-Newton update steps. In the future, we intend to apply the methods also to volume-coupled scenarios, where these gains can be decisive for the feasibility of the coupling approach.

Keywords: fluid-structure interaction; quasi-Newton; multiphysics coupling

1. Introduction

Multiphysics simulations have shown immense usefulness in the engineering design sector, and are increasingly being applied to more complex problems, ranging from biomedical devices [1] and wind loads on structures [2], to hydraulic fracture simulation [3]. The rise in challenging applications of multiphysics simulations has led to an increased focus on developing flexible, efficient, and scalable multiphysics coupling software. A practical and user friendly approach is to develop partitioned-coupling software, which couples existing standalone physics simulation solvers together to solve new types of simulation problems. Here, the physics solvers themselves are treated as black boxes. This is in contrast to monolithic methods, where all of the equations from each physics domain are solved together in a single system. Partitioned coupling requires an additional piece of software taking care of the actual numerical and technical coupling of the separate solvers. Amongst various such coupling software packages available is preCICE [4]. The key features of preCICE are the minimally invasive library approach, sophisticated numerical methods, parallel scalability, and a strong focus on usability, maintainability, and extensibility. In this paper, we present enhancements and robust parameter choices for numerical equation coupling with preCICE, substantially improving performance, robustness and usability.

Various other general-purpose coupling software exist that are able to perform partitioned coupling for multiphysics (including fluid-structure interaction) and multi-scale problems. Software coupling packages similar to preCICE are DTK [5] and

OpenPALM [6], which both offer a slightly different approach to simulation coupling than preCICE. DTK's application programming interface (API) offers lower-level features compared to preCICE, allowing more flexibility regarding the coupling logic, but at a greater development effort for the user, whereas OpenPALM offers a higher-level approach, with built-in coupling logic and a graphical user interface. A comparison of preCICE with the mentioned libraries, as well as many others coupling software solutions, is provided in [7].

In addition to numerical coupling of separate solvers, a coupling software has to provide communication between solvers, data mapping between non-matching meshes at the interface between solvers, and interpolation in time (if higher order time stepping shall be achieved). We focus on the numerical coupling in this paper, with a specific focus on parallel quasi-Newton schemes. Quasi-Newton schemes for partitioned multi-physics coupling were introduced in 2009 [8] and have been improved since [9–13]. Similar methods have been developed in a different community in the context of acceleration of fixed-point solvers under the name Anderson mixing or Anderson acceleration [14–17]. Quasi-Newton acceleration schemes have been shown to provide fast and stable coupling between various physics solvers for a variety of problems [3,9,18]. However, to provide fast and stable coupling, additional numerical techniques are implemented in preCICE [10]. These additional interface operations have often been considered to have a negligible computational cost compared to the solvers, as the interface degrees of freedom are assumed to be much fewer than those of the coupled solvers themselves. However, this might not be true if the interface becomes large or in the case of volume coupling, where the entire domain is essentially the coupling interface.

The aim of this work is to implement minor enhancements to the existing implementation of quasi-Newton methods in preCICE, which, however, give major improvements in terms of efficiency, robustness, and usability. The enhancements allow easier access to good input parameter choices for standard users, while also offering maximal flexibility for expert users without having to consider the computational cost of the coupling library. We approach this goal by achieving a detailed understanding on how the implementation of quasi-Newton schemes affects computational performance.

The remainder of the paper is structured as follows. In Section 2, we provide an overview of multisecant quasi-Newton methods for fixed-point problems, as well as general existing methods used to improve the quasi-Newton performance in preCICE. In Section 3, we show how these additional general methods can be further enhanced to reduce the number of coupling iterations, on the one hand, and the runtime of the actual acceleration, on the other hand. In Section 4, we present different test cases to analyse the improvements. In Section 5, we present results for these test cases, followed by a discussion of the results in Section 6.

2. Methods—Introduction to Quasi-Newton Coupling

In this section, we present variants of iterative quasi-Newton coupling as implemented for multiphysics simulation coupling. We start with the different versions of fixed-point equations that are solved in iterative partitioned coupling approaches in Section 2.1. In Section 2.2, we present the basic ideas of the quasi-Newton approaches used, followed by enhancements of these methods improving convergence and robustness in Section 2.3. The contributions of this paper, enhancements and rules that further improve the efficiency, robustness and usability of quasi-Newton coupling for multiphysics simulations, are presented in Section 3.

2.1. Partitioned Coupling

For time-dependent problems, partitioned coupling, i.e., coupling of separate simulation codes, can be divided into two types: explicitly (loosely) coupled or implicitly (strongly) coupled. In explicit coupling, each solver performs its time step only once and proceeds with the next time step after exchanging data with the other solvers. In implicit coupling, all solvers iterate their time steps exchanging data after each iteration until a fixed-point problem describing the coupling conditions is solved. Solving this fixed point problem requires the introduction of sophisticated methods, such as quasi-Newton methods, to achieve acceleration of the convergence of the respective fixed-point iterations.

To explain the derivation of the fixed-point formulation, we consider two coupled solvers for simplicity, represented by mapping functions S_1 and S_2 operating on data defined at the coupling interface Γ. A common example is a fluid-structure interaction (FSI) model, where S_1 is the fluid solver that maps interface displacements or velocities x_1 to forces x_2 exerted at the structure, whereas S_2 is the structure solver that maps forces x_2 to interface displacements or velocities x_1. The following description of the coupling, however, generalises to any multi-physics problem. The mapping S_1 requires the output of S_2 and vice-versa such that

$$S_1 : x_1 \mapsto x_2 \quad \text{and} \quad S_2 : x_2 \mapsto x_1.$$

For the considered time-dependent problems, x_1 and x_2 are the respective interface values at the new time step. Strong or implicit coupling between S_1 and S_2 is formulated as a fixed point problem. Two mathematically equivalent variants of this fixed-point problem can be formulated in matrix-like notation as

$$x_1 = S_2 \circ S_1(x_1) \quad \text{(Gauss-Seidel type coupling) and} \tag{1}$$

$$\begin{pmatrix} x_1 \\ x_2 \end{pmatrix} = \begin{pmatrix} 0 & S_2 \\ S_1 & 0 \end{pmatrix} \begin{pmatrix} x_1 \\ x_2 \end{pmatrix} \quad \text{(Jacobi type coupling).} \tag{2}$$

These correspond to two different types of fixed-point iterations as depicted in Figure 1. In this work, we focus on the second, so-called Jacobi type system as it results in higher parallelism and, thus, more efficient usage of computational resources on large compute clusters and supercomputers [11], and allows for an arbitrary number of solvers to be coupled [19].

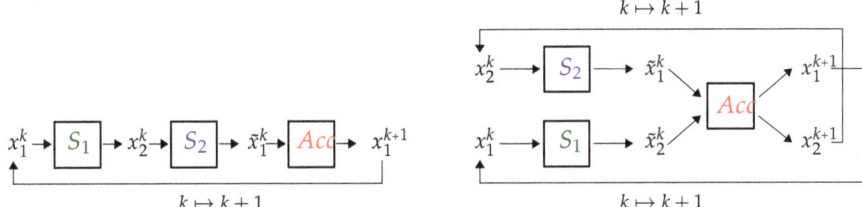

Figure 1. General coupling options, Gauss-Seidel (**left**) and Jacobi (**right**), for partitioned coupling of two solvers S_1 and S_2.

The unmodified fixed-point iterations $x^{k+1} = H(x^k)$, where

$$x = \begin{pmatrix} x_1 \\ x_2 \end{pmatrix} \quad \text{and} \quad H = \begin{pmatrix} 0 & S_2 \\ S_1 & 0 \end{pmatrix} \tag{3}$$

in matrix-like notation, may be slow to converge or not converge at all [9]. We therefore apply an acceleration scheme, $Acc()$, such that

$$x^{k+1} = Acc(\tilde{x}^k) \quad \text{with} \quad \tilde{x}^k = H(x^k).$$

In the following, we explain how the acceleration operator Acc is realised based on quasi-Newton approaches.

2.2. Introduction to Quasi-Newton Coupling Methods

The partitioned coupling schemes described above require solving a non-linear problem

$$R(x) := H(x) - x = 0. \tag{4}$$

To solve this for problem sizes typically encountered in multi-physics simulations, full Newton methods are undesirable due to large computational costs and memory requirements and, in addition, infeasible for black-box coupling as the derivatives of the fixed-point operator H are inaccessible. Quasi-Newton methods are able to approximate a Newton step and in general accelerate the convergence compared to the pure fixed-point iteration [8,9,12]. At each iteration, we compute the next iteration via

$$x^{k+1} = \tilde{x}^k - J^{-1} r^k, \tag{5}$$

where $r^k = R(x^k) = \tilde{x}^k - x^k$ and J^{-1} is an approximation of the inverse Jacobian of the mapping $\tilde{R} : \tilde{x} \mapsto \tilde{x} - H^{-1}(\tilde{x})$, i.e., the function that maps the result of the fixed point iteration to the residual ($\tilde{R}(H(x)) = R(x)$). Equation (5) represents a sequence of a fixed-point iteration and a modified approximated Newton step. As multiple iterations are performed in each time step for implicitly coupled systems, we introduce a convergence criteria to define when we proceed to the next time step. Implicitly coupled solvers are considered to have converged if

$$\frac{\|x_1^k - \tilde{x}_1^k\|_2}{\|x_1^k\|_2} < \epsilon_{\text{conv}}. \tag{6}$$

For solver S_1, and similarly for solver S_2. Critical to the performance of the quasi-Newton method is the manner in which we approach the approximation of the inverse Jacobian J^{-1}. We use the input/output differences collected from previous iterations collected in matrices

$$\begin{aligned} W_k^\eta &= \left[\Delta \tilde{x}^k, \Delta \tilde{x}^{k-1}, \ldots, \Delta \tilde{x}^{k-\eta}\right] \quad \text{with} \quad \Delta \tilde{x}^k = \tilde{x}^k - \tilde{x}^{k-1}, \\ V_k^\eta &= \left[\Delta r^k, \Delta r^{k-1}, \ldots, \Delta r^{k-\eta}\right] \quad \text{with} \quad \Delta r^k = r^k - r^{k-1}, \end{aligned} \tag{7}$$

where $W_k^\eta \in \mathbb{R}^{N \times \eta}$, $V_k^\eta \in \mathbb{R}^{N \times \eta}$, N is the number of degrees of freedom (DoF) at the coupling interface, and η is the maximum number of previous iterations that are retained. For transient coupled problems, this can also include iterations from previous time steps. We, however, drop iterations older than ζ time steps. The approximation of the inverse Jacobian is required to fulfil the multi-secant equation

$$J^{-1} V_k = W_k, \tag{8}$$

a strongly under-determined system as $\eta \ll N$. We, thus, have to add a norm minimisation

$$\min \|J^{-1} - J_{\text{prev}}^{-1}\|_F. \tag{9}$$

where J_{prev}^{-1} is a previous approximation. Depending on the choice of J_{prev}^{-1}, we get two types of quasi-Newton methods:

IQN-ILS. The Interface Quasi-Newton Inverse Least-Squares (IQN-ILS) method is a popular and frequently used multiphysics coupling acceleration scheme. It was first introduced in [8]. For IQN-ILS, we choose $J_{\text{prev}}^{-1} = 0$, i.e., to determine J^{-1}, we solve

$$J^{-1} V_k = W_k \quad \text{with} \quad J^{-1} = \mathrm{argmin} \|J^{-1}\|_F, \tag{10}$$

yielding

$$J^{-1} = W_k \left(V_k^T V_k\right)^{-1} V_k^T. \tag{11}$$

The benefit of this classic *least-squares* approach is the option for a matrix-free implementation of the quasi-Newton step (5):

$$x^{k+1} = \tilde{x}^k + W_k \alpha \quad \text{with} \quad \alpha = \mathrm{argmin} \|V_k \alpha + r^k\|_2. \tag{12}$$

To solve this least-squares problem, we compute a QR-decomposition of $V_k = QR$ and solve the small $\eta \times \eta$ system $R\alpha = -Q^T r^k$.

The IQN-ILS method builds the approximation of J^{-1} exclusively from the retained input/output vectors stored in W_k and V_k. Therefore, the amount and quality of the information stored in V_k and W_k is decisive for the convergence rate and robustness of the IQN-ILS method, and many problems typically require storing many previous iterations, η, over many previous time steps, ζ.

IQN-IMVJ. The Interface Quasi-Newton Inverse Multi-Vector Jacobian method [12,13,20] implicitly retains information from previous time steps using J^{-1}_{prev} as the previous time step's inverse Jacobian approximation and, thus, allows us to implicitly use information on J^{-1} already collected in previous time steps. In the IQN-IMVJ method, we also perform a QR-decomposition of $V_k^\eta = QR$ to determine the pseudo-inverse $V_k^+ = \left(V_k^T V_k\right)^{-1} V_k^T$ and get J^{-1} from

$$J^{-1} = J^{-1}_{prev} + \left(W_k - J^{-1}_{prev} V_k\right)\left(V_k^T V_k\right)^{-1} V_k^T. \tag{13}$$

The disadvantage of this method is that we need an explicit representation of J^{-1}_{prev} which requires $\mathcal{O}(N^2)$ both in memory and computational complexity. By smart approximations, this cost can, however, be reduced to $\mathcal{O}(N)$ [12,21]. In this paper, we focus on improvements to the IQN-ILS method, as the most commonly used quasi-Newton variant.

2.3. IQN-ILS Enhancement Techniques

When implementing the IQN-ILS method into a software library, various numerical methods need to be utilised to further improve numerical stability and reduce runtime and memory requirements. The two methods we highlight here are filtering of columns in V_k [10] and pre-scaling of the interface values [22]. These techniques are discussed below, and the advantages and disadvantages are highlighted.

2.3.1. Filtering

For IQN-ILS, there is no guarantee that all columns in V_k are linearly independent. Thus, to aid convergence for all acceleration schemes, filtering columns of the matrices V_k and W_k is performed to remove any linearly dependent columns [10]. We discuss two filtering variants: QR1 and QR2 filters (referred to as *old* QR and *new* QR filter, respectively, in Haeltemann et al. [10]). Both methods begin by performing a QR-decomposition $V_k = QR$, where V_k is decomposed into an orthogonal matrix $Q^{N \times \eta}$ and an upper triangular matrix $R^{\eta \times \eta}$.

In every iteration, a new column is added to the left of both V_k and W_k (see Equation (7)). Therefore the leftmost columns represent newest information and the rightmost columns older information. The QR-decomposition is performed column-wise from right to left, and is realised by a QR update procedure as shown in Figure 2. Taking an already computed decomposition QR from the previous iteration, a new column on the left $v = (V_k)_{:,1}$ (the subscript indicates the row and column number in a Python-like notation. :,1 refers to the leftmost column of V_k—this is the most recent column added to V_k) is orthogonalised against Q via a modified Gram–Schmidt procedure. The orthogonalised v is then added as additional (rightmost) column in Q (column q in Figure 2). A new (leftmost) column is also added to R (column r in Figure 2) along with a bottom row of zeros. A series of Givens rotations eliminate any non-zero sub-diagonal entries in R. A detailed explanation of the procedure can be found in [23]. Note that this seems to be unnecessarily complicated compared to adding columns from the right of V_k, which would only require orthogonalisation of the new column q with a standard Gram–Schmidt algorithm, and adding a new column r on the right of R. However, having the oldest information in the rightmost columns of V_k and W_k offers the advantage that deleting old information, possibly multiple columns at a time, is cheap. Deleting columns from V_k and W_k typically occurs at least every time step.

Deleting the rightmost (oldest) column of V_k only requires removing the rightmost column from Q and the rightmost column as well as the last row from R. This does not introduce non-zero sub-diagonal elements in R and, thus, does not require additional Givens rotations. Removing an arbitrary column from V_k requires (1) removing the corresponding column from R, (2) removing any sub-diagonal elements from R using Givens rotations, (3) applying the corresponding Givens-rotations to Q from the right, (4) removing the last column from Q as well as the bottom (zero) row from R.

If a complete QR decomposition is required, Q and R are discarded completely, and new Q and R matrices are rebuilt using each column from V_k, adding one column at a time.

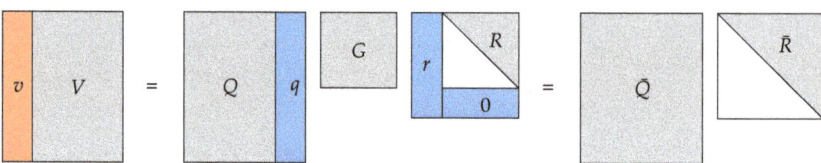

Figure 2. QR-decomposition updating procedure. For every new column added to V_k on the left, the QR decomposition can be updated by adding the orthogonalised version of this column as new (**rightmost**) column q in Q, adding an additional row of zeros to R and subsequently adding a new (**leftmost**) column r to R representing the respective orthogonalisation factors. To compute the final decomposition, we have to apply Given rotations (represented by the matrix G in the figure) to eliminate sub-diagonal entries in R. $\bar{Q}$ and $\bar{R}$ represent the final decomposition result.

QR1. In the QR1 filter, the impact of a column on the condition of R is estimated by comparing the diagonal elements of R to the complete norm of R as a metric for the norm of orthogonalised columns of V_k before normalisation. This means, we delete a column i if

$$R_{ii} < \epsilon_f \cdot \|R\|_F, \tag{14}$$

where ϵ_f is the filtering limit, a user specified parameter. This filter has a potential drawback as the QR-decomposition is built from the oldest column of information to the newest and, thus, has a tendency to delete new columns instead of (potentially outdated) old ones. However, [22] found that this was not problematic with a well selected filter limit. The main advantage of the QR1 filter is that it does not enforce a complete re-computation of the QR-decomposition in each coupling iteration, but allows updating the decomposition by inserting the newest column v into QR only. Any filtered column can be removed with a single column deletion step.

QR2. The QR2 filter (Algorithm 1) was introduced in [10] as a means to quantify the amount of new information a column in V_k adds to the QR-decomposition, and to filter the columns during the construction of Q and R itself. Importantly, the QR2 method re-constructs the QR-decomposition beginning with the newest column of information and, thus, has to rebuild the QR decomposition in each iteration, and tends to remove the oldest columns from V_k. This is favoured as older columns might no longer be relevant to the current dynamics of the physical system.

Algorithm 1 : QR2 Filter [10]

$R_{11} = \|(V_k)_{:,1}\|_2$ ▷ This is the newest column
for $i = 1, ..., \eta$ **do**   ▷ Starts from newest column
    $\bar{v} = (V_k)_{:,i}$
    for $j = 1, ..., i-1$ **do**
        $R_{ji} = Q_{:,j}^T \cdot \bar{v}$
        $\bar{v} = \bar{v} - R_{ji} \cdot Q_{:,j}$
    end for
    if $\|\bar{v}\|_2 < \epsilon_f \|(V_k)_{:,i}\|_2$ **then**
        delete column i
    end if
    $R_{ii} = \|\bar{v}\|_2$ and $Q_{:,i} = \bar{v}/R_{ii}$
end for

2.3.2. Pre-Scaling

When using the Jacobi-type fixed point equation as introduced in Equation (2), largely different orders of magnitude of the two parts of the vector x, forces and displacements/velocities, and their residuals $R(x)$ may cause numerical issues as only the field with the larger magnitude is *seen* in the approximation of J^{-1}. For example, in case of a stiff solid structure, the surface pressure may be in the order of 10^5, whereas the structural deformation may be in the order of 10^{-3}. Additionally, the solution of the stiff structure might not change much between time steps, and therefore have a small residual value $R(x)$, while the residual of the fluid solver may still be several order of magnitude larger. Therefore, additional scaling of the sub-vectors of x and the respective residual components is required. We apply a pre-scaling to $V_k = \Lambda_k V_k$ and $r^k = \Lambda_k r^k$, where $\Lambda_k = \text{diag}(\lambda_{k,1} \ldots \lambda_{k,1} \lambda_{k,2} \ldots \lambda_{k,2})^T \in \mathbb{R}^{N \times N}$. The two pre-scaling weights $\lambda_{k,1}$ and $\lambda_{k,2}$ are selected to normalise the values of each sub-vector in r^k. Using the so-called residual-sum pre-scaling introduced in [24],

$$\lambda_{k,1} = \left(\sum_{j=1}^{k} \frac{\|S_2(x_2^j) - x_1^j\|_2}{\|R(x^j)\|_2} \right) \quad \text{and} \quad \lambda_{k,2} = \left(\sum_{j=1}^{k} \frac{\|S_1(x_1^j) - x_2^j\|_2}{\|R(x^j)\|_2} \right). \qquad (15)$$

Uekermann [22] showed that this provided a suitable scaling for multiphysics simulations. In this pre-scaling, we use the norms of the fixed point equation residuals in each sub-field divided by the norm of the complete residual as scaling factors. This is then summed over all iterations, k, within one time step. The summation over all previous iterations in one time-step prevents a zig-zag convergence behaviour found in [22].

By updating the pre-scaling weights of Λ_k in each iteration, a complete QR-decomposition of V_k is required in every iteration instead of just a cheap update as all columns of V_k change. A complete QR-decomposition is required as each row in V_k is not scaled by the same value. This does not further affect the QR2 filter, but now a complete QR-decomposition in each iteration is also required for the QR1 filter.

2.4. Current Good Practice for IQN-ILS

Even though quasi-Newton methods for multiphysics coupling problems is advanced, tuning both convergence speed and runtime requires a careful understanding of the method in combination with the enhancements described above. Therefore, we describe some important aspects of IQN-ILS in combination with pre-scaling and filtering and the choice of other methodological parameters.

The pseudo-code of the IQN-ILS method with additional pre-scaling (Section 2.3.2) and filtering (Section 2.3.1) steps is shown in Algorithm 2. As the IQN-ILS method builds the approximation of J^{-1} exclusively from V_k and W_k, the problem dependent parameters η and ζ can have a large impact on the robustness and convergence as mentioned above. Whereas the general idea is to improve the approximation of J^{-1} by using more information than generated in a single time step, a large number of columns also has obvious negative impact: In addition to increasing the risk of rank deficiency of V_k (which can be resolved by filtering), retaining a large number of columns in W_k and V_k also leads to (i) larger memory consumption for matrix storage and (ii) larger computational effort in performing the QR-decomposition of V_k. A choice of between $\eta = 100$ and $\eta = 200$ previous iterations over the previous $\zeta = 10$ to $\zeta = 20$ time steps is a good choice if combined with a well-tuned filter. The filter QR2 is preferred as it tends to delete older columns and only if they add little information to the QR-decomposition. The filter limit should be chosen such that some, but not too many columns are deleted. A good filter limit is typically between $\epsilon_f = 0.1$ and $\epsilon_f = 0.001$.

Algorithm 2 : IQN-ILS

---

initial value x^0
$\tilde{x}^0 = H(x^0)$ and $r^0 = \tilde{x}^0 - x^0$
$x^{k+1} = x^0 + \omega(\tilde{x}^0 - x^0)$
for $k = 1, 2, \ldots$ do
    $\tilde{x}^k = H(x^k)$ and $r^k = \tilde{x}^k - x^k$
    if converged then
        break
    end if
    $V_k = \left[\Delta r^k, \ldots, \Delta r^1\right], \Delta r^k = r^k - r^{k-1}$
    $W_k = \left[\Delta \tilde{x}^k, \ldots, \Delta \tilde{x}^1\right], \Delta \tilde{x}^k = \tilde{x}^k - \tilde{x}^{k-1}$
    Determine pre-scaling weights Λ_k
    Compute $\Lambda_k V_k = QR$
    Filter columns in QR
    solve $R\alpha = -Q^T \Lambda_k r^k$
    $x^{k+1} = \tilde{x}^k + W_k \alpha$
end for

---

3. Computational Improvements for the Quasi-Newton Method

The complex manner in which input parameters interact with each other makes it difficult to find the optimal input settings for a given problem, while still maintaining fast simulation runtimes. In the following section, we discuss two methods developed to improve the computational runtime while maintaining the robustness of the quasi-Newton methods.

3.1. Pre-Scaling Weight Monitoring

As the pre-scaling weights $\lambda_{k,1}$ and $\lambda_{k,2}$ change in each iteration, a complete QR-decomposition is performed in each iteration, adding a significant computational expense to the quasi-Newton update. We introduce a new pre-scaling weight monitoring method to freeze the pre-scaling weights after the first time step. In each iteration in the first time step, we recompute and apply the pre-scaling weights as usual. Starting from the second time step,, we only compute the *theoretical* new pre-scaling weight values Λ_k^* from the newest residual values using Equation (15). We update the actual pre-scaling weights $\Lambda_k \mapsto \Lambda_k^*$ if these *theoretical* weights change by more than one order of magnitude (i.e., if $\lambda_{k,i}^* > 10 \cdot \lambda_{k,i}$ or $\lambda_{k,i}^* < 0.1 \cdot \lambda_{k,i}$ for each solver i). This allows us to implement further methods to reduce the computational runtime for iterations where we keep pre-scaling weights constant:

1. We alter the QR-decomposition strategy for the QR1 filter, such that it only recomputes the QR-decomposition if pre-scaling weights change. Previously, the full recomputation was done in every iteration if pre-scaling was enabled.
2. We develop a new faster QR filter, QR3, that can mimic the behaviour of the QR2 filter. The previous QR2 inherently required to recompute the QR-decomposition in each iteration independent on whether pre-scaling was enabled. We describe the new filter below in Section 3.2).

3.2. Fast Alternative QR Filter

The current QR2 filtering technique relies on performing a complete QR-decomposition in every iteration. We typically assume that the number of DoF on the coupling interface, N, is much smaller than the number of DoF within the solvers. Therefore, the computational time for the QR-decomposition should be negligible. However, this may not always be the case as

1. the number η of columns in V_k can grow very large and the cost of inserting a column into QR has a computational complexity of $\mathcal{O}(\eta^3)$,
2. in volume coupling, the number of coupling DoF is equal to the number of all DoF in the domain, and is not negligible.

It is unnecessary for the QR2 filter step to perform a complete QR-decomposition if actually no column is deleted. In this case, a single column insertion step could have been performed. In this work, we introduce a new QR3 filter. A requirement for this filter is that the pre-scaling weights remain constant, and therefore the pre-scaling weight monitoring (Section 3.1) is required to use the QR3 filter. If the weights are updated during a coupling iteration, then a normal QR2 filter step is performed as the QR-decomposition is then rebuilt ($\Lambda_k V_k = QR$) in the process. Otherwise, the new filter computes these three steps:

1. the newest column of V_k is inserted into an existing QR decomposition (see Section 2.3.1),
2. a check is performed to tag any column that should be removed according to the same criteria as QR2,
3. only if any one column is tagged to be removed, then a normal QR2 filter step is performed instead, that is, a complete QR-decomposition is performed and columns are removed in this step.

In step 2, the check begins from the oldest column, $R_{:,\eta}$ (the subscript refers to the row and column number. $:,\eta$ refers to the right most column of R), and moves towards the previous newest column, $R_{:,2}$. We do not check if the first column $R_{:,1}$ should be removed as we want to keep the latest information. A column i is tagged for deletion if $R_{ii} < \epsilon_f \|V_{k,(:,i)}\|_2$. This criteria aims to mimic the behaviour of the QR2 filter by tending to find older columns to delete similar to the regular QR2 filter, but without reconstructing QR. If at least one column is tagged for deletion, then the QR2 filter is applied to rebuild QR starting from newest information, ensuring that a good quality of QR is maintained. Further runtime improvements could be found if this criterion is made stricter, for instance if at least two or three columns must be tagged for deletion before a QR2 filter step is performed. For simplicity, we restrict our tests, however, to only a single column. The QR3 pseudo-code is shown in Algorithm 3.

Algorithm 3 : QR3 Filter

Add newest column $\bar{v} = (V_k)_{:,1}$ to QR
$filter \leftarrow$ `false`
for $i = \eta, ..., 2$ **do** ▷ Starts from oldest column and works forwards
    if $R_{ii} < \epsilon_f \|(V_k)_{:,i}\|_2$ **then**
        $filter \leftarrow$ `true`
        break
    end if
end for
if *filter* **then**
    Compute QR2 Filter Step
end if

---

4. Numerical Setup

To test the effectiveness of our suggested enhancements, we introduce three fluid-structure interaction test scenarios in this section, followed by descriptions of the used software and hardware.

4.1. Test Cases

We selected three test scenarios that feature fluid and solid density values known to be challenging in terms of stability due to the added mass effect [25,26]: a 3D elastic tube scenario and a breaking dam scenario in 2D and 3D. Whereas the elastic tube scenario represents an outer elastic structure, the breaking dam scenarios feature elastic structures immersed in the fluid and involve free surface flow, such that we cover the main types of fluid-structure interaction scenarios. In addition, the breaking dam scenarios are dynamically changing, which challenges the reuse of past information of quasi-Newton methods. To ensure reproducibility of our results, the complete test setups are available under an open-source license at https://github.com/KyleDavisSA/IQN-test-cases accessed on 15 February 2022.

Elastic-Tube-3D. The Elastic-Tube-3D problem, proposed for instance in [27] and used in many other studies [8,13,20], is a simplified heamodynamic FSI test case. The test case geometry consists of a cylindrical tube with elastic walls and an inner fluid domain. A time-dependent pressure boundary condition is applied at the inlet and at the outlet of the fluid domain. At the inlet, a pressure of 1.3332 kPa is applied for 3 ms followed by 0 Pa for another 7 ms. At the outlet, the pressure is kept constant at 0 Pa. The simulation is run for 10^{-2} s, with a time step size of $dt = 10^{-4}$ s, for a total of 100 time steps. The fluid flow causes the solid domain to expand, and a pressure pulse travels through the tube. The domain geometry and material properties are shown in Figure 3.

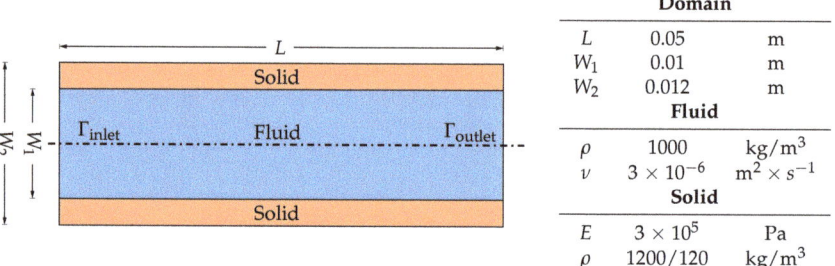

Figure 3. Domain geometry—not to scale—(**left**) and dimension and material parameters (**right**) of the Elastic-Tube-3D test case.

We compare two different densities for the structural solver to examine the added mass effect: (i) $\rho = 1200$ kg/m^3 and (ii) $\rho = 120$ kg/m^3, referred to as Elastic-Tube-3D-Heavy and Elastic-Tube-3D-Light, respectively. The structural solver has 11,735 elements in the domain and 1816 vertices on the coupling interface. The fluid domain contains 32,691 cells, with 1860 vertices on the coupling interface.

Breaking-Dam-2D. The Breaking-Dam-2D test case is a free surface problem, where a large body of water comes into contact with a flexible barrier [28,29]. This test case may pose problems for the quasi-Newton method: Firstly, the past information retained in the matrices V_k and W_k may not be entirely relevant once the water impacts the coupling interface and, thus, the character of the interaction between fluid and solid changes. Secondly, also the pre-scaling weight values may change dramatically at the moment of the impact. The domain and the material properties are shown in Figure 4.

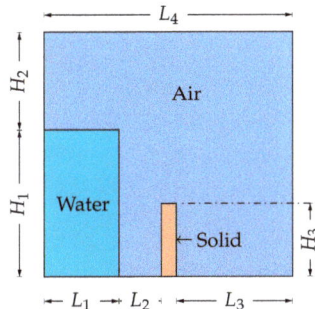

	Domain				
L_1	0.146	m	H_1	0.292	m
L_2	0.14	m	H_2	0.073	m
L_3	0.286	m	H_3	0.08	m
L_4	0.548	m			
Fluid - Water			**Fluid - Air**		
ρ	1000	kg/m³	ρ	1	kg/m³
v	1×10^{-6}	m²×s⁻¹	v	1×10^{-5}	m²×s⁻¹
Solid					
E	3×10^5	Pa			
ρ	2500	kg/m³			

Figure 4. Domain geometry—not to scale—(**left**) and dimension and material parameters (**right**) of the Breaking-Dam-2D test case.

A no slip boundary condition is applied at the bottom, the left, and the right boundary, and a zero pressure outlet at the top. The test case was run for 1 s with a time step size of $dt = 0.005$ s, for a total of 200 time steps. The fluid domain contains 1382 cells in the domain, with 44 vertices on the interface, and the structural domain uses 325 quadratic finite elements in the domain, and 282 vertices on the coupling interface.

Breaking-Dam-3D. The Breaking-Dam-3D test case is a more complex test case inspired by the Breaking-Dam-2D example. A new, larger domain was created with larger bodies of water placed on either side of the wall, and a heavier solid wall was placed between the water columns. The water bodies are offset in the third dimension such that they hit the wall at opposite ends and at different times, resulting in a non-symmetrical movement of the dam wall (Figure 5 (left)). The solid domain is fixed only at the bottom and the sides are free to move in-plane. The test case was run for 0.75 s with a time step of $dt = 0.005$ s, for a total of 150 time steps. The domain geometry is shown in Figure 6, with the dimension and material properties given in Figure 5 (right). The fluid domain contains 25,712 cells in the domain, with 714 vertices on the interface. The solid domain is simulated using 319 linear elements with 387 vertices on the coupling interface.

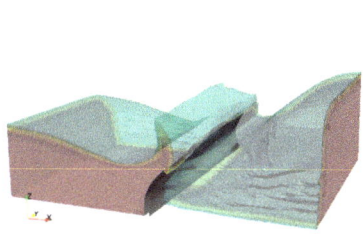

	Domain				
L_1	2	m	H_2	2	m
L_2	0.95	m	H_3	1	m
L_3	1.95	m	H_4	0	m
L_4	1	m	H_5	4	m
L_5	6	m	H_6	1	m
H_1	2	m	H_7	4	m
Fluid - Water			**Fluid - Air**		
ρ	1000	kg/m³	ρ	1	kg/m³
v	1×10^{-6}	m²×s⁻¹	v	1×10^{-5}	m²×s⁻¹
Solid					
E	1×10^7	Pa			
ρ	7850	kg/m³			

Figure 5. Breaking-Dam-3D waters striking the flexible wall at 0.75 s (**left**) and dimensions and material parameters (**right**).

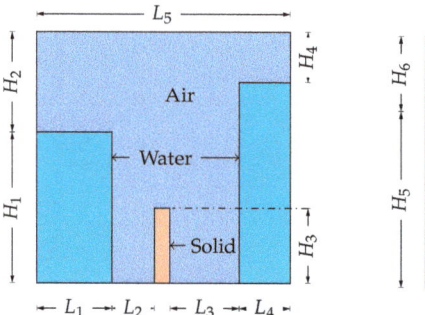
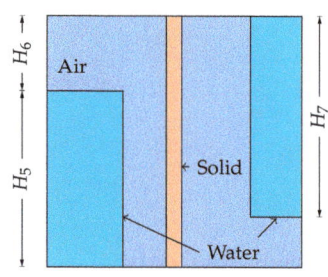

Figure 6. Domain geometry—not to scale—of the Breaking-Dam-3D test cases: front view (**left**), and top view (**right**).

4.2. Quasi-Newton Configuration

A few numerical parameters are the same for each test case. The following values are used unless stated otherwise:

- number of time steps reused: $\zeta = 10$ or $\zeta = 20$;
- maximal number of previous iterations: $\eta = 100$ or $\eta = 200$;
- convergence threshold for IQN-ILS: $\epsilon_{conv} = 10^{-3}$;
- maximum number of iterations allowed per time step before proceeding to the next time step, even if ϵ_{conv} is not reached: 30 or 50 (Breaking-Dam-3D);
- limit for QR2 or QR3 filter: $\epsilon_f = 0.001$, $\epsilon_f = 0.01$, or $\epsilon_f = 0.1$;
- type of pre-scaling: residual-sum pre-scaling as defined in Equation (15);
- initial under-relaxation value: $\omega = 0.1$.

A simulation run is denoted as diverged only if one of the physics solvers crashed, not if the convergence threshold was not reached in one or several time steps.

4.3. Software

We only use open-source software to test the quasi-Newton enhancements. We use the following versions:

- Fluid solver: https://www.openfoam.com/news/main-news/openfoam-v20-12, accessed on (OpenFOAM v2012) [30]; 15 February 2022
- Fluid solver adapter: https://github.com/precice/openfoam-adapter/releases/tag/v1.0.0, accessed on OpenFOAM-preCICE Adapter v1.0.0; 15 February 2022
- Solid solver: http://www.calculix.de/, accessed on CalculiX v2.17 [31]; 15 February 2022
- Solid solver adapter: https://github.com/precice/calculix-adapter/tree/5d42fb6160ede35926a59786ef8ae25dd71d7cdb, accessed on CalculiX-preCICE Adapter, commit 5d42fb6, 15 February 2022.

All quasi-Newton methods described in Sections 2 and 3 have been implemented in the coupling library preCICE [4,7]. We use two different versions of preCICE:

- https://github.com/precice/precice/releases/tag/v2.3.0 accessed on 15 February 2022, as baseline without the enhancements presented in Section 3
- https://github.com/precice/precice/tree/3fb3d8d465e45e1eadba766a8ce5f1f96c138b20 accessed on 15 February 2022, for the enhancements presented in Section 3.

4.4. Hardware

All simulations are run on a single core of a Lenovo T480, with an Intel Core i5–8250U CPU, 1.60 GHz × 4, and 16 GB main memory.

5. Results and Discussion

In this section, we present the results and discussion for the pre-scaling weight monitoring and the new QR3 filtering procedure.

5.1. Pre-Scaling Weight Monitoring

We introduced pre-scaling weight monitoring in order to reduce the number of weight updates throughout the simulation and, as a consequence, to be able to reduce the computational cost of QR-decompositions and filtering. Therefore, we examine the impact of different variants of pre-scaling weight updating on both, the number of quasi-Newton iterations and on the computational cost for QR-decompositions and filtering.

Table 1 shows the impact of different versions of pre-scaling on the number of quasi-Newton iterations. First, we provide a baseline set of results where we update the pre-scaling weights in each iteration. Second, we freeze the pre-scaling weights after the first time step. Finally, we use the pre-scaling weight monitoring.

Table 1. Comparison of the average number of quasi-Newton iterations per time step for IQN-ILS with QR2 filter ($\epsilon_f = 0.01$) for (i) pre-scaling update in every quasi-Newton iteration (column "baseline"), (ii) pre-scaling with freezing of weights after the first time step (column "freeze"), (iii) the new pre-scaling weight monitoring approach (column "monitoring"). For each test case, the first row uses a maximum of $\eta = 100$ iterations from previous $\zeta = 10$ time steps, and the second row $\eta = 200$ and $\zeta = 20$. Braces indicate how many time steps did not converge within 30 iterations before moving to the next time step.

Test Case	ζ	η	Baseline	Freeze	Monitoring
Elastic-Tube-3D Heavy	10	100	4.51	4.45	4.48
Elastic-Tube-3D Heavy	20	200	3.84	3.96	3.94
Elastic-Tube-3D Light	10	100	7.23	7.30	7.16
Elastic-Tube-3D Light	20	200	5.84	6.00	5.83
Breaking-Dam-2D	10	100	7.08 (2)	8.59 (8)	7.9 (3)
Breaking-Dam-2D	20	200	8.06 (6)	9.16 (10)	8.14 (2)
Breaking-Dam-3D	10	100	4.34	4.77	4.33
Breaking-Dam-3D	20	200	4.3	5.19	4.15

From the comparison of quasi-Newton iteration counts in Table 1, we observe that, in general, our new pre-scaling weight monitoring approach does not lead to an increase of the number of iterations compared to the more expensive baseline approach, where weights are updated in every iteration. All test cases apart from the Breaking-Dam-2D case offer comparable results in terms of iteration counts. For the Breaking-Dam-2D case, a slight increase can be observed for our weight monitoring approach. However, overall, the results show that the weight monitoring is effective in the sense that it detects when updates are necessary.

On the other hand, comparing our new approach to simulation runs where we freeze the pre-scaling weights after the first time step shows that freezing the weights substantially increases the number of iterations. It yields comparable results only for both Elastic-Tube-3D cases. There is a significant increase in the number of iterations for both Breaking-Dam cases, which shows that simply freezing the weights is sub-optimal if scenarios undergo sudden changes (fluid hitting the obstacle in the Breaking-Dam case). We conclude, that the weight monitoring is not only sufficient, but also necessary to ensure fast quasi-Newton convergence in general.

To further analyse the impact of the pre-scaling weight monitoring on the overall computational cost, we present more detailed results for the fluid and the solid domain in Table 2, in particular including the number of weight updates, the most important factor for computational cost as each update requires a complete re-computation of the QR-decomposition of V_k. Here, we use the pre-scaling weight monitoring with the new

QR3 filter. The combined impact of this new filter and the pre-scaling weight monitoring is presented in Section 5.2.

The more detailed analysis of these tests in Table 2 leads to further insights: (i) The large difference between the fluid and solid pre-scaling weights is immediately apparent. This large difference necessitates using pre-scaling in the first place. Completely removing the pre-scaling results in either a large increase in coupling iterations, or divergence of the solvers, for all test cases. For brevity, we do not explicitly list these results. (ii) In addition, we observe a large difference between the minimum and the maximum value of the weights for each of the fields over the entire simulation—fluid and solid. This indicates that updating the pre-scaling weights is necessary to ensure suitable scaling.

Table 2. Performance details of pre-scaling weight monitoring with QR3 filter ($\epsilon_f = 0.01$). For each test case, the first row uses a maximum of $\eta = 100$ iterations from previous $\zeta = 10$ time steps, and the second row $\eta = 200$ and $\zeta = 20$. We show the total number of quasi-Newton iterations for the entire simulation with the average number of quasi-Newton iterations per time step in braces (column "Its"), the numbers of updates of the pre-scaling weights during both the first time step (column "Upd 1") and all other time steps (column "Upd 2-"), as well as the range of weights $\lambda_{f/s}$ for both solvers.

Test Case	ζ	η	Its	Upd 1	Upd 2-	λ_s	λ_f
Elastic-Tube-3D-Heavy	10	100	448 (4.48)	14	1	17.38–3155.7	0.071–1.0
Elastic-Tube-3D-Heavy	20	200	394 (3.94)	14	1	17.38–3155.7	0.071–1.0
Elastic-Tube-3D-Light	10	100	716 (7.16)	22	6	4.68–2365.8	0.045–1.0
Elastic-Tube-3D-Light	20	200	583 (5.83)	22	7	2.13–2401.5	0.045–1.0
Breaking-Dam-2D	10	100	1525 (7.63)	4	54	0.98–1018.5	0.09–1.0
Breaking-Dam-2D	20	200	1651 (8.23)	4	75	0.98–621.3	0.09–1.0
Breaking-Dam-3D	10	100	649 (4.33)	10	11	9.67×10^4–3.23×10^7	0.1–1.0
Breaking-Dam-3D	20	200	623 (4.15)	10	6	2.19×10^5–3.23×10^7	0.1–1.0

During the first time step, the pre-scaling weight monitoring updates the weights in each iteration. Therefore, a large number of weight updates is observed for the first time step in Table 2 (column "Upd 1") compared to the rest of the simulation (column "Upd 2-"). For all test cases, the number of updates after the first time step is low compared to the total number of iterations, with the Elastic-Tube-3D-Light cases requiring only six and seven updates after the first time step, respectively. The Breaking-Dam-2D scenario requires 54 and 75 updates after the first time step. However, this is a low number compared to the total of 1525 and 1651 iterations for the entire simulation.

Comparing the amount of retained iterations η over a certain window of time steps ζ, we observe that the number of pre-scaling weight updates as well as the range of weight values do not appear to be very sensitive to these parameters. This was expected since converged solutions at the end of each time step should be rather independent of these parameters.

The range of pre-scaling weight values is also interesting. The Breaking-Dam-3D case features very large scaling values for the solid solver. This indicates that the solid solver residuals are far smaller than the fluid solver's residuals. For this test case, the structure is a heavy wall and "slow" to move. This could account for a high stability in the solid solver itself, but necessitates the use of pre-scaling. A change in the pre-scaling magnitudes is observed for the Elastic-Tube-3D scenarios, with slightly larger solid solver pre-scaling weight values for the Elastic-Tube-3D-Heavy scenario with a higher structure density, once again indicating smaller residual values compared to the fluid solver.

Summarising, our results show that the pre-scaling weights do not change considerably between successive iterations majority of the time, and that the factor of 10 used to determine when the pre-scaling weights is updated is a suitable choice. However, the weights can change significantly over time, and adjusting the weights is necessary.

5.2. QR3 Filter

We introduced the QR3 filter in order to reduce the number of complete QR-decompositions performed when filtering columns from QR in the quasi-Newton update. The new QR3 filter mimics the previous QR2 filter method in the way it detects which columns are to be deleted in V_k and W_k. However, it does not require a complete QR-decomposition in each quasi-Newton iteration, but only cheap QR-updates as long as (i) pre-scaling weights are unchanged and (ii) no column is tagged to be deleted. In the experiments in this section, we analyse the impact of the new filter in combination with pre-scaling weight monitoring on both the number of required quasi-Newton iterations and on the computational runtime. We compare the number of quasi-Newton iterations in Table 3 and the number of deleted columns in Table 4 for the QR2 filter while computing new pre-scaling weights in each iteration, and the new QR3 filter with pre-scaling weight monitoring.

Table 3. Comparison of the average number of quasi-Newton iterations per time step for (i) pre-scaling update in every quasi-Newton iteration with the QR2 filter (column "QR2"), and (ii) the new pre-scaling weight monitoring approach with the new QR3 filter (column "QR3"). Three different filter limits $\epsilon_f = 0.001$, $\epsilon_f = 0.01$, and $\epsilon_f = 0.1$ are compared. Values in brackets indicate how many time steps did not converge within 30 iterations before moving to the next time step.

Test Case	ζ	η	QR2 $\epsilon_f = 0.001$	0.01	0.1	QR3 $\epsilon_f = 0.001$	0.01	0.1
Elastic-Tube-3D-Heavy	10	100	4.26	4.51	5.12	4.59	4.48	5.24
Elastic-Tube-3D-Heavy	20	200	4.09	3.84	4.91	4.05	3.94	3.92
Elastic-Tube-3D-Light	10	100	7.27	7.23	8.83	7.18	7.16	8.69
Elastic-Tube-3D-Light	20	200	5.78	5.84	7.88	5.83	5.83	7.67
Breaking-Dam-2D	10	100	div	7.08 (2)	5.87 (3)	12.5 (25)	7.63 (3)	5.76 (2)
Breaking-Dam-2D	20	200	12.12 (19)	8.10 (6)	6.11 (2)	div	8.26 (2)	5.74 (2)
Breaking-Dam-3D	10	100	4.45	4.34	4.21	4.8	4.33	4.31
Breaking-Dam-3D	20	200	4.51	4.30	4.15	4.51	4.15	4.33

Table 4. Average number of columns deleted per time step for (i) pre-scaling update in every quasi-Newton iteration with the QR2 filter (column "QR2"), and (ii) the new pre-scaling weight monitoring approach with the new QR3 filter (column "QR3"). Three different filter limits $\epsilon_f = 0.001$, $\epsilon_f = 0.01$, and $\epsilon_f = 0.1$ are compared.

Test Case	ζ	η	QR2 $\epsilon_f = 0.001$	0.01	0.1	QR3 $\epsilon_f = 0.001$	0.01	0.1
Elastic-Tube-3D-Heavy	10	100	0.01	0.1	2.01	0.01	0.03	1.61
Elastic-Tube-3D-Heavy	20	200	0.01	0.04	2.19	0.01	0.03	0.31
Elastic-Tube-3D-Light	10	100	0.01	0.04	2.67	0.01	0.01	2.88
Elastic-Tube-3D-Light	20	200	0.01	0.03	2.8	0.01	0.03	2.65
Breaking-Dam-2D	10	100	div	3.45	3.95	4.42	3.82	3.8
Breaking-Dam-2D	20	200	8.25	5.81	4.64	div	5.81	4.19
Breaking-Dam-3D	10	100	1.99	2.06	2.19	2.14	1.88	2.17
Breaking-Dam-3D	20	200	2.03	2.27	2.45	2.26	2.03	2.38

The number of quasi-Newton iterations as presented in Table 3 does not increase when using the new QR3 filter instead of the QR2 filter. It even decreases for 13 of the 24 simulations performed. The Breaking-Dam-2D case diverges for $\epsilon_f = 0.001$ for both filters. This is in-line with previous observations, that for scenarios that are prone to linear dependencies in V_k, the filter limit cannot be chosen too small [10,22]. Changing from the QR2 to the QR3 filter is, thus, not the source of this issue.

Table 4 shows that also the number of deleted columns is comparable for both filters. The Elastic-Tube-3D-Heavy test case has only one and three columns deleted for $\epsilon_f = 0.001$ and $\epsilon_f = 0.01$, respectively. The Breaking-Dam test cases have more columns deleted, approximately half of all iterations. Both filters are able to address this with similar efficiency.

For $\epsilon_f = 0.001$, the filtering criterion tends to remove fewer columns. This seems to be somewhat contradicted by the Breaking-Dam-3D case with the QR3 filter, where the number of deleted columns is larger for $\epsilon_f = 0.001$. However, the total number of iterations for $\epsilon_f = 0.001$ is also larger than for $\epsilon_f = 0.01$. Therefore, more QR filter checks are performed and more columns are deleted over the entire simulation runtime. The same happens for the Breaking-Dam-2D case, where the number of removed columns for $\epsilon_f = 0.01$ is larger than for $\epsilon_f = 0.1$.

The runtime of the QR filter (filtering of columns including QR decomposition time) as a percentage of the total simulation runtime is shown in Table 5. The total simulation runtime includes the total runtime of the solvers and preCICE including initialisation.

Table 5. QR decomposition and filtering time as a percentage of the total simulation runtime for (i) pre-scaling update in every quasi-Newton iteration with the QR2 filter (QR2), and (ii) the new pre-scaling weight monitoring approach with the new QR3 filter (QR3). Three filter limits $\epsilon_f = 0.001$, $\epsilon_f = 0.01$ and $\epsilon_f = 0.1$ were tested. The test cases used a maximum of $\eta = 100$ iterations from previous $\zeta = 10$ time steps, and $\eta = 200$ and $\zeta = 20$.

Test Case	ζ	η	QR2 $\epsilon_f = 0.001$	0.01	0.1	QR3 $\epsilon_f = 0.001$	0.01	0.1
Elastic-Tube-3D-Heavy	10	100	1.99%	2.48%	1.51%	0.06%	0.06%	0.62%
Elastic-Tube-3D-Heavy	20	200	6.67%	4.9%	3.18%	0.08%	0.09%	0.36%
Elastic-Tube-3D-Light	10	100	7.41%	7.23%	6.60%	0.15%	0.15%	2.42%
Elastic-Tube-3D-Light	20	200	12.99%	13.12%	13.20%	0.38%	0.43%	5.04%
Breaking-Dam-2D	10	100	div	4.69%	1.27%	3.21%	2.52%	0.85%
Breaking-Dam-2D	20	200	27.98%	9.72%	2.33%	div	5.75%	1.65%
Breaking-Dam-3D	10	100	0.61%	0.59%	0.54%	0.08%	0.03%	0.04%
Breaking-Dam-3D	20	200	1.75%	1.57%	1.44%	0.23%	0.09%	0.07%

Comparing the relative runtimes shown in Table 5, we see that, for every test case, the runtime of the new QR3 filter is significantly smaller than the runtime of the QR2 filter. More noticeable improvements occur for $\zeta = 20$ and $\eta = 200$ than for $\zeta = 10$ and $\eta = 100$, which is expected since the larger number of columns in V_k increases the cost of each complete QR-decomposition. The filtering accounts for a large percentage of the simulation runtime for the Breaking-Dam-2D case, as the solver meshes are rather small. Especially for $\zeta = 20$ and $\eta = 200$ and the QR2 filter with limit $\epsilon_f = 0.001$, the QR-decompositions are relatively expensive. The Elastic-Tube-3D-Light scenario spends up to 13.20% of the simulation runtime filtering the QR2 filter with $\epsilon_f = 0.1$. The largest improvement in runtime performance is approximately 12.5% of the simulation runtime for the Elastic-Tube-3D-Light with $\zeta = 20$ and $\eta = 200$.

Also in terms of runtime and runtime gains, we observe large differences between Breaking-Dam-2D and Breaking-Dam-3D scenarios, which can, however, be easily explained as the 2D problem has a much smaller domain in terms of size and number of elements in the domain. Each fluid solver call is relatively "cheap" computationally, and therefore the relative cost of the QR decomposition increases.

Note that these runtime comparisons represent the overall gain of both improvements, pre-scaling weight monitoring and the QR3 filter as the QR2 filter is not able to exploit the advantage of pre-scaling weight monitoring, that is, it always re-computes the complete QR-decomposition of V_k independent of whether weight updates are required or not.

6. Conclusions

We provide an overview of current quasi-Newton acceleration methods commonly used for partitioned fluid-structure interaction. We also discuss numerical techniques that improve the computational efficiency and enhance convergence of quasi-Newton acceleration. From this, two new methods are introduced to further improve the computational efficiency of the acceleration: pre-scaling weight monitoring and a faster QR filtering procedure. The combination of both methods allows us to significantly reduce the necessary number of computationally expensive QR-decompositions. Instead, we can simply update existing QR decompositions in most acceleration steps—a rather cheap operation. We study the effectiveness of these new methods with three common fluid-structure interaction test cases, each offering a unique coupling difficulty. The new methods reduce the runtime of the QR filter significantly, while not decreasing the convergence speed. A small, but already significant speed up for complete fluid-structure interaction simulations is observable. This is true despite the fact that there are drastically fewer degrees of freedom at the coupling interface compared to the solver domains for such surface-coupled problems. For volume-coupled problems, we expect an even higher impact of the newly introduced methods.

Author Contributions: Conceptualization, K.D.; sata curation, K.D.; formal analysis, K.D.; funding acquisition, M.S. and B.U.; investigation, K.D.; methodology, K.D.; project administration, M.S.; software, K.D. and B.U.; supervision, M.S. and B.U.; validation, K.D.; visualization, K.D.; writing—original draft, K.D.; writing—review and editing, K.D., M.S. and B.U. All authors have read and agreed to the published version of the manuscript.

Funding: This research was funded by Deutsche Forschungsgemeinschaft (DFG, German Research Foundation) under grant number ME 2067/2-1 and under Germany's Excellence Strategy—EXC 2075—390740016.

Data Availability Statement: Publicly available datasets were analyzed in this study. This data can be found at: preCICE Repository: https://github.com/precice/precice 15 February 2022, preCICE Tutorials Repository: https://github.com/precice/tutorials 15 February 2022, Test cases and Configuration Files: https://github.com/KyleDavisSA/IQN-test-cases 15 February 2022.

Acknowledgments: We would like to thank the preCICE development team, and all those who have helped develop preCICE. We acknowledge the support by the Stuttgart Center for Simulation Science (SimTech).

Conflicts of Interest: The funders had no role in the design of the study, in the collection, analyses, or interpretation of data; in the writing of the manuscript, or in the decision to publish the results.

References

1. Grognuz, J. A New Heart Valve Replacement Procedure Modeled with Multiphysics Simulation Could Eliminate the Need for Open-Heart Surgery. Available online: https://www.enginsoft.com/expertise/a-new-heart-valve-replacement-procedure.html (accessed on 2 February 2021).
2. Jain, R.K.; Saha, P. *Fluid-Structure Interaction Simulations Prove Ability of Solar Artifacts to Withstand Wind Gusts*; ANSYS Inc.: Canonsburg, PA, USA, 2021. Available online: https://www.ansys.com/content/dam/product/3d-design/aim/csir-cmeri-cs.pdf (accessed on 15 February 2022).
3. Schmidt, P.; Jaust, A.; Steeb, H.; Schulte, M. Simulation of flow in deformable fractures using a quasi-Newton based partitioned coupling approach. *Comput. Geosci.* **2022**, *26*, 381–400. [CrossRef]
4. Bungartz, H.J.; Lindner, F.; Gatzhammer, B.; Mehl, M.; Scheufele, K.; Shukaev, A.; Uekermann, B. preCICE—A fully parallel library for multi-physics surface coupling. *Comput. Fluids* **2016**, *141*, 250–258. [CrossRef]
5. Slattery, S.; Wilson, P.P.H.; Pawlowski, R.P. The data transfer kit: A geometric rendezvous-based tool for multiphysics data transfer. In Proceedings of the International Conference on Mathematics and Computational Methods Applied to Nuclear Science and Engineering (M&C 2013), Sun Valley, ID, USA, 5–9 May 2013.
6. Duchaine, F.; Jauré, S.; Poitou, D.; Quémerais, E.; Staffelbach, G.; Morel, T.; Gicquel, L. Analysis of high performance conjugate heat transfer with the OpenPALM coupler. *Comput. Sci. Discov.* **2015**, *8*, 015003. [CrossRef]
7. Chourdakis, G.; Davis, K.; Rodenberg, B.; Schulte, M.; Simonis, F.; Uekermann, B. preCICE V2: A Sustainable and User-Friendly Coupling Library. *arXiv* **2021**, arXiv:2109.14470.
8. Degroote, J.; Bathe, K.J.; Vierendeels, J. Performance of a new partitioned procedure versus a monolithic procedure in fluid-structure interaction. *Comput. Struct.* **2009**, *87*, 793–801. [CrossRef]

9. Bogaers, A.E.J.; Kok, S.; Reddy, B.D.; Franz, T. Quasi-Newton methods for implicit black-box FSI coupling. *Comput. Methods Appl. Mech. Eng.* **2014**, *279*, 113–132. [CrossRef]
10. Haelterman, R.; Bogaers, A.; Uekermann, B.; Scheufele, K.; Mehl, M. Improving the performance of the partitioned QN-ILS procedure for fluid-structure interaction problems: Filtering. *Comput. Struct.* **2016**, *171*, 9–17. [CrossRef]
11. Mehl, M.; Uekermann, B.; Bijl, H.; Blom, D.; Gatzhammer, B.; van Zuijlen, A. Parallel coupling numerics for partitioned fluid–structure interaction simulations. *Comput. Math. Appl.* **2016**, *71*, 869–891. [CrossRef]
12. Scheufele, K.; Mehl, M. Robust multisecant Quasi-Newton variants for parallel fluid-structure simulations and other multiphysics applications. *SIAM J. Sci. Comput.* **2016**, *39*, 404–433. [CrossRef]
13. Spenke, T.; Hosters, N.; Behr, M. A Multi-Vector Interface Quasi-Newton Method with Linear Complexity for Partitioned Fluid-Structure Interaction. *Comput. Methods Appl. Mech. Eng.* **2020**, *361*, 112810. [CrossRef]
14. Anderson, D.G. Iterative procedures for nonlinear integral equations. *J. ACM* **1965**, *12*, 547–560. [CrossRef]
15. Miller, K. Nonlinear krylov and moving nodes in the method of lines. *J. Comput. Appl. Math.* **2005**, *183*, 275–287. [CrossRef]
16. Ni, P. Anderson Acceleration of Fixed-Point Iteration with Applications to Electronic Structure Computations. Ph.D. Thesis, Worcester Polytechnic Institute, Worcester, MA, USA, 13 November 2009. Available online: https://www.semanticscholar.org/paper/Anderson-Acceleration-of-Fixed-point-Iteration-with-Ni/8ca4703c5ec5c4580950a9c5c806604a595db3cb (accessed on accessed on 15 February 2022).
17. Oosterlee, C.W.; Washio, T. Krylov subspace acceleration of nonlinear multigrid with application to recirculating flows. *SIAM J. Sci. Comput.* **2000**, *21*, 1670–1690. [CrossRef]
18. Risseeuw, D. Fluid Structure Interaction Modelling of Flapping Wings. Master's Thesis, Delft University of Technology, Delft, The Netherlands, 2019.
19. Bungartz, H.J.; Lindner, F.; Mehl, M.; Uekermann, B. A plug-and-play coupling approach for parallel multi-field simulations. *Comput. Mech.* **2015**, *55*, 1119–1129. [CrossRef]
20. Lindner, F.; Mehl, M.; Scheufele, K.; Uekermann, B. A Comparison of various Quasi-Newton Schemes for Partitioned Fluid-Structure Interaction. In Proceedings of the VI International Conference on Computational Methods for Coupled Problems in Science and Engineering, Venice, Italy, 18–20 May 2015; pp. 477–488.
21. Scheufele, K. Coupling Schemes and Inexact Newton for Multi-Physics and Coupled Optimization Problems. Ph.D. Thesis, University of Stuttgart, Stuttgart, Germany, 2019.
22. Uekermann, B. Partitioned Fluid-Structure Interaction on Massively Parallel Systems. Ph.D. Thesis, Technical University of Munich, Munich, Germany, 2016.
23. Daniel, J.W.; Gragg, W.B.; Kaufman, L.; Stewart, G. Reorthogonalization and stable algorithms for updating the Gram–Schmidt QR factorization. *Math. Comput.* **1976**, *30*, 772–795. [CrossRef]
24. Marks, L.; Luke, D. Robust mixing for ab initio quantum mechanical calculations. *Phys. Rev. B* **2008**, *78*, 075114. [CrossRef]
25. Förster, C.; Wall, W.A.; Ramm, E. Artificial added mass instabilities in sequential staggered coupling of nonlinear structures and incompressible viscous flows. *Comput. Methods Appl. Mech. Eng.* **2007**, *196*, 1278–1293. [CrossRef]
26. Van Brummelen, E.H. Added Mass Effects of Compressible and Incompressible Flows in Fluid-Structure Interaction. *J. Appl. Mech.* **2009**, *76*, 021206. [CrossRef]
27. Gerbeau, J.F.; Vidrascu, M. *A Quasi-Newton Algorithm Based on a Reduced Model for Fluid-Structure Interaction Problems in Blood Flows*; [Research Report] RR-4691, INRIA; Cambridge University Press: Cambridge, UK, 2003.
28. Bogaers, A.E.J.; Kok, S.; Reddy, B.D.; Franz, T. An evaluation of quasi-Newton methods for application to FSI problems involving free surface flow and solid body contact. *Comput. Struct.* **2016**, *173*, 71–83. [CrossRef]
29. Walhorn, E.; Kölke, A.; Hübner, B.; Dinkler, D. Fluid–structure coupling within a monolithic model involving free surface flows. *Comput. Struct.* **2005**, *83*, 2100–2111. [CrossRef]
30. Weller, H.G.; Tabor, G.; Jasak, H.; Fureby, C. A tensorial approach to computational continuum mechanics using object-oriented techniques. *Comput. Phys.* **1998**, *12*, 620–631. [CrossRef]
31. Dhondt, G. *The Finite Element Method for Three-Dimensional Thermomechanical Applications*; John Wiley and Sons: Hoboken, NJ, USA, 2004.

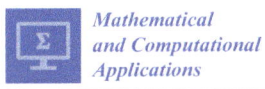

Article

A Computational Fluid Dynamics-Based Model for Assessing Rupture Risk in Cerebral Arteries with Varying Aneurysm Sizes

Rohan Singla [1], Shubham Gupta [1] and Arnab Chanda [1,2,*]

[1] Centre for Biomedical Engineering, Indian Institute of Technology (IIT), Delhi 110016, India
[2] Department of Biomedical Engineering, All India Institute of Medical Sciences (AIIMS), Delhi 110029, India
* Correspondence: arnab.chanda@cbme.iitd.ac.in

Abstract: A cerebral aneurysm is a medical condition where a cerebral artery can burst under adverse pressure conditions. A 20% mortality rate and additional 30 to 40% morbidity rate have been reported for patients suffering from the rupture of aneurysms. In addition to wall shear stress, input jets, induced pressure, and complicated and unstable flow patterns are other important parameters associated with a clinical history of aneurysm ruptures. In this study, the anterior cerebral artery (ACA) was modeled using image segmentation and then rebuilt with aneurysms at locations vulnerable to aneurysm growth. To simulate various aneurysm growth stages, five aneurysm sizes and two wall thicknesses were taken into consideration. In order to simulate realistic pressure loading conditions for the anterior cerebral arteries, inlet velocity and outlet pressure were used. The pressure, wall shear stress, and flow velocity distributions were then evaluated in order to predict the risk of rupture. A low-wall shear stress-based rupture scenario was created using a smaller aneurysm and thinner walls, which enhanced pressure, shear stress, and flow velocity. Additionally, aneurysms with a 4 mm diameter and a thin wall had increased rupture risks, particularly at specific boundary conditions. It is believed that the findings of this study will help physicians predict rupture risk according to aneurysm diameters and make early treatment decisions.

Keywords: cerebral; hemodynamics; aneurysm; rupture; artery; CFD

Citation: Singla, R.; Gupta, S.; Chanda, A. A Computational Fluid Dynamics-Based Model for Assessing Rupture Risk in Cerebral Arteries with Varying Aneurysm Sizes. *Math. Comput. Appl.* **2023**, *28*, 90. https://doi.org/10.3390/mca28040090

Academic Editor: Fábio Fernandes

Received: 18 May 2023
Revised: 3 July 2023
Accepted: 11 July 2023
Published: 2 August 2023

Copyright: © 2023 by the authors. Licensee MDPI, Basel, Switzerland. This article is an open access article distributed under the terms and conditions of the Creative Commons Attribution (CC BY) license (https://creativecommons.org/licenses/by/4.0/).

1. Introduction

A cerebral aneurysm is a weak or thin region of an arterial section in the human brain that balloons or bulges out when filled up with blood. The bulging aneurysm may compress the nerves or brain tissue. Additionally, this may lead to rupture or bursting, causing blood to flow into adjacent tissues (called a hemorrhage). A ruptured aneurysm can cause serious health problems such as hemorrhagic stroke, brain damage, unconsciousness, and even death [1,2]. The adult population within the range of 30 to 60 years of age is the most vulnerable to aneurysms, and women are more likely than men to have one. A brain aneurysm rupture affects roughly 6.7 million people in the USA each year, and the most serious aneurysm complication is rupture, which affects 2–4% of the global population [3]. In addition to this, 500,000 deaths related to aneurysm rupture are reported worldwide [4]. These problems are also prevalent in Europe, southern Asia, and Eastern Asia, with death rates of approx. 41%, 46%, and 60%, respectively [5]. In a recent study by Bechstein et al. [6], it was reported that Mongolia reported a crude incidence of 14.53 ruptures per 100,000 persons.

Hence, understanding aneurysm development and its progression is essential, which is the aim and rationale for this study.

Fukazawa et al. [7] performed a computational study to investigate the rupture risk of middle cerebral artery (MCA) aneurysms. They estimated the points with a high risk of rupture on the basis of the dynamic blood flow. Using three-dimensional CT scans and computational fluid dynamics (CFD), the aneurysm shapes were reconstructed, and it was found that the wall shear stress had a substantial impact on aneurysm rupture. Geers et al. [8] performed steady-flow simulations of the hemodynamics of cerebral aneurysms to evaluate wall shear stress (WSS) and examined the time-averaged (TA), peak systole (PS), and end-diastole (ED) WSS fields to compare simulations of steady flow and pulsatile flow. Cebral et al. [9] investigated pulsatile flow CFD models with patient-specific models. They studied 210 consecutive cerebral aneurysms and observed that concentrated inlet jets, narrow impingement zones, and complex and unstable flow patterns were all related to a clinical history of previous aneurysm rupture. This was performed using image-based, patient-specific geometry and qualitative hemodynamic analysis. These qualitative evaluations serve as a springboard for even more complex quantitative analyses intended to calculate the likelihood that an aneurysm will collapse later.

John et al. [10] characterized the various flow types found in anterior communicating artery aneurysms. Individual patient computational models were created, and simulations were performed to evaluate the risk of rupture. The findings demonstrated that aneurysms with smaller impaction regions and more blood entering the aneurysm were more likely to rupture. The results showed that aneurysms with smaller impaction areas and greater amounts of blood entering the aneurysm had a higher probability of rupture. Jeong et al. [11] compiled and discussed the probability of cerebral aneurysm initiation, growth, and rupture. The results demonstrated that the highest stress and displacement values were obtained by aneurysms that ruptured close to artery walls. Similarly, Luckrajh et al. [12] used CT angiograms of individuals with anterior communicating artery (ACoA) aneurysms and these statistics may act as baseline information of the morphological and morphometric features of ACoA aneurysms which should also be understood when organizing and carrying out aneurysm treatments. Using finite element analysis, Foutrakis et al. [9] investigated the hemodynamics of cerebral saccular aneurysm formation. Models of the curved artery section and arterial splitting at various stages were created to assess the development of aneurysms. The results demonstrated that pressure and shear stress generated across an artery's outer wall and at the leading edge of a capillary bifurcation increased the formation of saccular aneurysms. Torii et al. [13] created two brain aneurysms for the purpose of conducting fluid–structure interaction (FSI) simulations with hypertensive and normal blood pressure parameters. They looked at the distribution of wall shear stress (WSS) to completely comprehend an aneurysm. They noticed that the larger distribution of high WSS reduced wall weakening while increasing wall deformation-related alterations in flow patterns. Therefore, knowledge of pressure, wall shear, and blood flow is required to analyze an aneurysm rupture.

It is well known that one of the key causes of aneurysm rupture is hemodynamics. Aneurysm rupture is indicated by flow impingement, increased pressure, and unusual wall shear stress. Understanding the function of wall shear stress in cerebral aneurysms at comparable anatomic sites may be possible. These results indicate that CFD may be a key factor in the clinical assessment of aneurysm risks. There is an increasing need for accurate prediction of aneurysm growth and rupture in order to select the most appropriate and effective endovascular treatment. Concerning the possibility of aneurysm rupture, there is a huge knowledge gap. This study characterized the rupture risk by evaluating aneurysm growth and wall thickness during initiation in a real anterior cerebral artery (ACA). The formation of the aneurysm throughout its five stages was modeled by adjusting the aneurysm's diameter. Following an assessment of the rupture risk for each stage under hypertension settings and consideration of two wall thicknesses (0.075 mm and 0.15 mm), an analysis of both flow and pressure distributions on the artery-aneurysm models was conducted. The findings of this study are anticipated to open up novel possibilities for understanding the development of aneurysms and their risk of perforation under situations of variable aneurysm diameters and wall thicknesses.

The novel aspects of our work include the consideration of different aneurysm sizes on actual MRI-scanned artery models to study the behavior of fluids (i.e., blood), compared to the plethora of previous studies that over-simplified the artery geometry as a hollow cylindrical tube. Our study showcases economical modeling methods using realistic scanned geometries.

2. Materials and Methods

2.1. Selection of Artery and Aneurysm

As compared to the posterior lobes, the frontal lobes receive more blood from the anterior cerebral arteries. Due to this, anterior arteries are more vulnerable to induced pressures hence, it is important to understand the disorders that are associated with the anterior region. One of the most frequent locations is the anterior cerebral artery (ACA) of intracranial aneurysms which has been reported as a significant site for ruptures. Over 92% of the total rupture cases have been reported due to an aneurysm at this location [2,12,14]. Data shows that brain arteries with curvature or bifurcation are prone to develop or spread aneurysms [11]. Hence, according to statistical data and literature review, we found that ACA is more prominent and vulnerable to diseases. Hence our study focused on this area as discussed in the following sections.

2.2. Geometrical Modeling

The initial steps in the geometric modeling of the arteries included recreation of the segmented MRI data. To develop geometric models of aneurysm development, the cerebral artery was further divided and altered. The thorough technique is explained in the ensuing subsections.

2.2.1. Preparing Arterial 3-D Geometry and Modeling

To create a three-dimensional model of the human cerebral arteries, MRI scans of the brain were taken from the NIH Visible Human database. The 3-D geometry of the cerebral artery network was obtained by segmenting and reconstructing the MRI data using TurtleSeg software (Figure 1A). It is evident from the literature review that aneurysm genesis or progression is more frequent in cerebral arteries that curve or split. As a result, specific parts of the anterior cerebral artery, which are reported to be at risk of aneurysm development, were dissected from the full arterial network using the MeshLab program (Figure 1B). The surface features of the cut-out cerebral arteries were then smoothened, and all the discrepancies were resolved using the 'stitch and remove' command. The ACA sections were then assigned a diameter of 2 mm and a thickness of 0.2 mm (i.e., 10% of the diameter [15]).

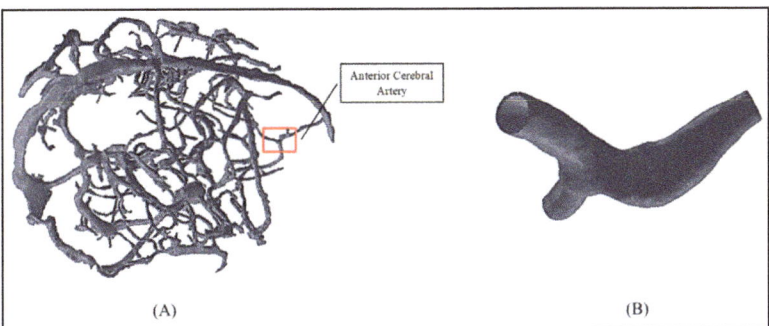

Figure 1. (**A**) Network of cerebral veins and arteries; (**B**) ACA section.

2.2.2. Modeling of Aneurysms

In order to simulate the aneurysm on the cut-out portion of the cerebral arteries, the optimal aneurysm positions on the arterial segment were determined. After that, a cavity was built where the aneurysm was to be placed. The geometry was then exported in '.STEP' format using Solidworks (Dassault Systèmes, Vélizy-Villacoublay, France). and then imported to ANSYS Workbench 2020R1 (Canonsburg, PA, USA) for analysis using the computational fluid dynamics (CFD) technique.

In the current work, the shape of the aneurysms was considered spherical due to their commonality. In a previous study, MRI scans from 84 patients reported a spherical shape of the aneurysms [14]. Hence, on the basis of these data, we considered the shape of the aneurysm to be spherical. In this work, we simulated aneurysm growth through the development of five different models with different diameters, where each model with a fixed size represented a particular aneurysm growth stage. Additionally, two different wall thicknesses were simulated for all five models. The five models were then used to understand and evaluate the nature of blood flow and differences in wall shear stresses and pressures among all the stages of aneurysm growth [16]. For the ACA component, aneurysms with sizes of 2 mm, 4 mm, 6 mm, 8 mm, and 10 mm were modeled. In a previous study, the aneurysms were reported to have walls that ranged in thickness from 0.01 to 0.216 mm [17]. As a result, in this work, two aneurysm thicknesses—0.075 mm and 0.15 mm—were taken into account for the arterial sections [18]. The models developed for the ACA segment of the cerebral arteries have different aneurysm dimensions, as shown in Figure 2A–E. As our study also includes the effect of wall thickness, a fluid domain was created in each of the respective artery models using both Fusion 360 and SolidWorks. As the artery was developed using MRI technology, it had a large number of faces and edges. The STL format was used to resolve the small edges and faces, which added complexity to the computational domain. So, to overcome this, an optimal STL was obtained using Fusion 360 by reducing the model's number of faces and edges in several steps and then exporting the file in STEP format. The STEP file was imported into SolidWorks for the development of the fluid domain and was finally assembled.

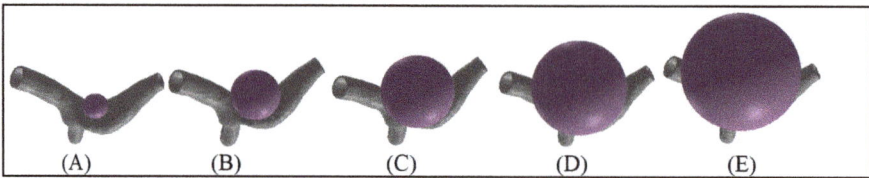

Figure 2. Aneurysms of different diameters that simulate growth in ACA: (**A**) 2 mm, (**B**) 4 mm, (**C**) 6 mm, (**D**) 8 mm, and (**E**) 10 mm.

2.3. Finite Element Modeling

The mesh convergence analysis of the models was conducted using ANSYS Workbench. The Solid-187 elements were used to combine the aneurysm and artery models. Higher-order, three-dimensional, ten-node elements with quadratic displacement behavior were chosen, similar to prior computational studies [19,20]. Surface-to-surface contact was applied between aneurysm and artery models. The method involving bonding contact pairs at all times was adopted [21,22]. A mesh adaptive technique was used to ensure an optimal mesh with low skewness and high orthogonality. Six meshes were produced for each artery model by altering the element size and spacing; 30,867, 40,560, 45,993, 78,831, 123,634, and 420,125 were the generated number of mesh elements for the artery model. Additionally, the residuals also converged to a convergence value of 10^{-6}. Figure 3 represents the meshed model of a single artery, including the fluid domain.

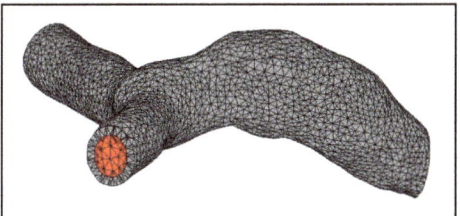

Figure 3. Meshed artery model of the ACA section.

Fluid Properties, Loads, and Boundary Conditions

From the literature, it was determined that the blood density was 1070 kg/m^3 and the viscosity was 0.004 kg/m·s [23]; hence, these values were used in the current work. We presumptively used Newtonian laminar flow, which is an incompressible fluid flow. Furthermore, the lumen's wall was considered rigid. Inlet boundary conditions were applied to the elements on the face of the inlet. At the inlet, the blood velocity was adopted from the literature [23]. A constant velocity of 0.10 m/s was assigned at the inlet. The operating pressure was set to atmospheric pressure, i.e., 101,325 Pa, and the reference pressure value was set to zero. The rigid wall's properties were modeled with a Young's modulus of 10 MPa, a Poisson's ratio of 0.49, and a density of 10 MPa. These values were based on a recent study by Khe et al. [24]. As per a previous study [25], the zero-pressure boundary condition indicated zero pressure as the bleeding was reported without considering any impacts from the vessels outside the simulation area. Therefore, in our study, the outlet gauge pressure was given the static pressure value of 0.024 MPa, corresponding to 180 mmHg, and was used to simulate the pressure circumstances associated with hypertension. Figure 4 shows the inlet and outlet flows.

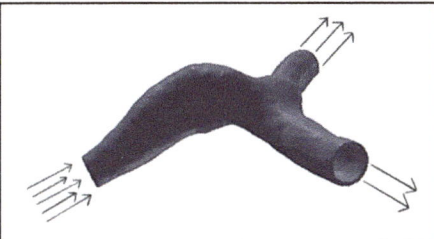

Figure 4. Annotation of the inlet and outlet section for artery ACA.

3. Results

3.1. Results of Mesh Convergence

For the mesh convergence investigation, the produced meshes were subjected to an input flow velocity of 0.1 m/s. The highest pressure introduced in the mesh-based designs was evaluated across consecutive meshes in order to determine the ideal number of mesh elements. The ACA section's maximum pressure variation was within a tolerable limit (i.e., 5%) for the pressure after the mesh elements crossed 123,634 elements (Figure 5A). As a result, the mesh with 123,634 elements was found to be optimal for the ACA segment. Figure 5B shows the change in area-weighted average pressure with respect to the number of iterations. A constant line was reported, which showed converged results for the corresponding meshed models. The mesh independence study revealed that considering a more refined mesh does not show any significant differences as compared to coarsely meshed structures. By adopting a number of elements 123,634, the net mass flow was achieved to a convergence value of 10^{-7}. Additionally, the residuals also converged to a convergence value of 10^{-6}. The developed computational framework took 70 iterations to reach convergence.

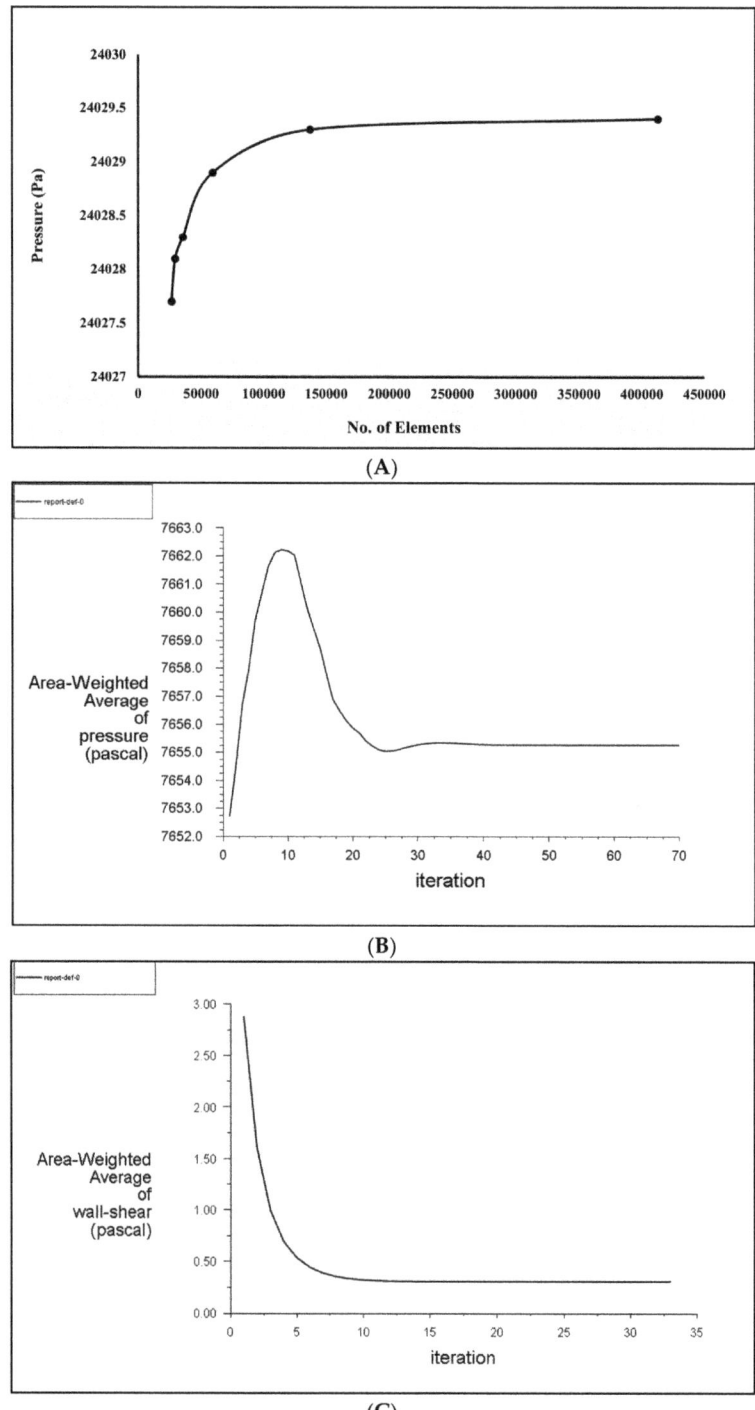

Figure 5. (**A**) Mesh convergence results for ACA; (**B**) CFD result convergence; (**C**) plot of surface report for wall shear stress result vs. iteration.

3.2. Distributions of Stress at Various Aneurysm Progression Levels

Figure 6 shows the pressure distributions for the ACA's 0.075 mm (Figure 6a) and 0.15 mm (Figure 6b) aneurysm thickness models. In line with previous computational studies on hemodynamics, our study showed similar static pressure values across all the models [7,11,23]. On the arterial and aneurysmal structures, it was found that the pressure concentration was not uniform in all the models. Near the artery-aneurysm's neck region, there was a concentration of pressure, and the aneurysm's neck and aneurysm showed homogeneous pressure ranges. According to the literature [26], the pressure range observed in the MCA was 22,000.00 Pa–26,287.00 Pa. Hence, the pressure results of our study are in the same order as the literature results. For various models with 0.075 mm thickness, the highest pressure varied between 24,005.00 Pa and 24,035.00 Pa, while for models with 0.15 mm thickness, it varied from 24,007.00 Pa to 24,032.00 Pa. Evaluation of the pressure data showed that variations in aneurysm sizes and in boundary thicknesses significantly affect the pressure distribution. In summary, the findings showed considerable influence on the aneurysm's structural characteristics, including its width and thickness. It was noticed that the pressure decreased from 24,031.97 to 24,009.75 as the aneurysm's size increased from 2 mm to 10 mm in the ACA artery with a wall thickness of 0.15 mm. In the ACA artery with a wall thickness of 0.075 mm, the pressure first increased from 24,029.24 to 24,034.5 for an aneurysm diameter of 2 mm to 4 mm, and then further decreased from 24,024.15 to 24,008.75 with the increase in diameter. Also, a strong correlation R^2 was found (i.e., 0.97 and 0.8294) between pressure and a changing diameter of the aneurysm for an artery with walls of thickness 0.15 mm and 0.075 mm, respectively.

3.3. Distributions of Wall Shear Stress at Varying Aneurysm Progression Stages

Figure 7 shows the wall shear stresses for the aneurysm with thicknesses of 0.075 mm (Figure 7a) and 0.15 mm (Figure 7b), respectively. In the figures, the red region signifies a wall shear stress of more than 2 Pa. A non-uniform WSS was observed throughout the artery. Overall, the findings showed considerable influence on the aneurysm's structural characteristics, including its diameter and thickness. As per previous studies, our study also exhibited similar results, i.e., in all model walls, shear stress was found to be zero at the aneurysm [9,27]. As the aneurysm's size rose from 2 mm to 10 mm in the ACA artery with a wall thickness of 0.15 mm, it was found that the wall shear stress dropped from 3.57 Pa to 0.5 Pa. With an aneurysm diameter of 2 mm to 4 mm, the wall shear stress in the ACA artery with a wall thickness of 0.075 mm first increased from 2.42 Pa to 5.83 Pa, and as the diameter increases, it continues to decrease from 1.68 Pa to 0.48 Pa.

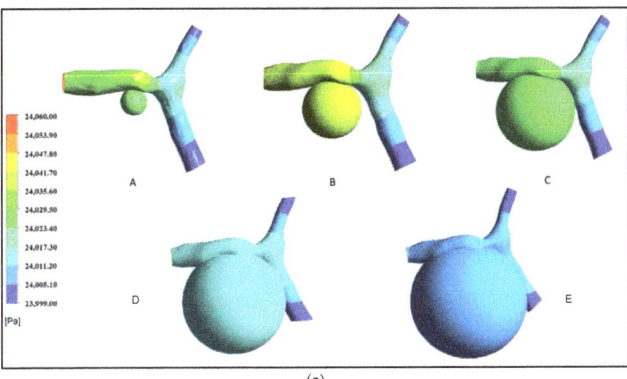

(a)

Figure 6. *Cont.*

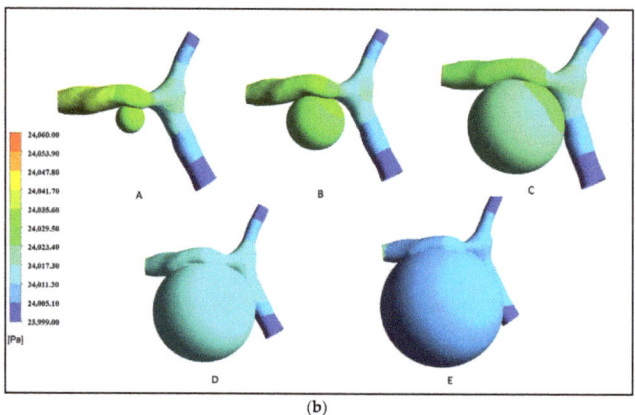

(b)

Figure 6. (**a**,**b**) Pressure contours of artery ACA of wall thickness 0.075 mm and 0.15 mm, respectively, at hypertension pressure conditions for different aneurysm diameters: (**A**) 2 mm, (**B**) 4 mm, (**C**) 6 mm, (**D**) 8 mm, and (**E**) 10 mm.

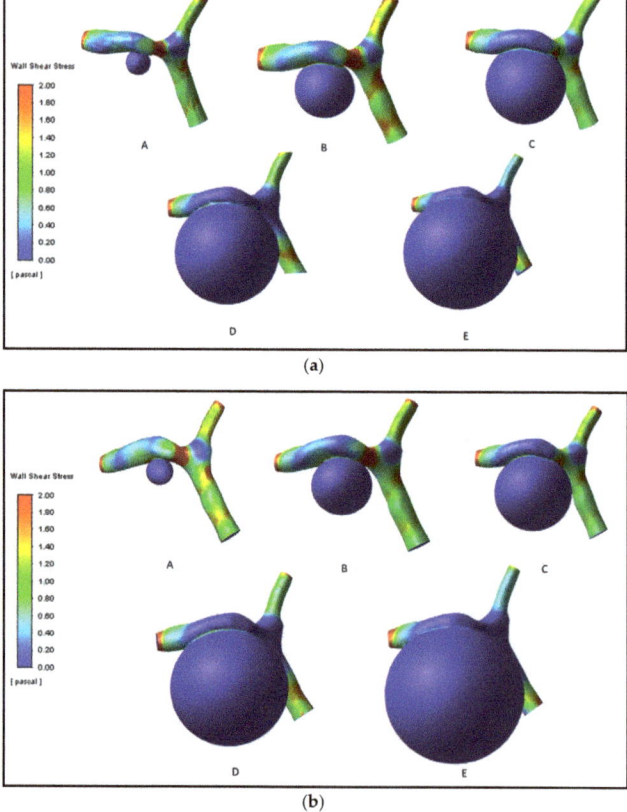

Figure 7. (**a**,**b**) Wall shear stress contours of artery ACA of wall thickness 0.075 mm and 0.15 mm, respectively, at hypertension pressure conditions for different aneurysm diameters: (**A**) 2 mm, (**B**) 4 mm, (**C**) 6 mm, (**D**) 8 mm, and (**E**) 10 mm.

3.4. Velocity Streamlines at Various Aneurysm Progression Stages

Our findings also showed that all rupture regions have low WSS magnitudes and slower flow rates. As per a previous study by Fukazawa et al. [7], it was reported that lower WSS could be a significant component contributing to the degenerative process ongoing in the aneurysm wall and aneurysm rupture, independent of the aneurysm morphology. Based on this, our study reported lower WSS at the apex of all the aneurysms. After the apex, the geometrical boundary joining the artery showed reduced WSS as compared to other regions of the artery. Hence, the low WSS and slow flow rates at these locations may suggest possible aneurysm ruptures. It has been suggested that the flow pattern may be a crucial element in the pathogenic process of aneurysm rupture [7,9,27]. In our study, the maximum streamline velocity was observed near the bifurcation of the arteries (Figure 8). The flow velocity in the ACA artery, which has a wall thickness of 0.075 mm, initially increased from 0.14 m/s to 0.15 m/s (Figure 8a) as the aneurysm diameter increased from 2 mm to 4 mm. However, as the diameter expanded further, it continued to decrease up to 0.01 m/s. The velocity also decreased from 0.15 m/s to 0.01 m/s when the aneurysm's diameter was increased from 2 mm to 10 mm in the ACA artery with a 0.15 mm thick wall (Figure 8b). Furthermore, there was a significant association (i.e., R^2 of 0.96 and 0.85) between pressure and aneurysm diameter for an arterial wall thickness of 0.150 mm and 0.075 mm, respectively.

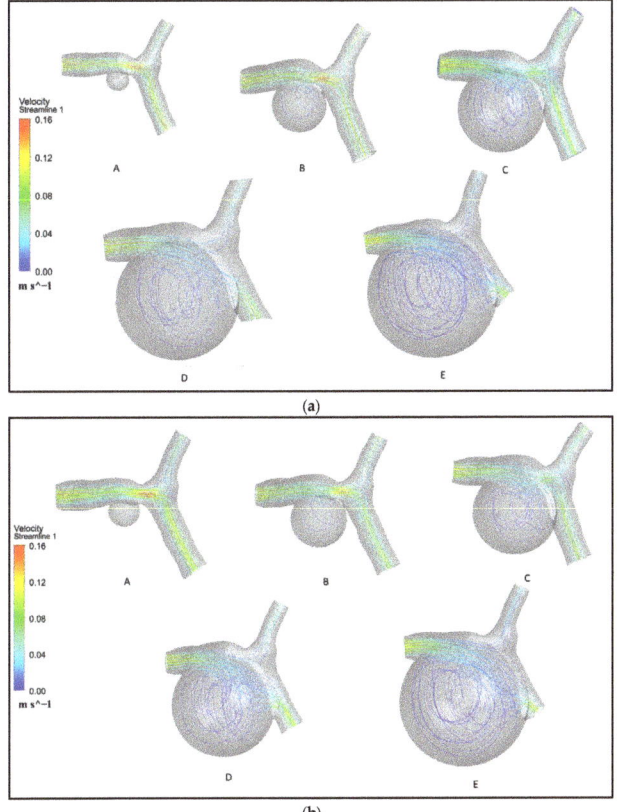

Figure 8. (**a**,**b**) Velocity streamlines contours of artery ACA of wall thickness 0.075 mm and 0.15 mm, respectively, at hypertension pressure conditions for different aneurysm diameters: (**A**) 2 mm, (**B**) 4 mm, (**C**) 6 mm, (**D**) 8 mm, and (**E**) 10 mm.

4. Discussion

This study has demonstrated that steady-flow simulations can accurately approximate the wall shear stress, pressure, and velocity streamline field of an aneurysm as well as an artery. Figure 9 displays the changes in aneurysm pressure and average neck pressure as well as the correlation between pressure and diameter for aneurysms having thicknesses of 0.15 mm and 0.075 mm. When wall tension is higher than the tissue's mechanical strength, aneurysms rupture. Pathological wall remodeling is directly related to the localized weakening of the aneurysm wall, which is characterized by the narrowing of the media and a lack of collagen fibers [28]. An artery with a thickness of 0.15 mm and the smallest diameter of 2 mm, as well as an artery with a wall thickness of 0.075 mm and a diameter of 4 mm, both experienced the highest pressure. These results are in line with previous studies. This research showed that an artery with a 4 mm diameter and a 0.075 mm wall thickness was the most vulnerable to aneurysm rupture. Additionally, the fluctuation in pressure with thickness is consistent with the literature [11], which asserts that the wall tension of a spherical aneurysm is inversely proportional to the wall thickness and directly proportional to the internal pressure and radius. As a result, high pressure, and a thin wall increase wall tension.

One of the important hemodynamic factors is WSS. The current analysis showed that the WSS at the rupture locations had a low magnitude. Therefore, these findings imply that reduced WSS would substantially lead to the continuous deterioration of the aneurysm wall and its rupture [7]. A 4 mm aneurysm demonstrated the peak wall shear stress in an artery with a 0.15 mm wall thickness, while a 4 mm aneurysm demonstrated the maximum wall shear stress in an artery with a 0.075 mm wall thickness. The opposite side of the aneurysm in both situations had a critical area with wall thicknesses of 0.075 mm and 0.15 mm in the artery and a 4 mm aneurysm diameter. This region had a high value of WSS and pressure close to the aneurysm bifurcation, raising the risk of aneurysm development at that location. Additionally, our observations and results were in line with those of previous research [11], which demonstrated that low WSS and modified flow patterns might have a long-term influence on aneurysm wall deterioration via wall reconfiguration.

A flow pattern analysis based on 3D streamlines of intra-aneurysmal flow and cross-sectional flow was performed, which revealed a complex flow structure and pattern. We discovered that the artery with a wall thickness of 0.15 and the lowest diameter, or one with a wall thickness of 0.075 and a diameter of 4 mm, produced the highest pressures when observing the streamline velocity in the artery. Based on this study, the most vulnerable aneurysm that can rupture was observed to have a diameter of 4 mm and a wall thickness of 0.075 mm. These results are in line with previous studies [9]. A ruptured aneurysm is more likely to have complicated and unstable flows, concentrated inputs, and tiny impingement zones. Figure 10 represents the correlation of velocity with diameter for wall thicknesses of 0.075 mm and 0.15 mm, respectively. These results are in line with previous studies, according to the statistical analysis, which also shows a substantial correlation between these qualitative hemodynamic features and aneurysm rupture. Even though diffuse inflows characterized the majority of unruptured aneurysms, many of them also featured complicated flows, unstable flows, and/or tiny impingement zones.

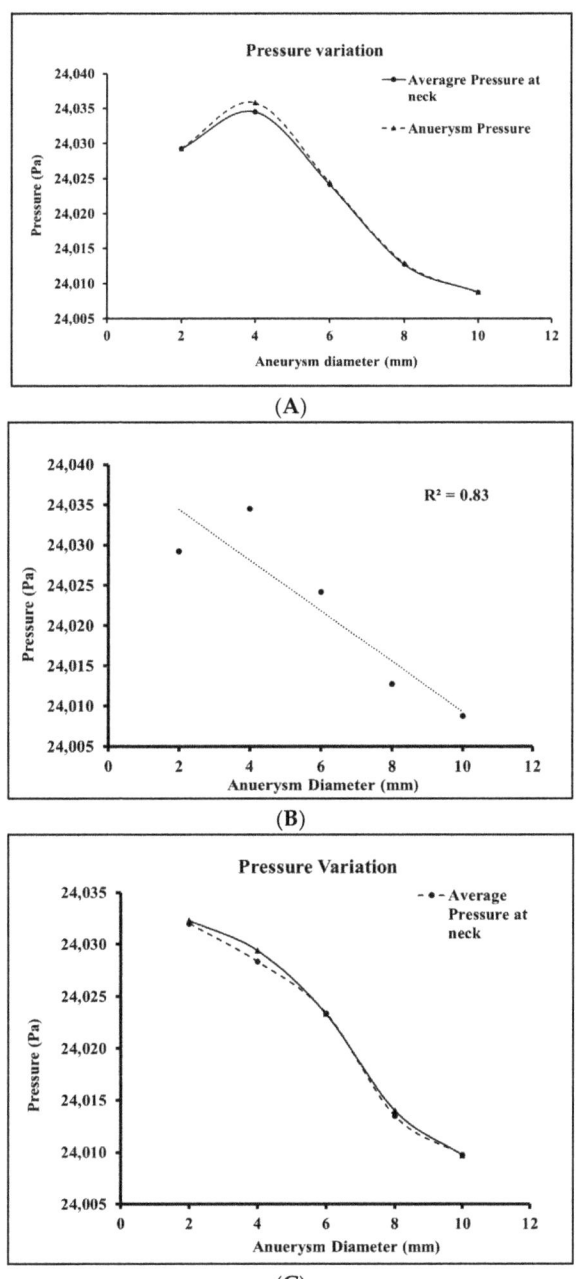

Figure 9. *Cont.*

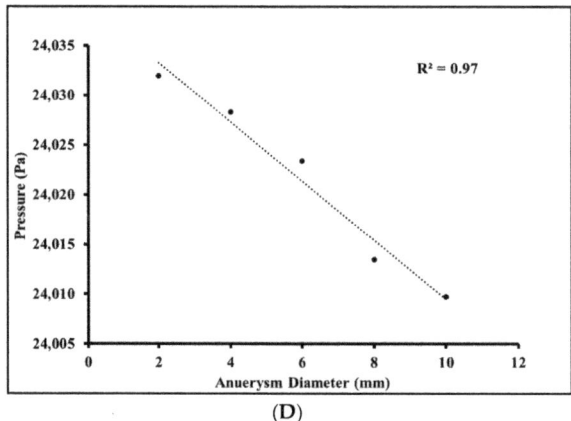

(D)

Figure 9. (**A**,**C**) Pressure at the aneurysm and average pressure at the neck variation with diameter of wall thickness 0.075 mm and 0.15 mm, respectively. (**B**,**D**) Correlation of pressure with diameter for wall thicknesses of 0.075 mm and 0.15 mm, respectively.

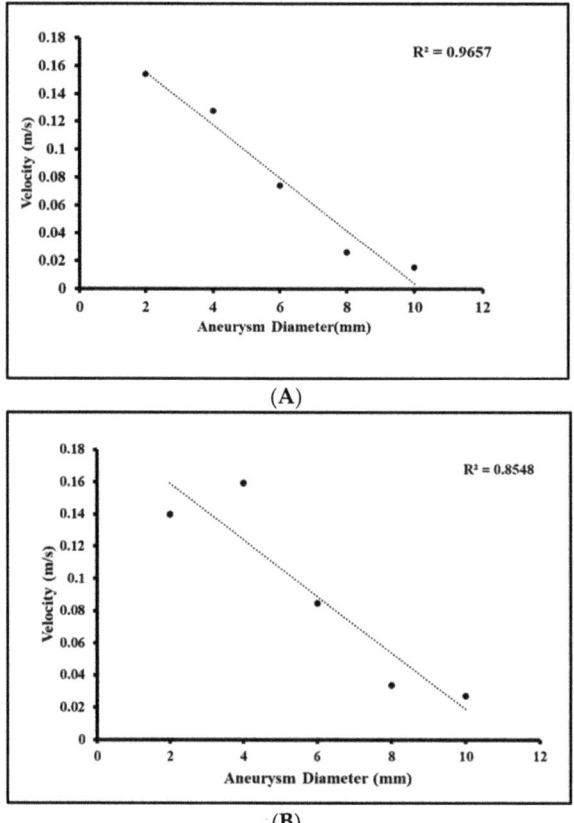

Figure 10. Correlation of velocity with diameter for wall thicknesses of (**A**) 0.075 mm and (**B**) 0.15 mm.

In past arterial flow studies, wall shear stresses have been extensively studied and reported as an important metric in determining thrombosis, aneurysms, minor clots, etc. In a previous study by Shojima et al. [29], 20 middle cerebral artery aneurysms were evaluated using CFD, and it was concluded that the WSS of the aneurysm region may be of some help for the prediction of rupture. In another study by Cebral et al. [9], a qualitative hemodynamic analysis of cerebral aneurysms was performed using images from 210 patients. The specific geometry, concentrated inflow jets, small impingement zones, complicated flow patterns, and unstable flow patterns were all found to be associated with a clinical history of past aneurysm rupture. The findings demonstrate the possibility that CFD could have a significant impact on the clinical assessment of aneurysm risks. Our study is in line with previous studies; it follows the same trend and also conforms to the range of values for WSS, pressure, and velocity. A total of 12 intracranial aneurysms (IAs), 8 ruptured and 4 unruptured, at the middle cerebral artery bifurcation were studied using FSI to better identify the characteristics of ruptured IAs [30]. In conclusion, they found that ruptured IAs had a larger low WSS area and more complex, concentrated, and unstable flow [31]. Computational models of six ruptured middle cerebral artery aneurysms with intraoperative confirmation of rupture points were constructed from three-dimensional rotational angiography images, and they were able to find the rupture points in all cases. With those findings, the local hemodynamics of ruptured aneurysms were quantitatively investigated. The rupture point was located in a low WSS region of the aneurysm wall [7]. Twelve ruptured middle cerebral artery bifurcation aneurysms were analyzed by three-dimensional computed tomographic angiography and CFD, and it was found that CFD may determine the rupture point of aneurysms using the feature of markedly low WSS [32]. It was found that ruptured aneurysms tend to have complex and/or unstable flow patterns, concentrated flow jets, and small impingement regions.

These assumptions were carefully taken into consideration by studying the past literature in this area. One of the assumptions, such as the consideration of human blood as Newtonian, was based on a previous study by Berger et al. [33]. The blood flow mechanics, along with their impact on the artery walls, were reviewed. It was mentioned that the consideration of blood as a Newtonian and laminar fluid was appropriate to understand the flow mechanics if the viscosity and density were accurately modeled. In larger arteries, the blood was treated as a Newtonian fluid. This assumption was found to be acceptable for most regions except those with a small strain rate; hence, no significant differences were reported for non-Newtonian flows in these arteries. Based on this study, we considered the viscosity and density of the blood to model the fluid computationally. Our work consisted of understanding the effect of different aneurysm sizes on fluid pressures, wall shear stresses, and blood velocities. Hence, the modeling of water, ions, proteins, nutrients, red and white blood cells, and platelets was not considered as it could have led to a much more complex flow system. The complex flow system could lead to difficulties in model convergence and, hence, reduced accuracy and precision of the CFD model. Moreover, the results from our work were comparable with a recent study by Berger et al. [33], which was based on the actual clinical setting.

Another assumption, such as the consideration of a lumen's wall as rigid, was based on a recent study by Humphrey et al. [34]. The lumen's wall was considered rigid, and modeling the multiple layers was reported as computationally expensive. These considerations were found to be significant at peak systolic pressures. Also, based on a previous study by Jou et al. [35], the aneurysm's mechanical properties were reported to be twice as stiff (due to a lack of elastin) as compared to other vessels; hence, it was considered rigid. In other studies, such as those by Shojima et al. [29] and Gao et al. [21], similar assumptions were considered to investigate the hemodynamics of an aneurysm by calculating the wall shear stress, wall pressures, and velocity streamlines. Other novel aspects of our work include the consideration of different aneurysm sizes on the actual MRI-scanned artery models to study the behavior of fluid (i.e., blood), whereas a plethora of previous studies oversimplified the artery geometry, considering it as a hollow cylindrical tube. Our study showcases economical modeling methods using realistic scanned geometries. To the best of our knowledge, these modeling methods and results have not been reported to date.

5. Conclusions

This study presented the development of a computational framework for modeling cerebral aneurysm growth and estimating rupture risk. By altering the aneurysm width and wall thickness in vulnerable arterial sections, such as the anterior cerebral artery, aneurysm growth was simulated. It was determined how blood pressure, aneurysm wall thickness, and aneurysm diameter affect the wall shear stresses at the aneurysm wall. According to this study, fluid–structure interaction can locate aneurysm rupture spots by utilizing distinctive flow dynamic properties, including complicated flow and noticeably low WSS. To evaluate the risk of rupture, a computational fluid dynamics-based model was incorporated to determine the effect of geometrical characteristics on the pressure gradients of an aneurysm. The results of this study may be used to estimate the risk of aneurysm rupture under varied aneurysm sizes and wall thicknesses. It is believed that this research will offer new perspectives on cerebral aneurysms and assist physicians and surgeons in formulating preventative methods to lessen aneurysm ruptures.

There are a few limitations to this work that must be acknowledged. Only spherical aneurysms with constant wall thicknesses were taken into account to develop artery-aneurysm models. It is known that certain aneurysms in real-world situations have irregular geometries and differing wall thicknesses at various aneurysm sites. The computational modeling of the artery and aneurysm structures in this work also made use of the isotropic material properties. Because of the fibers that are present in the skin layers, arteries exhibit anisotropic material characteristics. The consideration of the mechanical properties of the blood as Newtonian is a simplification based on studies by Berger et al. [33]. It was mentioned that the consideration of blood as a Newtonian and laminar fluid was appropriate to understand the flow mechanics if the viscosity and density were accurately modeled. In larger arteries, the blood was treated as a Newtonian fluid. This assumption was found to be acceptable for most regions, except in those with a small strain rate; hence, no significant differences were reported for non-Newtonian flows in these arteries. The lumen's wall was considered rigid since modeling the multiple layers was reported as computationally expensive, but this assumption was found to be significant at peak systolic pressures. Also, based on a previous study by Jou et al. [35], the aneurysm's mechanical properties were reported to have twice the stiffness due to a lack of elastin as compared to other vessels, and, hence, it was considered rigid. The artery wall was assumed to be rigid, and it was assigned as a solid domain with its material properties, such as density, Young's modulus, and Poisson ratio, in the Ansys Fluent solver. As per the Fluent Theory Guide [36], it has an intrinsic FSI function that takes into account the effect of changing thickness on the wall stiffness to some extent. Future studies considering FSI, actual aneurysm forms, wall thicknesses, and anisotropic material-based models can help enhance the overall accuracy and precision of the developed framework.

Author Contributions: Conceptualization, A.C. and S.G.; Data curation, R.S. and S.G.; Formal analysis, R.S., S.G. and A.C.; Investigation, S.G. and A.C.; Methodology, R.S., S.G. and A.C.; Project administration, S.G. and A.C.; Validation, R.S., S.G. and A.C.; Visualization, R.S., S.G. and A.C.; Writing—original draft, R.S. and S.G.; Writing—review and editing, R.S., S.G. and A.C. All authors have read and agreed to the published version of the manuscript.

Funding: This research received no external funding.

Data Availability Statement: The datasets generated and/or analyzed during the current study are not publicly available due to their large size but are available from the corresponding author on reasonable request.

Acknowledgments: This paper and the research would not have been possible without the exceptional support of A.C. and the facilities provided by the Indian Institute of Technology Delhi (IIT-D), India.

Conflicts of Interest: The authors declare no conflict of interest.

References

1. Han Seok, J.; Jang Han Seok, C. Current Science Management of Cerebral Aneurysm. *J. Mol. Pathophysiol.* **2021**, *10*, 1–2.
2. Mura, J.M.; Degaspari, S.; Spetzler, R.F.; Fennell, V.S.; Yashar, M.; Kalani, S.; Atwal, G.; Martirosyan, N.L. Biology of Saccular Cerebral Aneurysms: A Review of Current Understanding and Future Directions. *Front. Surg.* **2016**, *3*, 43. [CrossRef]
3. Vlak, M.H.M.; Algra, A.; Brandenburg, R.; Rinkel, G.J.E. Prevalence of unruptured intracranial aneurysms, with emphasis on sex, age, comorbidity, country, and time period: A systematic review and meta-analysis. *Lancet Neurol.* **2011**, *10*, 626–636. [CrossRef] [PubMed]
4. Statistics and Facts—Brain Aneurysm Foundation. Available online: https://www.bafound.org/statistics-and-facts/ (accessed on 25 June 2023).
5. Etminan, N.; Chang, H.S.; Hackenberg, K.; De Rooij, N.K.; Vergouwen, M.D.I.; Rinkel, G.J.E.; Algra, A. Worldwide Incidence of Aneurysmal Subarachnoid Hemorrhage According to Region, Time Period, Blood Pressure, and Smoking Prevalence in the Population: A Systematic Review and Meta-analysis. *JAMA Neurol.* **2019**, *76*, 588–597. [CrossRef]
6. Bechstein, M.; Gansukh, A.; Regzengombo, B.; Byambajav, O.; Meyer, L.; Schönfeld, M.; Kniep, H.; Hanning, U.; Broocks, G.; Gansukh, T.; et al. Risk Factors for Cerebral Aneurysm Rupture in Mongolia. *Clin. Neuroradiol.* **2022**, *32*, 499–506. [CrossRef] [PubMed]
7. Fukazawa, K.; Ishida, F.; Umeda, Y.; Miura, Y.; Shimosaka, S.; Matsushima, S.; Taki, W.; Suzuki, H. Using Computational Fluid Dynamics Analysis to Characterize Local Hemodynamic Features of Middle Cerebral Artery Aneurysm Rupture Points. *World Neurosurg.* **2015**, *83*, 80–86. [CrossRef] [PubMed]
8. Geers, A.J.; Larrabide, I.; Morales, H.G.; Frangi, A.F. Approximating hemodynamics of cerebral aneurysms with steady flow simulations. *J. Biomech.* **2014**, *47*, 178–185. [CrossRef]
9. Cebral, J.R.; Mut, F.; Weir, J.; Putman, C.M. Association of Hemodynamic Characteristics and Cerebral Aneurysm Rupture. *AJNR Am. J. Neuroradiol.* **2011**, *31*, 264–270. [CrossRef]
10. Barile, J.P.; Mitchell, S.A.; Thompson, W.W.; Zack, M.M.; Reeve, B.B.; Cella, D.; Smith, A.W. Patterns of Chronic Conditions and Their Associations with Behaviors and Quality of Life, 2010. *Prev. Chronic Dis.* **2015**, *12*, 150179.
11. Jeong, W.; Rhee, K. Hemodynamics of cerebral aneurysms: Computational analyses of aneurysm progress and treatment. *Comput. Math. Methods Med.* **2012**, *2012*, 782801. [CrossRef]
12. Luckrajh, J.S.; Harrichandparsad, R.; Satyapal, K.S.; Lazarus, L. A clinical investigation of the anatomy of the proximal anterior cerebral artery and its association with anterior communicating artery aneurysm. *Transl. Res. Anat.* **2022**, *27*, 100200. [CrossRef]
13. Torii, R.; Oshima, M.; Kobayashi, T.; Takagi, K.; Tezduyar, T.E. Numerical investigation of the effect of hypertensive blood pressure on cerebral aneurysm-Dependence of the effect on the aneurysm shape. *Int. J. Numer. Meth. Fluids* **2007**, *54*, 995–1009. [CrossRef]
14. Dehdashti, A.R.; Chiluwal, A.K.; Regli, L. The Implication of Anterior Communicating Complex Rotation and 3-Dimensional Computerized Tomography Angiography Findings in Surgical Approach to Anterior Communicating Artery Aneurysms. *World Neurosurg.* **2016**, *91*, 34–42. [CrossRef] [PubMed]
15. Valencia, A.; Burdiles, P.; Ignat, M.; Mura, J.; Bravo, E.; Rivera, R.; Sordo, J. Fluid structural analysis of human cerebral aneurysm using their own wall mechanical properties. *Comput. Math. Methods Med.* **2013**, *2013*, 293128. [CrossRef]
16. Liu, Z.; Ajimu, K.; Yalikun, N.; Zheng, Y.; Xu, F. Potential Therapeutic Strategies for Intracranial Aneurysms Targeting Aneurysm Pathogenesis. *Front. Neurosci.* **2019**, *13*, 1238. [CrossRef]
17. Macdonald, D.J.; Finlay, H.M.; Canham, P.B. Directional Wall Strength in Saccular Brain Aneurysms from Polarized Light Microscopy. *Ann. Biomed. Eng.* **2000**, *28*, 533–542. [CrossRef]
18. Singh, G.; Yadav, P.N.; Gupta, S.; Chanda, A. Biomechanical modeling of aneurysm in posterior cerebral artery and posterior communicating artery: Progression and rupture risk. *Brain Multiphysics* **2023**, *4*, 100069. [CrossRef]
19. Gupta, S.; Singh, G.; Chanda, A. Prediction of diabetic foot ulcer progression: A computational study. *Biomed. Phys. Eng. Express* **2021**, *7*, 065020. [CrossRef]
20. Gupta, V.; Chanda, A. Expansion potential of skin grafts with novel I-shaped auxetic incisions. *Biomed. Phys. Eng. Express* **2021**, *8*, 015016. [CrossRef]
21. Gupta, S.; Gupta, V.; Chanda, A. Biomechanical modeling of novel high expansion auxetic skin grafts. *Int. J. Numer. Methods Biomed. Eng.* **2022**, *38*, e3586. [CrossRef]
22. Singh, G.; Gupta, S.; Chanda, A. Biomechanical modelling of diabetic foot ulcers: A computational study. *J. Biomech.* **2021**, *127*, 110699. [CrossRef] [PubMed]
23. Gao, B.L.; Hao, H.; Hao, W.; Ren, C.F.; Yang, L.; Han, Y. Cerebral aneurysms at major arterial bifurcations are associated with the arterial branch forming a smaller angle with the parent artery. *Sci. Rep.* **2022**, *12*, 5106. [CrossRef]
24. Khe, A.K.; Cherevko, A.A.; Chupakhin, A.P.; Bobkova, M.S.; Krivoshapkin, A.L.; Orlov, K.Y. Haemodynamics of giant cerebral aneurysm: A comparison between the rigid-wall, one-way and two-way FSI models. *J. Phys. Conf. Ser.* **2016**, *722*, 012042. [CrossRef]
25. Chul Suh, D.; Sang Lee, J.; Young Moon, J.; Sang Lee, Y.; Woo Kim, Y. Considerations of Blood Properties, Outlet Boundary Conditions and Energy Loss Approaches in Computational Fluid Dynamics Modeling. *Neurointervention* **2014**, *9*, 1–8. [CrossRef]

26. Hariri, S.; Mirzaei Poueinak, M.; Hassanvand, A.; Barzegar Gerdroodbary, M.; Faraji, M. Effects of blood hematocrit on performance of endovascular coiling for treatment of middle cerebral artery (MCA) aneurysms: Computational study. *Interdiscip. Neurosurg.* **2023**, *32*, 101729. [CrossRef]
27. Russin, J.; Babiker, H.; Ryan, J.; Rangel-Castilla, L.; Frakes, D.; Nakaji, P. Computational Fluid Dynamics to Evaluate the Management of a Giant Internal Carotid Artery Aneurysm. *World Neurosurg.* **2015**, *83*, 1057–1065. [CrossRef]
28. Ma, Z.; Mao, C.; Jia, Y.; Fu, Y.; Kong, W. Extracellular matrix dynamics in vascular remodeling. *Am. J. Physiol. Cell Physiol.* **2020**, *319*, C481–C499. [CrossRef]
29. Shojima, M.; Oshima, M.; Takagi, K.; Torii, R.; Hayakawa, M.; Katada, K.; Morita, A.; Kirino, T. Magnitude and Role of Wall Shear Stress on Cerebral Aneurysm. *Stroke* **2004**, *35*, 2500–2505. [CrossRef]
30. Gao, B.; Ding, H.; Ren, Y.; Bai, D.; Wu, Z. Study of Typical Ruptured and Unruptured Intracranial Aneurysms Based on Fluid–Structure Interaction. *World Neurosurg.* **2023**, *175*, e115–e128. [CrossRef]
31. Omodaka, S.; Sugiyama, S.I.; Inoue, T.; Funamoto, K.; Fujimura, M.; Shimizu, H.; Hayase, T.; Takahashi, A.; Tominaga, T. Local hemodynamics at the rupture point of cerebral aneurysms determined by computational fluid dynamics analysis. *Cerebrovasc. Dis.* **2012**, *34*, 121–129. [CrossRef]
32. Cebral, J.R.; Mut, F.; Sforza, D.; Löhner, R.; Scrivano, E.; Lylyk, P.; Putman, C. Clinical application of image-based CFD for cerebral aneurysms. *Int. J. Numer. Methods Biomed. Eng.* **2011**, *27*, 977–992. [CrossRef]
33. Berger, S.A.; Jou, L.D. Flows in Stenotic Vessels. *Annu. Rev. Fluid Mech.* **2003**, *32*, 347–382. [CrossRef]
34. Humphrey, J.D.; Na, S. Elastodynamics and arterial wall stress. *Ann. Biomed. Eng.* **2002**, *30*, 509–523. [CrossRef] [PubMed]
35. Jou, L.-D.; Quick, C.M.; Young, W.L.; Lawton, M.T.; Higashida, R.; Martin, A.; Saloner, D. Computational Approach to Quantifying Hemodynamic Forces in Giant Cerebral Aneurysms. *Am. J. Neuroradiol.* **2003**, *24*, 1804–1810. [PubMed]
36. ANSYS. ANSYS FLUENT Theory Guide. Available online: https://www.afs.enea.it/project/neptunius/docs/fluent/html/th/main_pre.htm (accessed on 3 July 2023).

Disclaimer/Publisher's Note: The statements, opinions and data contained in all publications are solely those of the individual author(s) and contributor(s) and not of MDPI and/or the editor(s). MDPI and/or the editor(s) disclaim responsibility for any injury to people or property resulting from any ideas, methods, instructions or products referred to in the content.

MDPI
St. Alban-Anlage 66
4052 Basel
Switzerland
www.mdpi.com

Mathematical and Computational Applications Editorial Office
E-mail: mca@mdpi.com
www.mdpi.com/journal/mca

Disclaimer/Publisher's Note: The statements, opinions and data contained in all publications are solely those of the individual author(s) and contributor(s) and not of MDPI and/or the editor(s). MDPI and/or the editor(s) disclaim responsibility for any injury to people or property resulting from any ideas, methods, instructions or products referred to in the content.

www.ingramcontent.com/pod-product-compliance
Lightning Source LLC
LaVergne TN
LVHW070426100526
838202LV00014B/1538